Absorbing Boundaries for the Time-Domain Analysis of Dam-Reservoir-Foundation Systems

Glauco Feltrin

Institute of Structural Engineering

Swiss Federal Institute of Technology Zurich

November 1997

Institut für Baustatik und Konstruktion, ETH Zürich

Absorbing Boundaries for the Time-Domain Analysis of Dam-Reservoir-Foundation Systems

Glauco Feltrin

Springer Basel AG

IBK Bericht Nr. 232, November 1997

KEYWORDS: nonlinear transient analysis, finite element analysis, fluid-structure interaction, soil-structure interaction, absorbing boundaries in time domain, balanced truncation approximation, Hankel norm approximation

© 1997 Springer Basel AG
Originally published by Birkhäuser Verlag Basel in 1997

Gedruckt auf säurefreiem Papier

ISBN 978-3-7643-5869-3   ISBN 978-3-0348-7610-0 (eBook)
DOI 10.1007/978-3-0348-7610-0

987654321

Preface

This report deals with the modeling and analysis of the earthquake behavior of concrete gravity dams. This task becomes very challenging when the interaction with the reservoir and foundation has to be considered. In addition, a concrete gravity dam may get severely damaged by dynamically propagating cracks when subjected to a severe earthquake ground motion. This strongly nonlinear behavior of dams impedes the application of state-of-the-art frequency-domain methods for the analysis because the superposition principle is no longer valid. As a consequence, the model has to be formulated in the time domain complicating the analysis even more.

In this report, which is based on a doctoral thesis, a new method is presented that permits to develop absorbing boundary conditions for the nonlinear time-domain analysis of concrete gravity dams. These conditions are based on a approximation of dynamic stiffness matrices using series of orthogonal functions and advanced model reduction techniques of linear systems theory. The new absorbing boundary conditions are very accurate so that they are virtually equivalent to dynamic stiffness matrices obtained by a rigorous solution of a far field. This allows to reduce the size of near fields dramatically and permits an accurate and numerically efficient nonlinear time-domain analysis.

This work is an important contribution to the realistic modeling and numerically efficient dynamic analysis of the nonlinear earthquake behavior of concrete gravity dams. In addition, the very same method can be applied effectively to the analysis of other engineering structures interacting with an unbounded medium.

Zurich, October 1997   Prof. Hugo Bachmann

Contents

3   DtN-maps for fluids and solids

4   Approximation of scalar DtN-maps I

5   Linear time-invariant systems

Chapter 1

Introduction

1.1 Interaction and radiation damping

Most civil engineering structures are elements of much larger systems, called here the overall system, containing several other system components. When subjected to transient loads, these structures interact with the other components of the overall system such that a continuous transfer of energy is established between them. The effects of interaction on the dynamic behavior of these structures are determined by the mechanical properties of all the components of the overall system, the interaction mechanism and the type of dynamic loading. Dams belong exactly to this category of structures. They interact with the foundation rock and with the reservoir. The kind and intensity of the interaction depends on the physical processes that occur at the interfaces of the dam with the foundation rock and the reservoir. As an example, the effects of the interaction on the dam will be different if a dam is allowed to slip on the foundation or not.

Usually we speak of weak or strong interaction depending on whether the interaction has little or large effects on the behavior of the structure being analysed. This classification has obvious implications for the modeling. In the case of weak interaction, we may neglect the interaction effects in the analysis without serious consequences for the accuracy. In contrast, strongly interacting structures impose a sophisticated modeling of the interaction effects in order to capture the relevant features of the response. In many cases, say for moderate interaction, simple models may yield very good results. However, their range of application is usually limited. Examples of simple models in the field of earthquake analysis of dams are the added mass approach of Westergaard for fluid-structure interaction or the massless foundation method for structure-foundation interaction.

A class of interaction problems requiring special methods for its analysis is that involving overall systems which are much larger in size than the structure itself. Typical examples are foundations resting on rock or soil or dams interacting with the reservoir and the foundation rock. The size of the overall system is so large that its direct modeling, e.g. with finite elements, is virtually impossible because of the tremendous effort needed to compute a solution. In addition, usually engineers are mainly interested on the response of the structure and its close neighborhood so that an accurate modeling is only needed for a small subdomain of the overall system. Therefore, the overall system is subdivided by a suitable imposed boundary, the so-called artificial boundary Γ_B, in two sub-systems (Figure 1.1). One subsystem, the inner domain or near field Ω_N, contains the structure and its close neighborhood. This near field may behave nonlinearly. It is usually modeled in detail by finite elements or similar numerical methods.

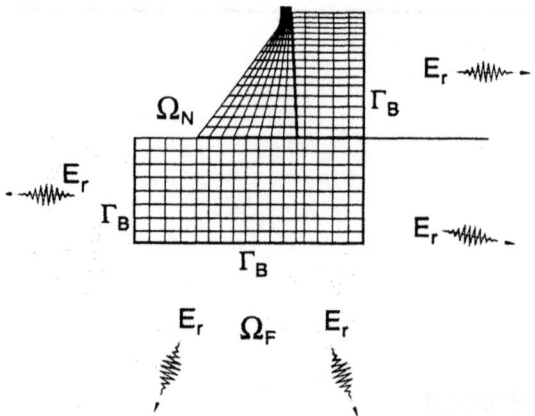

Figure 1.1: Near field Ω_N, far field Ω_F, artificial boundary Γ_B and energy E_r radiated to infinity because of radiation damping

The second sub-system, the outer domain or far field Ω_F, is modeled by methods that are usually quite different from the finite element method. Usually, the far field is simplified in the sense that it is assumed to be infinite in at least one of its dimensions. This model assumption is certainly justified if the distant parts of the physical boundaries have little or no appreciable effect on the response of the structure and its close neighborhood. Moreover, it is usually assumed that the far field has regular properties (e.g. isotropy, homogeneity etc.) and that it behaves linearly. These premises are introduced to provide a reasonably simple solution of the far field. However, they are even reasonable from a physical point of view. In fact, we may expect that isolated confined inhomogeneities situated far away from the structure have small effects on its response.

The unboundedness of the far field has an important consequence in wave dynamics: waves travelling in the unbounded direction are not reflected back to the near field. Because waves transport energy, we obtain a mechanism which irreversibly transfers energy from the near field to the far field. This mechanism is called radiation damping and extracts energy from the near field. Its effect on the response of structures is similar to that of viscous damping where a part of the mechanical energy of a structure is irreversibly transformed into heat. In both cases, the response of the structure is appreciably reduced. For this kind of problem, the accurate modeling of radiation damping becomes a central issue. This is exactly the case for dams interacting with the adjacent reservoir and foundation rock.

1.2   The modeling of radiation damping

In the past, many different models and methods have been proposed for the solution of interaction problems involving radiation damping. In this section, we give a brief review of this field of research. To keep its size reasonable, we only discuss the models and methods which have been relevant in civil engineering and in particular in the numerical analysis of dams. In addition, we refer to models and methods which are important for this work as well.

The usual way to include radiation damping in a model is to formulate special boundary conditions at the artificial boundaries Γ_B of the near field. The most frequent denominations of these boundary conditions are non-reflecting boundaries, silent boundaries, transmitting boundaries, radiation boundaries or absorbing boundaries. Throughout this work we will use the last denomination and the term absorbing boundary condition as synonyms.

Absorbing boundary conditions can be subdivided in two main classes: local and nonlocal absorbing boundaries. Local absorbing boundary conditions are formulated using differential operators with respect to space and time. In contrast, nonlocal absorbing boundary conditions are described through differential and integral operators with respect to space or time. In Figure 1.2 we try to give a picture of the concept of localness in space in the context of a discretization with finite elements. When the absorbing boundary condition is local in space, the direct coupling of the degrees of freedoms is limited to the portion of the boundary modeled by one element. In contrast, an absorbing boundary condition being nonlocal in space directly couples all degrees of freedom of the boundary.

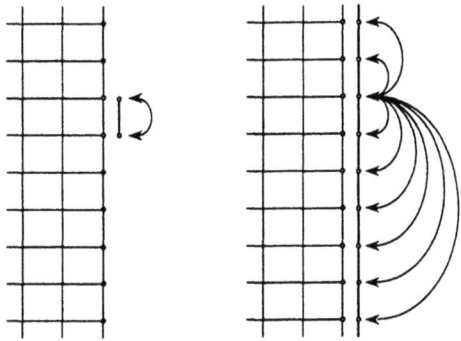

Figure 1.2: The coupling of local (left) and nonlocal (right) absorbing boundary conditions with respect to space

Generally, nonlocal absorbing boundary conditions are used in the context of frequency-domain analysis because the Fourier transform cancels the non-localness with respect to time. But in frequency-domain analysis, the near as well as the far field have to behave linearly. In contrast, local absorbing boundary conditions are used in the context of time-domain analysis. Hence, they can be applied in nonlinear finite element analysis (near field nonlinear, far field linear) as well. However, because the accurate absorbing boundary conditions of wave dynamics are nonlocal in space and in time, local absorbing boundary conditions are necessarily approximations.

1.2.1 Nonlocal absorbing boundaries in the frequency domain

Nonlocal absorbing boundaries in the frequency domain are generally accurate because they are nonlocal in space (the time is eliminated by the Fourier transform). Kausel [Kau88] refers to these absorbing boundaries as the nonlocal consistent boundaries. They are defined by a relationship between the generalized forces and displacements of the far field at the artificial boundary. This relationship is called dynamic stiffness matrix, transfer matrix, impedance matrix or DtN-map (Dirichlet to Neumann map). The last designation, proposed by Givoli [Giv92], will be used throughout this work.

DtN-maps are derived via exact or accurate approximate solutions of wave propagation problems of the far field. Generally, these solutions are factorized into two terms. The first term considers waves propagating towards infinity. This part of the solution fulfills the radiation condition such that only outgoing waves are considered. The second term describes the motion perpendicular to the wave propagation direction, the cross-section, and is usually described by a series of eigenfunctions. In simple situations the eigenfunctions are invariant with respect to frequency so that they describe propagation modes. The DtN-map may then be constructed using purely analytical methods. Because the solution in the cross-section is represented by an eigenfunction expansion, DtN-maps derived by this method are nonlocal in space. Finally, the DtN-map is discretized such that the displacement field agrees with that of the finite element mesh at the associated artificial boundary. This method is described in depth in the book of Givoli [Giv92] and was called by him the DtN-method.

However, the motion of the far field in the cross-section is generally too complicated to be expressed in a closed form. This part of the solution is then computed numerically applying the finite element method. Because the part of the solution containing the outgoing wave components is still given analytically, this technique is usually referred to as a semi-analytical method. This method is equivalent to the DtN-method of Givoli with the difference that the only discretization process is accomplished already at the level of the far-field solution. The semi-analytical method was first developed by Lysmer and Waas [LW72] to formulate DtN-maps of two-dimensional and axisymmetric solid-layers. Later, Chopra and co-workers, for references see section 1.3.2, used it to construct DtN-maps of semi-infinite prismatic fluid channels. Lotfi et al. [LTF86] extended this technique to two-dimensional coupled solid-fluid layers.

Another class of methods which allows to derive accurate DtN-maps is the boundary integral method. The basic idea of this method is to formulate the equation of motion of the far field in the form of an integral equation instead of a differential equation. Finally, this integral equation is solved numerically. Today, the most important numerical method for its solution is the boundary element method. As the name suggests, the far field is discretized only in a finite boundary domain. The interior of the domain is treated by a so-called fundamental solution.

However, accuracy for far field problems is only guaranteed if the fundamental solution satisfies the boundary conditions of those boundaries having an infinite extension. Otherwise, a modeling error occurs. In addition, the DtN-maps resulting from boundary element methods are usually non-symmetric. This spoils the symmetry of the system of equations when the far field is coupled to the near field. Nevertheless, the coupled finite element/boundary element method has been widely used in wave propagation problems. A critical overview with many references may be found in chapter 2 of the book of Givoli [Giv92].

Somewhat similar to the semi-analytical method but closer in spirit to the finite element method are the so-called infinite elements. These elements approximate the displacement field of the far field in the cross-section as well as in the unbounded direction. The form of the shape functions in the unbounded direction is usually derived from exact solutions of associated one-dimensional wave propagation problems. Contrary to the semi-analytical method, infinite elements capture only a few propagation modes so that they are generally not accurate even if the finite element mesh is very fine. In wave dynamics, infinite elements were first introduced by Bettess and Zienkiewicz [BZ77]. Medina and Penzien [MP82] applied infinite elements to axi-symmetric and three-dimensional elastodynamics problems of circular plates on a semi-infinite medium. Infinite elements for the analysis of strip foundations on a semi-infinite elastic medium were devised by Chuhan and Chongbin [CC87].

1.2.2   Nonlocal absorbing boundaries in the time domain: The convolution integral approach

Nonlocal absorbing boundaries in the time domain are usually derived from the associated frequency-domain DtN-map by way of the inverse Fourier transform. In time domain, the absorbing boundaries are defined by convolution integrals. Therefore, they are nonlocal in time. That is, the future response depends on the past time-history. Because these absorbing boundaries are nonlocal in space as well, the future response of a degree of freedom depends not only on its own past time-history but on the past time-histories of all degrees of freedom of the artificial boundary too. The coupling of these absorbing boundaries with the discretized near-field results in a system of Volterra integro-differential equations of convolution or hereditary type.

In numerical analysis, time-domain DtN-maps are usually defined on a regular grid of points on the time axis. To assure an adequate accuracy, the interval between the points has to be small enough. Hence, for large size problems, the storage requirement of time domain DtN-maps may become enormous because they must be computed in advance and stored on a disk. A mid size problem with 50 degrees of freedom and a grid of 2000 points on the time axis requires 20 MByte of data. Moreover, because of the repeated evaluation of the convolution integrals at every integration time step, fast execution can only be achieved if the complete DtN-maps are stored in the central memory unit. This requires computers with a sufficiently large central memory.

In addition, evaluating the convolution integrals becomes more and more time-consuming as the time advances because the complete past time history has to be considered. Therefore, Bennet [Ben76] concluded, that nonlocal time-domain absorbing boundaries are difficult to use in practise because of their large demand on storage capacity and computing time. Although today computers are much faster and their storage capacity is much bigger, the evaluation of convolution integrals is still a very demanding task. Nevertheless, several methods have been proposed which are based on the evaluation of convolution integrals. In some cases the numerical efficiency has been increased by evaluating the convolution integral with only a part of the past time history [Wag85], [Giv92]. This approach neglects the contribution of the tail of the past time history to the convolution integral.

On a spherical artificial boundary, Grote and Keller [GK95] derived an exact absorbing boundary condition for the scalar wave equation which is nonlocal in space but local in time. This boundary condition could be described by a system of differential equations of first order and is one of the very rare instances of exact absorbing boundary conditions which can be defined without a convolution integral.

1.2.3 Local absorbing boundaries in the time domain

To avoid the computational problems associated with the nonlocalness of the accurate absorbing boundary conditions many so-called local absorbing boundary conditions have been proposed. The great advantage of these boundary conditions is that they are local in space and in time such that they are numerically much more efficient than any nonlocal boundary condition. Local absorbing boundary conditions are derived by approximating the related exact absorbing boundary conditions. A great variety of asymptotic approximation techniques have been developed for this task. However, local absorbing boundary conditions are generally not accurate. Usually they show small spurious reflections for waves hitting the boundary at nearly normal incidence and large spurious reflection for waves hitting the boundary with tangent incidence. However, for many problems they provide reasonably accurate solutions with little numerical effort. In this section, we give a brief overview of the most important contributions of this field of research.

Remarkably, the first local absorbing boundary condition was proposed for the numerical solution of soil-structure interaction problems in civil engineering. Lysmer and Kuhlemeyer [LK69] developed the famous viscous boundary to account for an elastic half-space. Even today, nearly 30 years later, the viscous boundary is the most often used absorbing boundary in numerical engineering mechanics. It is implemented in many general purpose finite element programs like Abaqus, Adina, Ansys, Flowers etc. Its popularity is due to the fact, that it has a simple physical interpretation in the form of a dashpot so that it is easy to implement in a finite element environment. Later the viscous boundary condition was generalized by White et al. [WVL77].

Several years later a new set of absorbing boundary conditions was proposed by Engquist and Majda [EM77, EM79]. They developed absorbing boundary conditions of increasing complexity for acoustic and elastic waves described with respect to rectangular as well as cylindrical coordinate systems. Their technique, which was based on the rational approximation of pseudo-differential operators, differed considerably from that of Lysmer and Kuhlemeyer which was supported by simple physical considerations. The simplest boundary conditions for acoustic waves is equivalent to the viscous boundary condition of Lysmer and Kuhlemeyer. Notice, that it was the first time that the method of rational approximation appeared as a tool to derive local absorbing boundary conditions.

Using the equivalence between absorbing boundary conditions and one-way wave equations Trefethen and Halpern [TH86, HT88] formulated several absorbing boundary conditions for acoustic wave equation problems. Similarly to Engquist and Majda, they used rational approximations of increasing order of a pseudo-differential operator. Using different approximation techniques they obtained a set of associated absorbing boundary conditions. They found that the original Engquist-Majda boundary conditions are best at nearly normal incidence and worst at nearly tangent incidence.

Higdon [Hig86b] formulated absorbing boundary conditions for the acoustic wave equation using products of dispersionless one-way wave equations. He showed that all absorbing boundary conditions based on a rational approximation of a pseudo-differential operator can be cast in this form so that they are special cases of his absorbing boundary conditions. These boundary conditions can be made perfectly absorbing for plane waves hitting the artificial boundary at specific angles. Later, he used the same concept to formulate absorbing boundary conditions for the elastic wave equation [Hig91] and for wave equations in stratified media [Hig92].

Many local absorbing boundary conditions were derived for polar coordinate systems. The first were formulated by Engquist and Majda [EM77, EM79] using the technique of rational approximation of pseudo-differential operators. The most referenced are those of Bayliss and Turkel [BT80]. They were derived using an asymptotic expansion of an exact solution at large distances. Bayliss and Turkel demonstrated with several numerical examples that the accuracy of the solution can be increased by increasing the order of the approximation. Feng [Fen83] derived four boundary conditions by approximating an exact nonlocal boundary condition at large distances. He discussed these boundary conditions in the context of the finite element method but did not implement them.

All these local absorbing boundary conditions have found much interest in the context of finite difference methods. Many papers deal with their implementation, accuracy and stability. A detailed overview with supplementary references may be found in [Giv92]. But these absorbing boundary conditions had much less impact in the context of the finite element method. The first of the few finite element implementations was attempted by Cohen and Jennings [CJ83] for elastic wave dynamics problems. Using an asymptotic approximation technique they derived boundary conditions which they called paraxial boundary conditions. In the two-dimensional case, the paraxial boundary conditions are equivalent to those of Engquist and Majda. The finite element implementation was very involved because they employed special upwind elements to avoid spurious oscillations originated by non-symmetric terms. The numerical tests showed that the paraxial boundary conditions were not significantly better than the much simpler viscous boundary conditions of Lysmer and Kuhlemeyer.

Liao and Wong [LW84] developed the so-called extrapolation algorithm to account for radiation damping. This algorithm, which works in connection with an explicit time integration scheme, extrapolates the past motion of the interior of the near field to the artificial boundary. That is, the motion at the artificial boundary is predicted assuming a wave motion with constant speed propagating in a definite direction (usually perpendicular to the boundary). Kausel [Kau88] showed that the extrapolation algorithm for acoustic waves can be interpreted as a time and space discretization of the Engquist-Majda boundary conditions.

Barry, Bielak and McCamy [BBM88] derived local absorbing boundary conditions for stress waves in nonhomogeneous rods and for acoustic waves in two dimensional domains. They used techniques from geometrical optics. The absorbing boundary conditions were implemented using a mixed element formulation. Those of highest order had to be modified because they turned out to be unstable. A different formulation of these boundary conditions using auxiliary degrees of freedom was given later by Kallivokas, Bielak and McCamy [KB93].

The absorbing boundary conditions of Bayliss and Turkel were implemented by Pinsky and Abboud [PA90, PA91]. They obtained a symmetric equation of motion of fourth order in time. These could be reduced to second order by using the limitation principle of the mixed finite element formulation. However, the resulting equations of motion were nonsymmetric. Later Pinsky and Thompson [PT92] obtained local absorbing boundary conditions for two-dimensional acoustic wave propagation by an asymptotic approximation of an exact solution. Their finite element implementation was standard, i.e. by means of a natural boundary condition which was weakly enforced. Because of nonsymmetric terms in the absorbing boundary conditions the equations of motion turned out to be nonsymmetric.

1.2.4 Nonlocal absorbing boundaries in the time domain: The rational approximation approach

The computational problems associated with the evaluation of convolution integrals initiated new and innovative research in the field of interaction problems with unbounded domains. The aim of this field of research is to construct formulations which are approximately equivalent to convolution integrals but local in time. The idea stems from the simple observation that the response of a linear single degree of freedom system (mass-dashpot-spring) can be described by a convolution integral as well as by a differential equation with initial conditions. For numerical computation, the last approach is much more appealing because its solution can be obtained by standard incremental time step algorithms. These algorithms need the data of at most a few past time steps so that they are virtually local in time.

Because the transfer function of a linear mass-dashpot-spring system is given by a rational function we might approximate the DtN-map with rational functions too. Using the Laplace transform, the numerator and denominator polynomials of the rational function can be interpreted as differential operators in time. Therefore, systems described by convolution integrals may be approximated by systems of ordinary differential equations with initial conditions (initial value problem).

Simple applications of this rational approximation concept have been around since the seventies. Typical examples are the mass-dashpot-spring models for foundations resting on an elastic half space [MV74]. These models were inspired by a simple engineering approach without addressing the problem of a systematic rational function approximation. The rational function was a by-product of the mass-dashpot-spring models used to approximate the transfer function. Using essentially the same basic idea, Wolf and Somaini [WS86] constructed much more complicated mass-dashpot-spring models of embedded foundations on an elastic half space. They first defined a mass-dashpot-spring model. This was used to approximate an impedance function in a finite frequency range.

The first systematic approaches to tackle the approximation of convolution integrals in a general context were given by Wolf and Motosaka [WM89a], [WM89b]. They developed four different methods. Two of them consider the rational approximation of a scalar DtN-map in the frequency domain. The approximation was obtained with a least square approximation of DtN-maps in a finite frequency range, decomposed in a partial fraction expansion and then transformed back to the time domain and finally sampled at a constant time interval. This time series was transformed to a complex rational function applying the Z-transform. Finally, interpretating the numerator and denominator as difference operators, they obtained a recursive equation in the form of an ARMA model. Taking another route the numerator and denominator of the rational approximation in the frequency domain were interpreted as differential operators acting in the time domain. By this approach they obtained a linear ordinary differential equation with initial conditions. That is, a continuous time model for the evaluation of the interaction forces. Both methods were applied to the DtN-map of a rod on a elastic foundation. These methods were only used in the context of scalar DtN-maps.

Weber [WHB90] developed a rational approximation scheme for matrix valued DtN-maps of semi-infinite prismatic far fields governed by the acoustic wave equation. The DtN-map given in the frequency domain was approximated using a model reduction technique of linear system theory (truncated balanced realization). Contrary to the approximation scheme of Wolf and

Motosaka [WM89b], that of Weber approximates the DtN-map in the complete frequency range from zero to infinity. The rational approximation was obtained via the singular value decomposition of a block Hankel matrix. The blocks of the Hankel matrix were defined by the Markov parameters of a mapped discrete time system associated with the DtN-map by a so-called bilinear transform. In [Web90] the rational approximants of the matrix valued DtN-map was realized by a symmetric MDOF-system and coupled to the MDOF-system of the near field. The resulting enlarged MDOF-system comprising near and far field was symmetric and could be solved by standard numerical integration schemes. The boundary condition developed by Weber was called rational transmitting boundary and was devised to consider the fluid-structure interaction in the earthquake analysis of dams. A detailed description of the method with many computational examples with large finite element models of arch dams can be found in [Web94]. The rational transmitting boundaries have been implemented in the computer program DANAID.

Another realization of scalar rational approximants as symmetric MDOF-systems was developed by Wolf [Wol91a, Wol91b]. These realizations are called consistent lumped parameter models and can be visualized by a coupled system of mass-damper-spring elements. However, because the parameters of the mass-damper-spring elements are sometimes negative and because the systems are usually very complicated the physical insight of these models is very limited. In fact, the lumped parameter model representing a diagonal term of the DtN-map of a semi-infinite fluid-channel given by Paronesso and Wolf in [PW94] is too complicated for practical use. In addition, realizations with lumped parameter models are available only for scalar DtN-maps so that they can only be used for matrix valued DtN-maps when combined with a diagonalization procedure [PW95].

1.3 Models for the analysis of dams including interaction effects

1.3.1 First steps

The modeling of interaction effects in the field of earthquake analysis of dams has a long tradition. The first to study these types of problem was Westergaard in 1933 [Wes33]. He considered the problem of fluid-structure interaction of a two-dimensional dam-reservoir system subjected to horizontal earthquake ground motion. The dam was assumed to be rigid and the reservoir was supposed to be semi-infinite and of constant depth. With analytical methods he derived the pressure distribution in the fluid at the dam-reservoir interface. His finding was that the interaction forces are proportional to the acceleration of the earthquake ground motion such that they may be approximated by a mass density distributed parabolically over the height of the dam. This technique is called added mass approach. A key argument used to motivate this solution was that the dominant energy content of the earthquake ground motion is related to the spectral components being significantly smaller than the first eigenfrequency of the reservoir. Clearly, this argument cannot be seriously advocated today: The spectra of registered earthquake ground motions records show significant amplitudes over a wide range of frequencies. But at that time very little was known about earthquake ground motion.

The added mass approach of Westergaard allows to model an important effect of fluid-structure interaction which is in close agreement with more elaborated models. Because of the additional mass, the eigenfrequencies of the coupled dam-reservoir system relevant for the earthquake response of the dam are significantly lower than those of the dam alone. However, the added mass approach doesn't consider any radiation damping so that energy dissipation is only due to the structural damping of the dam. Nevertheless, because of its simplicity, it has been one of the most frequently used models for the numerical analysis of dams in engineering practice.

Using a model similar to Westergaard's (rigid dam, semi-infinite reservoir of constant depth) Kotsubo [Kot59] developed an fluid-structure interaction model in which radiation damping was fully considered. The interaction forces due to transient excitations were computed with convolution integrals. Kotsubo also considered three-dimensional prismatic reservoir models with different cross-sectional geometries. He showed that the added mass approach of Westergaard is only justified if the first eigenfrequency of the reservoir is significantly larger than the dominant frequency range of the earthquake ground motion.

In his first of a long series of publications devoted to interaction effects on dams, Chopra [Cho67] studied the effects of compressibility on the fluid-structure interaction forces. The physical model corresponded to that of Westergaard. He found that significant errors may occur if the compressibility of the fluid is neglected. The assumed incompressibility of water was the subject of many studies because it greatly simplifies the modeling of fluid-structure interaction. In incompressible water, waves propagate with infinite speed such that the interaction forces are proportional to the accelerations imposed to the fluid. Therefore, no convolution integrals must be evaluated.

In the references discussed so far the interaction forces during earthquake ground motion were applied as external forces to the dam. The response of the dam didn't contribute to the interaction forces because it was assumed to be rigid. Chopra [Cho68] studied the effects of a flexible dam on the interaction forces by modeling the dam as a mass-dashpot-spring system. The parameters of this system were defined by considering the first eigenmode of a triangular shaped dam. He showed that the natural frequencies of the coupled dam-reservoir system are different from those of the two uncoupled subsystems. He found that the compressibility of water can be neglected if the ratio of the first eigenfrequency of the dam to the first eigenfrequency of the reservoir is larger than two. In [Cho70] this dam-reservoir model was used to compute the response of the dam subjected to a horizontal earthquake ground acceleration. He found that the response of the dam significantly increases if the fluid-structure interaction is considered.

1.3.2 Frequency-domain models

Chopra and Chakrabarti [CC81] developed a model for concrete gravity dams in which the fluid-structure as well as the soil-structure interaction was considered. The dam was modeled with finite elements. The fluid-structure interaction was considered by the DtN-map of a semi-infinite channel of constant depth. Soil-structure interaction was modeled by a DtN-map of an elastic half-space. Reservoir-foundation interaction in the neighborhood of the dam was considered as well. To reduce the computational effort, the whole system was reduced to a few Ritz vectors. The model of Chopra and Chakrabarti [CC81] was extended by Fenves and Chopra [FC81a] to include the effects of an absorptive reservoir bottom. This absorption was modeled with dashpots (viscous boundary condition) and considered the radiation damping due to the interaction of the reservoir with the foundation. This model was implemented in the computer program EAGD84 [FC84c] and was utilized for extensive parametric studies [FC84b, FC85]. The interaction with the foundation lowers the first eigenfrequency of the system and significantly reduces the response of the system because of the radiation damping induced by the foundation.

Along a slightly different line Hall and Chopra [HC82a] developed a model of concrete gravity dams which allows to consider irregularly-shaped reservoirs. The near field is composed by the dam and a finite part of the reservoir was modeled with finite elements. This system was coupled with the DtN-map of a semi-infinite channel of constant depth which considered the energy absorption at the bottom too. The DtN-map was constructed using a semi-analytical solution of the far field. This model was used to study the effects of irregularly-shaped reservoirs on the response of dams [HC82b] and was generalized for the analysis of the fluid-structure interaction of arch dams [HC83]. The model was implemented in the computer program EACD-3D [FHC86]. Recently, this model was extended to include the effects of dam-foundation interaction [TC95].

Lotfi, Roesset and Tassoulas [LTF86] developed a two-dimensional model of dam-reservoir-foundation systems where the far fields were considered by DtN-maps constructed from semi-analytical solutions. The near field comprising the dam and finite portions of the reservoir and the foundation was modeled with finite elements. In the downstream direction the far field was considered by a solid layer whereas in the upstream direction the far field was composed by a coupled solid-fluid layer. The bottom of the foundation was assumed to be fixed.

Besides the models based on analytical or semi-analytical solutions of the far fields, several models have been developed using the boundary element technique. Wepf [Wep87, WWB88] developed a model where the arbitrarily shaped near field of the reservoir was discretized with boundary elements. This finite reservoir domain was coupled with the DtN-map of a semi-infinite reservoir of constant depth to consider radiation damping. Because this model was devised for time-domain analysis, the frequency-domain solution was only used to construct a DtN-map in the time domain via inverse Fourier transform. This DtN-map was coupled at the fluid-structure interface with a dam modeled with finite elements.

A similar model but exclusively for the analysis in the frequency domain was developed by Humar and Jablonski [HJ88]. This model was extended to include the effects of the interaction with the foundation by Humar and Chandrashaker [CH93].

1.3.3 Time-domain models

While the frequency-domain models have been improved continuously, very few time-domain models which rigorously consider radiation damping have been developed so far. Many recent time-domain models consider radiation damping with standard viscous dampers [VF89, EH89, AK92, DMS+95] or infinite elements [SBZ78, VZ92]. These models need large near fields to obtain an adequate accuracy.

A dam-reservoir model that accurately considered fluid-structure interaction effects in the time domain was proposed by Feltrin, Wepf and Bachmann [FWB90]. The interaction forces were computed evaluating the convolution integrals of the DtN-map developed by Wepf [WWB88] (see section before). The number of convolution integrals being evaluated was equal to the square of the number of boundary elements at the dam-reservoir interface. Because an explicit time integration scheme was used, the interaction forces could be applied as external forces.

Tsai, Lee and Ketter [TLK90] developed a two-dimensional dam-reservoir model in which the near field (the dam and a finite portion of the reservoir) were discretized with finite elements. The far field was considered by a DtN-map of a semi-infinite reservoir of constant depth. The DtN-map was represented as a finite sum of convolution integrals with matrix-valued coefficients. These were obtained from the normal mode expansion of the acoustic wave solution. By this approach the number of convolution integrals could be reduced to the number of modes considered in the model. By the same general approach (the DtN-map was based on a semi-analytical solution) they developed a model for three-dimensional dam-reservoir interaction [TL90].

The work of Weber [Web94] was already addressed in section 1.2.4. The rational transmitting boundaries allow for an accurate and efficient linear and nonlinear transient analysis of fluid-structure interaction problems on arch and concrete gravity dams.

Guan, Moore and Lin [GML94] proposed a time-domain model for the transient analysis of concrete gravity dams which includes fluid- as well as soil-structure interaction. The reservoir is modeled as semi-infinite channel of constant depth. The foundation being modeled as an elastic half-space is coupled to the bottom of the gravity dam. The interaction forces are defined by convolution integrals and act as external forces on the interfaces of the dam with the reservoir and the soil. The computational effort due to the evaluation of the convolution integrals was reduced by considering only part of the past time history.

Darbre [Dar93] and Chávez and Fenves [CF95] used the so-called hybrid frequency time-domain procedure to evaluate the fluid-structure interaction forces. In this procedure the response of the system is computed in the frequency domain. Because the dam is assumed to be nonlinear, many iterations are needed to compute the solution. In order to reduce the computational effort, the dam-reservoir system was reduced to a few degrees of freedom using Ritz vectors.

1.4   Objectives and scope of this work

The brief reviews of the foregoing sections indicate that there is a a need for accurate and numerically efficient absorbing boundary conditions for use in time-domain analysis procedures. In fact, highly complex models in the time-domain are extremely inefficient because of the repeated evaluation of convolution integrals when interaction effects are considered. In addition, local absorbing boundaries are of limited interest in earthquake analysis because they usually need large near fields to assure an adequate accuracy.

A promising approach for constructing accurate absorbing boundary conditions is the rational approximation of DtN-maps. Weber [Web94] has shown that this approach is very effective for fluid-structure interaction problems. So far, the generalization of this approach to soil-structure interaction problems has not been done. However, the approximation schemes developed so far suffer from several drawbacks. The least-squares method used by Wolf [WM89b] is apparently restricted to scalar DtN-maps and considers only a finite frequency range. Weber's approach [Web94] allows to approximate matrix-valued DtN-maps. However, this approach doesn't guarantee a priori that an arbitrary matrix-valued DtN-map can be approximated with sufficient accuracy with stable rational functions. In fact, possible unstable components of the approximation have to be removed. However, stability is of outstanding importance for a reliable rational approximation of arbitrary DtN-maps.

This is the background against which we formulate the objectives of this work. The main objective will be to develop a method for constructing accurate absorbing boundary conditions for the nonlinear transient finite element analysis of acoustic and elastic wave problems. These new absorbing boundary conditions should also be applicable in the nonlinear earthquake analysis of concrete gravity dams. We will assume that the infinite domain can be subdivided into a near and a far field. The near field will be modeled with finite elements. The far field is considered by matrix-valued DtN-maps being formulated in the frequency domain. The new absorbing boundaries should be based on the rational approximation of these DtN-maps. A new approximation procedure should be developed which guarantees a priori that the approximations are stable and uniformly converge to the DtN-maps. Furthermore, we should specify the conditions under which these approximations converge. The accuracy and applicability of the new absorbing boundary conditions should be demonstrated with computational examples. Moreover, for many problems the application of local boundary conditions may provide sufficiently accurate results. However, for most of these local boundary conditions no finite element implementation exists so far. In this work, we develop a novel method to develop such finite element implementations.

Because this work is devoted to the development of new absorbing boundary conditions, several other aspects which are of concern in nonlinear earthquake analysis of concrete gravity dams are not addressed herein. In particular we don't consider in detail the modeling of the earthquake input. This topic will be briefly discussed but no particular model will be implemented. Similarly, wave scattering problems are not considered in this work. Furthermore, the near field is supposed to behave linearly because the computational examples are mainly devoted to demonstrate the applicability and accuracy of the new absorbing boundary conditions for transient time-domain analysis. In addition, the computational examples should not be mistaken for a parametric study of a concrete gravity dam subjected to earthquake loading.

1.5 Outline of the work

In chapter 2 we introduce several fundamental concepts which will appear throughout this work. The exposition is based on the analysis of a simple model problem: the elastically restrained rod. Addressed are concepts like initial boundary value problem, harmonic and transient wave propagation, dispersion relation, energy flux, DtN-map, absorbing boundary conditions, convolution integral etc. In addition, it contains an overview of several existing methods to construct local time-domain absorbing boundary conditions and a detailed analysis of these boundary conditions.

Chapter 3 deals with the construction of DtN-maps of far fields governed by the scalar and the elastic wave equation. Most attention will be paid to DtN-maps of two-dimensional strata. Additionally, we will consider the half-space with a circular hole. The mathematical and physical structure of these DtN-maps will be investigated in detail. Finally, we will consider two-dimensional strata composed of solid and fluid layers.

In chapter 4 we make the first step towards the rational approximation of scalar DtN-maps. We analyze the structure of DtN-maps in detail. Approximation methods will be presented which allow to uniformly approximate DtN-map kernels in the complete frequency range. In addition, we establish the link to the theory of linear time-invariant systems by way of the Z-transform.

Chapter 5 deals with the theory of linear time-invariant systems. We will briefly introduce these concepts of the theory which will be needed in this work. The theory is focused on the state space representation of linear systems.

In chapter 6 we continue to analyze the problem of the rational approximation of scalar DtN-maps. The Padé approximation, the truncated balanced realization and the optimal Hankel norm approximation method are presented. The accuracy of these methods is discussed in detail.

Chapter 7 contains the generalization of the theory developed in chapter 7 for scalar DtN-maps to matrix valued DtN-maps. The truncated balanced realization and the sub-optimal and the optimal Hankel norm approximation methods will be presented in detail.

Chapter 8 deals with the time-domain implementation of the new absorbing boundaries in the context of the finite element method. First, we introduce the procedure by constructing the implementations of some local absorbing boundary conditions of the elastically restrained rod. Then, we systematically develop the implementation of general scalar- and matrix-valued absorbing boundary conditions. Finally, we use these methods to construct absorbing boundary elements for the Engquist-Majda and Bayliss-Turkel boundary conditions.

In chapter 9 we show how near and far field are coupled together. Furthermore, we discuss the problem of the well-posedness of initial boundary value problems with the absorbing boundaries constructed in this work.

In chapter 10 we show the absorbing boundary conditions at work. First we present several validation examples. Later we show several computational examples with concrete gravity dams.

Finally, in chapter 11 we present some conclusions and suggest possible extensions and improvements of the method.

Chapter 2

Some elementary concepts

The aim of the first part of this chapter is to introduce several fundamental concepts which will be frequently used throughout this work. We introduce these concepts analyzing a simple model problem: the propagation of longitudinal waves in a restrained elastic rod. Although apparently simple, the restrained elastic rod exhibits similar physical and mathematical structures of the much more complex two-dimensional strata we are going to analyse in the next chapter. This allows to discuss all relevant concepts with great transparency and with little mathematical effort. The second part of this chapter is devoted to the analysis of absorbing boundary conditions. Although nearly all absorbing boundary conditions are for application in two or three dimensional acoustic or elastic wave propagation problems, we discuss these boundary conditions in the context of the elastically restrained rod. This, because our main purpose is to give a brief overview of the methods used for their construction and analysis. For more comprehensive reviews of absorbing boundary conditions we refer to [Giv91] or [Giv92].

2.1 Longitudinal waves in an elastically restrained rod

2.1.1 An initial boundary value problem

Consider a straight prismatic rod as shown in Figure 2.1. The coordinate x refers to the cross-section of the rod. We assume that the longitudinal displacement of each section is given by $u(t, x)$ and is identical at all points of the section. We further assume that the rod is subjected to a dynamically varying stress field $\sigma(x, t)$. The rod is continuously restrained by two layers of linear springs and visco-elastic dampers characterized by the material constants k_e and e, respectively. Formulating the dynamic equilibrium on a differential element (Figure 2.2) yields the equation of motion.

$$-\sigma A + (\sigma + \sigma_{,x} dx)A - k_e u\, dx - e u_{,t}\, dx - \rho A u_{,tt}\, dx = 0 \qquad (2.1)$$

Let us assume that the material behaves linear elastically and is given by the constitutive equation: $\sigma = E u_{,x}$ (Hooke's law) where E is Young's modulus. Then, the equation of motion is

$$AE u_{,xx} = \rho A u_{,tt} + k_e u + e u_{,t}. \qquad (2.2)$$

Dividing the last equation by the stiffness AE we reduce the number of parameters

13

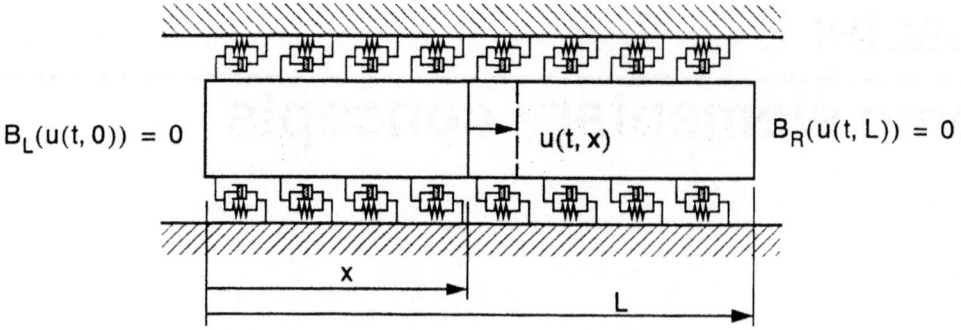

Figure 2.1: Model of an elastically restrained rod

$$u_{,xx} = \frac{1}{c^2} u_{,tt} + \kappa^2 u + 2\frac{\varepsilon}{c} u_{,t}, \tag{2.3}$$

where the new parameters c, κ and ε are defined according to

$$c = \sqrt{\frac{AE}{\rho}}, \quad \kappa = \sqrt{\frac{k_e}{AE}}, \quad \varepsilon = \frac{ec}{2AE}. \tag{2.4}$$

Equation (2.3) is a linear partial differential equation of second order. It is a hyperbolic partial differential equation (PDE) so that the solutions are wavetrains propagating in the positive and negative x-directions. The parameter c has the dimension of a velocity and is known as the wave speed.

In order to obtain a well-defined initial boundary value problem (IBVP), a set of equations describing the initial state of the rod and the boundary conditions must be added to the equation of motion (2.3). These additional equations are

$$u(x, 0) = f(x), \tag{2.5.a}$$

$$u_{,x}(x, 0) = g(x), \tag{2.5.b}$$

$$B_L(u) = b_L(t), \tag{2.5.c}$$

$$B_R(u) = b_R(t) . \tag{2.5.d}$$

The first two equations describe the initial deflections and the velocities on each section of the rod at the time $t = 0$. The last two equations define the boundary conditions at the left (index L) and at the right (index R) end of the rod. B_L and B_R are operators acting on the deflection u. The explicit form of the operators may change from problem to problem. The operators must be such that the IBVP defined by (2.3) and (2.5.a) to (2.5.d) is well-posed. That is, given bounded functions $f(x)$, $g(x)$, $b_L(t)$ and $b_R(t)$ there exists a unique and bounded solution of the IBVP. In this chapter we want restrict ourselves to problems where $f(x)$ and $g(x)$ vanish.

2.1.2 Harmonic wave propagation and the dispersion relation

The usual way to analyse the fundamental features of an equation of motion is to assume a solution in the form of free harmonic waves

$$u(x, t) = u_o e^{-ikx + i\omega t} , \tag{2.6}$$

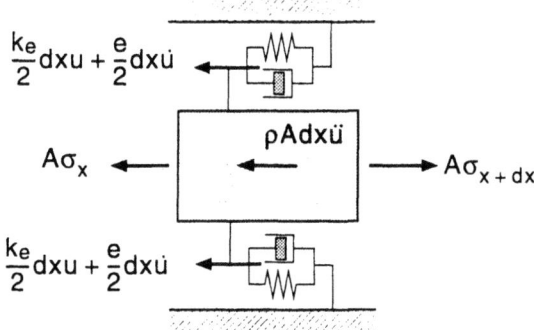

Figure 2.2:   Infinitesimal element of the elastically restrained rod

where ω is the circular frequency and k is the wave number. This approach deliberately ignores the initial and boundary conditions and emphasizes the intrinsic properties of the equation of motion. Substitution of (2.6) in (2.3) gives

$$\left(k^2 - \frac{\omega^2}{c^2} + \kappa^2 + i2\frac{\varepsilon}{c}\omega\right)u_o e^{-ikx + i\omega t} = 0.\qquad(2.7)$$

Expression (2.7) is fulfilled for arbitrary t and x only if the term in brackets vanishes. This defines a relationship between the wave number k and the circular frequency ω which is called the dispersion relation

$$k^2 = \frac{\omega^2}{c^2} - \kappa^2 - 2i\frac{\varepsilon}{c}\omega.\qquad(2.8)$$

The term dispersion relation stems from the fact that in a dispersive medium a propagating wavetrain continuously modifies its shape. The dispersion relation represents a description of the equation of motion in the frequency domain. This relation is completely equivalent to the differential equation (2.3). Indeed, performing the formal substitutions

$$k = i\partial u/\partial x \quad \text{and} \quad \omega = -i\partial u/\partial t \qquad(2.9)$$

in (2.8) we recover exactly the differential operator acting on $u(t, x)$ given in (2.3). Observe, that the wave number is related to the wave length by $\lambda = 2\pi/k$. Hence, the dispersion relation connects the wave length λ to an angular frequency ω.

The dispersion relation (2.8) may be represented in several ways. For the purposes of this work the appropriate form is the wave number k as a function of the circular frequency ω: $k = k(\omega)$. The quadratic character of the expression (2.8) allows two solutions $\{k_1, k_2\}$ corresponding to the two roots of the equation

$$k = \pm\sqrt{\frac{\omega^2}{c^2} - \kappa^2 - 2i\frac{\varepsilon}{c}\omega}.\qquad(2.10)$$

Let us analyse the implications of this expression on the propagation of wavetrains. First, the wave number k is a complex number and may be represented as $k = k_R + ik_I$ where k_R represents the real part and k_I the imaginary part of the wave number. Both are real functions of the circular frequency ω. Inserting this expression into (2.6) yields:

$$u(x, t) = u_o e^{-ik_R x + i\omega t + k_I x} = u_o e^{-ik_R x + i\omega t}e^{k_I x} \qquad(2.11)$$

The result is composed of a factor of a so-called phase function $\exp(i(-k_R x + \omega t))$, which represents a propagating harmonic wave, and the term $\exp(k_I x)$ which depends on the space variable x only. This term modulates the amplitude of the harmonic wave. k_I must be negative for all ω for waves propagating in the positive x-direction and positive for those propagating in the negative x-direction. Otherwise, the wavetrains would grow without limit which can't be correct because the distributed dashpots dissipate energy. Hence, wave propagation direction and modulation must be somehow interrelated.

From (2.11) we also deduce that waves propagating in the positive x-direction result only if $\omega / k_R > 0$. Because $\omega \geq 0$, we obtain the condition $k_R(\omega) > 0$ for all ω. Now let us verify this condition by computing the real and the imaginary part of the dispersion relation. Inserting $k(\omega) = k_R(\omega) + i k_I(\omega)$ into (2.10) yields

$$k_R^2(\omega) - k_I^2(\omega) = \left(\frac{\omega}{c}\right)^2 - \kappa^2, \tag{2.12.a}$$

$$k_R(\omega) k_I(\omega) = -\varepsilon \omega / c . \tag{2.12.b}$$

Inserting the last equation into the first one and resolving for $k_I(\omega)$ we obtain

$$k_I(\omega) = \pm \sqrt{\left(\kappa^2 - \left(\frac{\omega}{c}\right)^2\right)\!\Big/2 + \sqrt{\left(\left(\kappa^2 - \left(\frac{\omega}{c}\right)^2\right)\!\Big/2\right)^2 + \left(\frac{\varepsilon\omega}{c}\right)^2}} . \tag{2.13}$$

$k_I(\omega)$ is a strictly positive or a strictly negative real function of ω according to the sign in front of the outer root. In agreement with the requirement of a physically meaningful solution, we choose the negative root. Because of (2.12.b) this choice produces a strictly positive $k_R(\omega)$

$$k_R(\omega) = \frac{-\varepsilon\omega}{c k_I(\omega)} \tag{2.14}$$

since both the nominator and the denominator are negative functions for positive ω. The case $\varepsilon = 0$ is recovered from the case $\varepsilon > 0$ performing the limit $\varepsilon \to 0$ on both functions and requiring continuity at $\varepsilon = 0$. Hence, we find that the dispersive wave equation (2.3) has two solutions, a wavetrain travelling in the positive and one travelling in the negative direction. These wavetrains are associated with the two roots of the dispersion relation (2.8). The graphs of $k_R(\omega)$ and $k_I(\omega)$ are shown in Figure 2.3. For $\omega \to \infty$ the imaginary part converges to $k_I(\omega) = -\varepsilon$ as can be easily deduced by an asymptotic expansion of (2.10).

The dispersion relation (2.10) has a very interesting feature for $\varepsilon = 0$. At $\omega_c = \kappa c$ the wave number $k(\omega)$ changes from a purely imaginary to a purely real function (see Figure 2.3). This fact has also a physical interpretation. In the frequency range $|\omega| > \kappa c$, $k_I = 0$ and $k(\omega)$ reduces to the phase function (2.11): a propagating harmonic wave. In the frequency range $|\omega| < \kappa c$, $k_R = 0$ and $k(\omega)$ reduces to a spatially decaying function oscillating in phase with the circular frequency ω. This solution is called a non-propagating mode because there is no wave propagation or evanescent mode because the amplitude of the oscillation decays with increasing x.

The transition frequency ω_c is called the cut-off frequency and is characterised by $k(\omega_c) = 0$. The behaviour of the restrained rod at the cut-off frequency is apparent from (2.11): The x-dependency disappears in the harmonic wave solution so that it reduces to an oscillatory term $\exp(i\omega t)$. There is no propagating harmonic wave any more: each section of the rod has equal amplitude and phase. Hence, the flexible rod oscillates like a rigid rod. This interpretation is confirmed by setting $u_{,xx} = 0$ in the equation of motion (2.2) which transforms the partial differential equation to the ordinary differential equation (ODE)

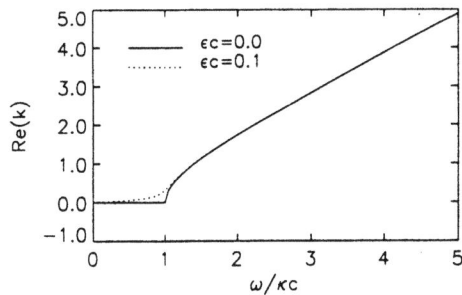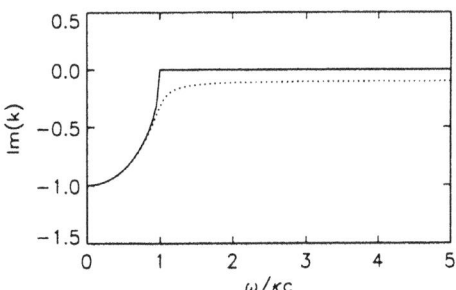

Figure 2.3: Real and imaginary part of the positive root of the dispersion relation.

$$m\frac{d^2u}{dt^2} + k_e u = 0, \qquad (2.15)$$

where $m_e = \rho A$. Equation (2.15) has exactly the structure of the equation of motion of a linear spring-mass system. Observe, that the eigenfrequency of (2.15) is exactly the cut-off frequency ω_c. For $\varepsilon > 0$, $k(\omega)$ is nowhere zero so that a cut-off frequency does not exist. Because of the dissipation of energy, the amplitude of the harmonic wave decreases with increasing values of x so that $u_{,xx} \neq 0$ for any frequency.

2.1.3 Reflection of waves at a boundary and exact absorbing boundary conditions

In the preceding sections we have derived many important properties of the wave equation considering the equation of motion only. In this section, we go a step further and analyse the steady state solutions of a finite rod of arbitrary finite length L. In this case, we have to impose boundary conditions at both ends of the rod to obtain a well-posed problem. Since the analysis is restricted to steady state solutions, the influence of the initial conditions can be neglected.

Consider the restrained rod of length L shown in Figure 2.1. We impose a Dirichlet boundary condition at the left end of the rod ($x = 0$):

$$u(0, t) = u_o e^{i\omega t} . \qquad (2.16)$$

At the right end of the rod ($x = L$) we impose the Neumann boundary condition:

$$\sigma(0, t) = Eu_{,x}(L, t) = \frac{F(t)}{A} . \qquad (2.17)$$

Because we will use the equation of motion (2.3) we rewrite the (2.17) as

$$u_{,x}(L, t) = \frac{F(t)}{AE} = f(t) . \qquad (2.18)$$

We impose that the function $f(t)$ is defined by a convolution integral

$$f(t) = -\int_{-\infty}^{t} G(t - \tau)u(L, \tau)d\tau, \qquad (2.19)$$

with convolution kernel $G(t)$. The boundary condition becomes

$$u_{,x}(L, t) + \int_{-\infty}^{t} G(t - \tau)u(L, \tau)d\tau = 0 . \qquad (2.20)$$

Of course (2.20) restricts the class of admissible boundary conditions. However, we will show later on that absorbing boundary conditions can be cast exactly in this form.

We assume that there exists a steady state solution in the form

$$u(t, x) = \hat{u}(x)e^{i\omega t} . \tag{2.21}$$

The equation of motion (2.3) and the boundary conditions (2.16) and (2.20) become

$$\hat{u}''(x) = -\left(\frac{\omega}{c}\right)^2 \hat{u}(x) + \kappa^2 \hat{u}(x) + i2\frac{\varepsilon}{c}\omega\hat{u}(x), \tag{2.22.a}$$

$$\hat{u}(0) = u_0 \tag{2.22.b}$$

$$\hat{u}'(L) + g(\omega)\hat{u}(L) = 0 . \tag{2.22.c}$$

where the prime indicates differentiation with respect to x. The function $g(\omega)$ is the Fourier-transform of the convolution kernel $G(t)$. The equations above define a boundary value problem (BVP) for a linear ordinary differential equation (ODE) of second order. The circular frequency ω is a parameter. We seek solutions of the ODE in the form

$$\hat{u}(x) = A(\omega)e^{ikx} . \tag{2.23}$$

Combining (2.23) with (2.22.a) we get a condition for the parameter k which corresponds exactly to the dispersion relation (2.7)

$$k^2 = \frac{\omega^2}{c^2} - \kappa^2 - i2\frac{\varepsilon}{c}\omega. \tag{2.24}$$

The two roots $\pm k(\omega)$ lead to two independent solutions so that the general solution is

$$\hat{u}(x) = A_1(\omega)e^{-ikx} + A_2(\omega)e^{ikx} . \tag{2.25}$$

The coefficients A_1 and A_2 are determined by the boundary conditions. Combining (2.25) with (2.22.b) and (2.22.c) yields

$$\hat{u}(x) = \frac{u_0}{1 + q(\omega)}(e^{-ikx + i\omega t} + q(\omega)e^{ikx + i\omega t}) , \tag{2.26}$$

where $q(\omega)$ is given by

$$q(\omega) = \frac{ik(\omega) - g(\omega)}{ik(\omega) + g(\omega)}e^{-2ikL} . \tag{2.27}$$

The solution (2.26) consists of a rightward and a leftward travelling wave. The leftward travelling wave is multiplied by the factor $q(\omega)$.

Since $q(\omega)$ is associated with the leftward travelling wave we may argue that it is related to the reflection of the rightward travelling wave at the right boundary. This is verified by computing the reflection coefficient for the boundary condition $g(\omega)$. For this purpose, we combine the solution (2.25) with the boundary condition (2.22.c) and compute the quotient $A_2(\omega)/A_1(\omega)$. Performing the limit $L \to 0$ yields

$$A_2(\omega)/A_1(\omega) = \frac{ik(\omega) - g(\omega)}{ik(\omega) + g(\omega)}, \tag{2.28}$$

which confirms $q(\omega)$ as the reflection coefficient. The exponent e^{-2ikL} in (2.27) doesn't depend on $g(\omega)$ and reflects the decay of the amplitude because of the nonpropagating mode and of the dissipation of energy in the rod. Now, we may ask which convolution kernel $g(\omega)$ gives no reflection at the right boundary. That is, which kernel leads to $q(\omega) \equiv 0$. From equation (2.27) we obtain

$$ik(\omega) - g(\omega) \equiv 0 \quad \text{or} \quad g(\omega) \equiv ik(\omega). \tag{2.29}$$

Hence, the transformed convolution kernel $g(\omega)$ must be identical to the positive root of the dispersion relation multiplied with the imaginary number i. Notice that this relation is unique. All other kernels $g(\omega)$ induce a reflection at the boundary.

If (2.29) holds, the solutions of the finite rod corresponds exactly to the solution of a semi-infinite rod. Hence, the finite rod with the boundary condition $\hat{u}_{,x}(L) + ik(\omega)\hat{u}(L) = 0$ is an exact finite model of the semi-infinite rod. This is a very significant result so that we will call a boundary condition which leads to $q(\omega) \equiv 0$ an exact absorbing boundary condition.

2.1.4 Exact absorbing boundary conditions and DtN-maps

In the previous section before, we have defined the exact absorbing boundary condition considering the reflection coefficient $q(\omega)$. A much easier procedure to define an exact absorbing boundary condition will be explained in this section. The central idea is that we can use the solution of the semi-infinite restrained rod to construct a relationship between the displacements and the stresses at the left end of the infinite rod. This relationship is used to formulate the boundary condition at the right end of the finite rod.

The harmonic wave solution of the semi-infinite rod is obtained from that of the finite rod given by (2.26) when we consider that the reflection coefficient is by definition $q(\omega) = 0$ (no waves can be reflected). Therefore, the solution is

$$u(x, t) = u_o e^{-ik(\omega)x + i\omega t}, \tag{2.30}$$

with the wave-number $k(\omega)$ defined by (2.10). In order to deduce a boundary condition having the structure defined in (2.20), we have to formulate a relation between the stresses and the displacements. With Hooke's law the stresses in every section of the rod become

$$\sigma_x(\omega, x) = E u_{,x}(\omega, x) = -ik(\omega) E u_o e^{-ik(\omega)x + i\omega t} = -ik(\omega) E u(\omega, x). \tag{2.31}$$

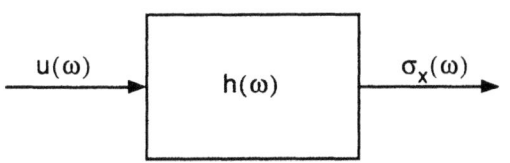

If (2.31) is applied at the left boundary of the semi-infinite rod it describes the relationship between imposed displacements and resulting stresses for any frequency ω. We call such a relation a DtN-map: a Dirichlet to Neumann map. The name was introduced by Givoli [Giv92] and reflects an aspect of the formal structure of the boundary condition (2.20): a

Figure 2.4: DtN-map as a linear input-output system defined by the transfer function $h(\omega)$

"Dirichlet item", the displacements $u(\omega)$, is mapped to a "von Neumann item", the stresses $\sigma_x(\omega)$. The mapping between input and output is defined by the DtN-map kernel $h(\omega)$. In (2.33) it is given by $h(\omega) = ik(\omega)E$. The DtN-map gives a description of the semi-infinite restrained rod by means of a so-called input-output system. The input, the displacements $u(\omega)$ are mapped to the output, the stresses $\sigma_x(\omega)$ (see Figure 2.4). Because (2.31) is a linear map in the frequency domain, a DtN-map is a linear time-invariant input-output system.

Impedance or transfer functions are other names which are frequently found instead of DtN-maps. Moreover, if (2.31) is expressed as a relationship between 'displacements' and 'forces'

$$f(\omega) = -ik(\omega)AEu(\omega) \tag{2.32}$$

then it is called a dynamic stiffness. This name, which is very common in engineering mechanics, has been chosen because of the formal analogy of (2.32) to the stiffness matrix in statics (frequency ω is fixed). Nevertheless, this name is somewhat misleading so that we prefer the more abstract term DtN-map. If we refer to the normalized equation of motion of the rod given in (2.3), the DtN-map is

$$u_{,x}(\omega, x) + ik(\omega)u(\omega, x) = 0, \tag{2.33}$$

so that the DtN-map kernel is $h(\omega) = ik(\omega)$. This is exactly the kernel $g(\omega)$ of the exact absorbing boundary condition defined in (2.29). Clearly, this is an obvious consequence of the very definition of a DtN-map.

2.1.5 Harmonic wave propagation and energy flux

As the name suggests, absorbing boundary conditions should absorb or dissipate energy carried by waves coming from the interior and hitting these boundaries. A concept with which we can quantify the energy absorbed or dissipated at the boundaries is the energy flux. The energy flux describes the amount of energy per time crossing a section. We derive this concept starting with the equation of motion (2.2). Multiplication with $\partial u / \partial t$ leads to

$$\frac{\partial}{\partial t}\left(\frac{EA}{2}\left(\frac{\partial u}{\partial x}\right)^2 + \frac{k}{2}u^2 + \frac{\rho A}{2}\left(\frac{\partial u}{\partial t}\right)^2\right) + \frac{\partial}{\partial x}\left(-AE\frac{\partial u}{\partial x}\frac{\partial u}{\partial t}\right) + e\left(\frac{\partial u}{\partial t}\right)^2 = 0 \quad . \tag{2.34}$$

Rewritten in a more concise form, (2.34) reveals the typical form of a conservation law

$$\frac{\partial W}{\partial t} + \frac{\partial S}{\partial x} + e\left(\frac{\partial u}{\partial t}\right)^2 = 0 \tag{2.35}$$

The energy density W has been identified by the first term in parenthesis in (2.34)

$$W = \frac{EA}{2}\left(\frac{\partial u}{\partial x}\right)^2 + \frac{k}{2}u^2 + \frac{\rho A}{2}\left(\frac{\partial u}{\partial t}\right)^2 . \tag{2.36}$$

The second term of (2.34) is identified by the energy flux S

$$S = -AE\frac{\partial u}{\partial x}\frac{\partial u}{\partial t} = -A\sigma_x\frac{\partial u}{\partial t} . \tag{2.37}$$

The last term in (2.34) accounts for the dissipation of energy because of the two layers of dashpots. Equation (2.35) states that the time variation of the energy density W is in equilibrium with the space variation of the energy flux S and the dissipation rate. This becomes even more evident if this energy relation is formulated on a finite part of the rod. We integrate equation (2.34) along the coordinate x

$$\int_{x_1}^{x_2}\frac{\partial W}{\partial t}dx = -S\big|_{x_1}^{x_2} - \int_{x_1}^{x_2}e\left(\frac{\partial u}{\partial t}\right)^2 dx \quad . \tag{2.38}$$

The time variation of the energy content in a portion of the rod equals the energy flux crossing the borders plus the local energy dissipation in the rod.

For harmonic wave solutions we can reformulate the energy density and the energy flux equations given in (2.34). Since the energy density and the energy flux must be real quantities we need real displacements. We define it as

$$u(x, t) = Re(ae^{-ikx + i\omega t}) = \frac{1}{2}(U(x)e^{-i\omega t} + U^*(x)e^{i\omega t}).$$ (2.39)

The asterisk denotes the operation of complex conjugation. Combining (2.39) with (2.36) and (2.37) we obtain

$$W = \frac{\rho A\omega^2}{2}\frac{|U|^2}{2} + \frac{\rho c^2 A\kappa^2}{2}\frac{|U|^2}{2} + \frac{\rho c^2 A}{2}\frac{|dU/dx|^2}{2} + F(x)e^{2i\omega t} + G(x)e^{-2i\omega t},$$ (2.40)

$$S = -\frac{\rho c^2 A\omega}{2} \cdot Im\left(\frac{dU}{dx}U^*\right) + f(x)e^{2i\omega t} + g(x)e^{-2i\omega t}.$$ (2.41)

More important than the instantaneous energy density and energy flux are their mean values over an oscillation period $T = 2\pi/\omega$: Integrating (2.40) and (2.41) over a period T and dividing the resulting expression by the same period T we obtain

$$W_m = \frac{\rho A\omega^2}{2}\frac{|U|^2}{2} + \frac{\rho c^2 A\kappa^2}{2}\frac{|U|^2}{2} + \frac{\rho c^2 A}{2}\frac{|dU/dx|^2}{2}$$ (2.42)

$$S_m = -\frac{\rho c^2 A\omega}{2} \cdot Im\left(\frac{dU}{dx}U^*\right).$$ (2.43)

The terms oscillating with twice the angular frequency drop out because their mean values vanish over a period. Now let us combine the harmonic wave solution of the semi-infinite restrained rod with (2.42) and (2.43). This yields

$$W_m = \left(\frac{\rho A\omega^2}{4} + \frac{\rho c^2 A\kappa^2}{4} + \frac{\rho c^2 A}{4}|k|^2\right)e^{2Im(k)x}$$ (2.44)

$$S_m = -\frac{\rho c^2 A\omega}{2} \cdot Im(-ik)e^{2Im(k)x} = \frac{\rho c^2 A\omega}{2}Re(k)e^{2Im(k)x}.$$ (2.45)

We analyse the simpler case of the nondissipative rod. Inserting $\varepsilon = 0$ in (2.44) and (2.45) yields

$$W_m = \frac{\rho A\omega^2}{2}e^{-2Im(\sqrt{(\omega/c)^2 - \kappa^2})x},$$ (2.46)

$$S_m = \frac{\rho c^2 A\omega}{2}Re(\sqrt{(\omega/c)^2 - \kappa^2})e^{-2Im(\sqrt{(\omega/c)^2 - \kappa^2})x}.$$ (2.47)

If the angular frequency is greater than the cut-off frequency ($|\omega| \geq \kappa c$) the imaginary part vanishes and the exponential factor becomes one. So both equations simplify to

$$W_m = \frac{\rho A\omega^2}{2} \quad \text{and} \quad S_m = \frac{\rho c^2 A\omega}{2}\sqrt{(\omega/c)^2 - \kappa^2} \quad \text{for } |\omega| \geq \kappa c.$$ (2.48)

The mean energy density increases with the square of the angular frequency. More interesting is the fact, that the mean energy flux is proportional to the real part of the dispersion relation. It vanishes at exactly the cut-off frequency and increases then steadily to the asymptotic value

$$S_m \sim \frac{\rho c A \omega^2}{2} = c W_m.$$ (2.49)

Beneath the cut-off frequency we obtain

$$W_m = \frac{\rho A \omega^2}{2} e^{-2\sqrt{\kappa^2 - (\omega/c)^2}\,x} \quad \text{and} \quad S_m = 0 \quad \text{for } |\omega| < \kappa c$$ (2.50)

The mean energy density is modulated in the x-direction by an exponential factor smaller than one. It tends to one if the angular frequency approaches the cut-off frequency. The mean energy flux vanishes. This is because the rod is in a nonpropagating state of motion and over a period no energy is exchanged between neighboring sections.

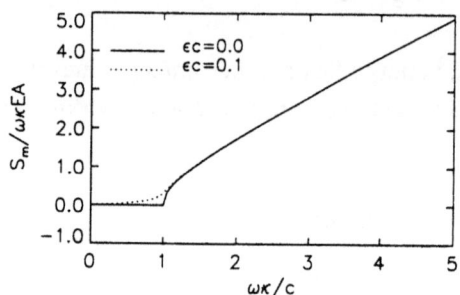

Figure 2.5: Mean energy flux of the rod

The graph of the mean energy flux is shown in Figure 2.5. When the rod has small energy dissipation, the shape of the mean energy flux doesn't change much. Below the cut-off frequency the mean energy flux is rather modest and becomes significant only above it. There, it approaches very quickly the shape of the energy flux of the nondissipative rod. That is, if ε is small, the nonpropagating mode state predominates at angular frequencies smaller than the cut-off frequency whereas for angular frequencies larger than the cut-off frequency the wave propagation mode dominates.

The left hand side of equation (2.49) suggests that for $\omega \to \infty$ the energy is carried with the velocity of the wave speed c. Hence, we may define a propagation speed of energy as $c_E = S_m / W_m$. With this definition, we obtain

$$c_E = c \frac{\sqrt{\omega^2 - (\kappa c)^2}}{\omega} \quad \text{for } |\omega| \geq \kappa c,$$ (2.51.a)

$$c_E = 0 \quad \text{for } |\omega| < \kappa c.$$ (2.51.b)

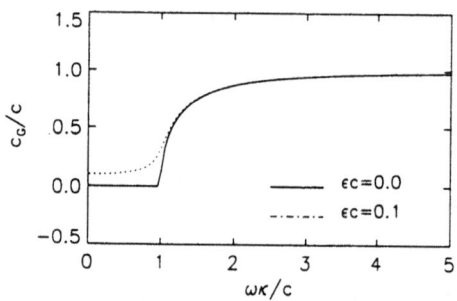

Figure 2.6: Graph of the group velocity of the nondissipative and dissipative rod.

This is exactly the formula describing the so-called *group velocity* defined by $c_g = d\omega/dk$. If the wave propagation is dispersive, the group velocity is a nonconstant function of the frequency. The group velocity - we always refer to its real part - describes the speed with which a narrow-band pulse propagates. However, it describes the speed of energy propagation as well. Hence, we have obtained a neat relationship between mean energy flux, mean energy density and group velocity. As shown in Figure 2.6 the group velocity is frequency-dependent. Below the cut-off frequency the group velocity is zero. Above the cut-off frequency the group velocity increases up to the wave speed $c = \sqrt{E/\rho}$. The relationship between group velocity or energy speed is strictly valid only for nondissipative media. However, if the dissipation is small, the group velocity is still a very good approximation.

To obtain more insight into the role of the kernel $g(\omega)$, we may have a short look at the mean energy flux at the right boundary of the finite rod. Before starting with the analysis, we will briefly discuss the exact meaning of the energy flux at a boundary. Since we are at the end of the rod there is obviously no further part of it where the energy can flow. But since a mean energy flux different from zero cannot be excluded a priori, we must assume that the boundary condition at the end of the rod is able to destroy or create energy. A positive value of the energy flux means that the boundary condition dissipates energy and a negative value that it creates energy. This point will be of great importance when we discuss the problem of well-posedness of IBVP with absorbing boundary conditions.

Next we want to compute the energy flux at the right boundary. We combine (2.26) with (2.43) and obtain the following relation for the mean energy dissipation rate

$$S_m = \rho c^2 A \omega |u(x, L)|^2 \mathrm{Im}(g(\omega)).\tag{2.52}$$

The mean energy flux is thus proportional to the imaginary part of $g(\omega)$. Equation (2.52) states that the boundary conditions dissipates energy if and only if the imaginary part of $g(\omega)$ is positive. This result suggests, that a real $g(\omega)$ gives no contribution to the energy dissipation. Hence, in this case we should obtain total reflection or $|q(\omega)| \equiv 1$ if the rod is nondissipative. Indeed, evaluating (2.27) with the assumption of a real $g(\omega)$ yields

$$|q(\omega)| = \sqrt{\frac{k_R^2 + (k_I + g)^2}{k_R^2 + (k_I - g)^2}}.\tag{2.53}$$

In the frequency range above the cut-off frequency we have harmonic wave solutions so that $k_I(\omega) = 0$ and $|q(\omega)| = 1$ for any $g(\omega)$. Since according to section 2.1.2 $k_I(\omega)$ is strictly negative for the dissipative rod we obtain $|q(\omega)| < 1$ if and only if $g(\omega) > 0$.

2.1.6 DtN-map, convolution operator and integro-differential equations

In this work, we will formulate approximated solutions of the IBVP by the method of finite elements. This transforms an IBVP in an initial value problem (IVP) described by a system of ordinary differential equations (ODE). We assume that the user is familiar with the standard concepts and methods of the finite element method. The semi-discretization with finite elements of the equation of motion of the restrained rod yields

$$M\ddot{u}(t) + Ku(t) = f(t) + u_{,x}(t, L),\tag{2.54}$$

where u is the vector of the nodal displacements, M and K are the mass and the stiffness matrices, respectively. $f(t)$ is the vector of the external forces and $u_{,x}(t, L)$ specifies the natural boundary condition at the right end of the rod. To complete the initial value problem we must add initial conditions to (2.54). We assume that the rod is at rest. That is,

$$u(0) = 0 \quad \text{and} \quad \dot{u}(0) = 0.\tag{2.55}$$

The natural boundary condition $u_{,x}(t, L)$ at the right end of the rod is specified by the convolution integral (2.20) with the kernel $g(\omega) = ik(\omega)$, where $k(\omega)$ is given by

$$k = \sqrt{\frac{\omega^2}{c^2} - \kappa^2 - i2\frac{\varepsilon}{c}\omega} = \left(\sqrt{\frac{\omega^2}{c^2} - \kappa^2 - i2\frac{\varepsilon}{c}\omega} - \frac{\omega}{c}\right) + \frac{\omega}{c}.\tag{2.56}$$

The first term in parenthesis is a bounded function of ω and may be called the *regular part* of the DtN-map kernel. The second term ω/c is unbounded as $\omega \to \infty$ and is called the *singular part* of the DtN-map kernel.

With the inverse Fourier transform the kernel is mapped in the time domain. The absorbing boundary condition of the nondissipative restrained rod is

$$u_{,x}(t, L) = -\int_{-\infty}^{\infty} \left(\kappa \frac{H(t-\tau)J_1(\kappa c(t-\tau))}{t-\tau} + \frac{1}{c}\delta'(t-\tau) \right) u(\tau, L)d\tau, \qquad (2.57)$$

where $\delta'(t)$ is the first derivative of Dirac's delta function, $H(t-\tau)$ is the Heaviside step function and $J_1(t)$ is the Bessel function of first kind and first order. Considering the initial conditions (2.55) equation (2.57) becomes

$$u_{,x}(t, L) = -\kappa \int_0^t \frac{J_1(\kappa c(t-\tau))}{t-\tau} u(\tau, L)d\tau - \frac{1}{c}\dot{u}(t, L). \qquad (2.58)$$

The left hand side has not exactly the form stated in (2.20) because in (2.58) the singular term of $g(\omega)$ for $\omega \to \infty$ has been extracted from the convolution integral kernel. This term induces a damping matrix C with only one nonzero component in the diagonal entry associated with the degree of freedom $u(t, L)$. Combining (2.58) with (2.54) yields

$$M\ddot{u}(t) + C\dot{u}(t) + Ku(t) + h_R(t) = f(t). \qquad (2.59)$$

$h_R(t)$ is a vector which contains a unique non-zero entry at the degree of freedom $u(t, L)$. This entry is given by the convolution integral

$$h_R(t) = \kappa \int_0^t \frac{J_1(\kappa c(t-\tau))}{t-\tau} u(\tau, L)d\tau. \qquad (2.60)$$

One observes that this term arises from the fact, that the medium is dispersive. A nondispersive medium would just induce a damping matrix C considerably simplifying the IVP. Combining (2.59) with (2.60) we obtain

$$M\ddot{u}(t) + C\dot{u}(t) + Ku(t) + \kappa \int_0^t \frac{J_1(\kappa c(t-\tau))}{t-\tau} u(\tau, L)d\tau = f(t) \qquad (2.61)$$

Equation (2.61) is called a Volterra integro-differential equation of convolution or hereditary type. Because of the convolution integral its solution depends on the past response history of the degree of freedom $u(t, L)$.

The convolution integral introduces the nonlocality in time. This fact has far-reaching consequences for the numerical solution of (2.59). Usually, an ODE-system is solved applying a numerical step by step integration scheme. Such an algorithm needs usually only very little information of the past history of the ODE-system. In the worst case, when a multistep algorithm is used, we have to store the data describing the state variables of the last few steps. Hence, the storage demand is limited and independent of time. In addition, the number of operations is invariant at each integration step.

This useful situation changes substantially when a convolution integral like (2.60) has to be evaluated. Its update at every incremental time step requires the computation of the past time history. Therefore, the numerical evaluation of convolution integrals demands additional memory resources and CPU-time. Its memory demand increases linearly with the number of time steps n_t. The number of operations grows proportional to n_t^2. This fact becomes really dramatic when the model has a large boundary. Because DtN-maps are nonlocal in space, all degrees of freedom are coupled together. Assuming n_Γ degrees of freedom the memory demand is of order $n_\Gamma^2 n_t$ and the CPU-time demand is of order $n_\Gamma^2 n_t^2$. Obviously, for large models, the demands on memory resources and CPU-time become prohibitive. This is why local absorbing boundaries are preferred in numerical analysis. The aim of this work is to develop a method which allows to avoid the direct computation of convolution integrals.

2.2 Absorbing boundary conditions

2.2.1 Asymptotic boundary conditions

In this section we shall derive several absorbing boundary conditions using the technique proposed by Engquist and Majda [EM77] which is based on the rational approximation of pseudo-differential operators. A precise definition of pseudo-differential operators is extremely involved and cannot be given in this context (see [Won91] for an introduction). Therefore, we give here an intuitive and necessarily imprecise description. In section 2.1.3 we found that the DtN-map of the nondissipative restrained rod is given by

$$u_{,x} + ik(\omega)u = u_{,x} + \sqrt{\kappa^2 - \left(\frac{\omega}{c}\right)^2} u = 0. \tag{2.62}$$

Instead of interpreting (2.62) as a boundary condition, we may consider it as an equation of motion being defined in the frequency domain. In fact, replacing $i\omega$ by the differential operator $\partial/\partial t$ we obtain the formal operator

$$\left(\frac{\partial}{\partial x} + \sqrt{\kappa^2 + \frac{1}{c^2}\frac{\partial^2}{\partial t^2}}\right)u = 0. \tag{2.63}$$

Although it looks very strange, the expression inside the brackets has a well defined meaning and can be considered as a sort of differential operator acting on the variable $u(x, t)$. It is a so-called pseudo-differential operator. Equation (2.63) obtains a precise meaning in the context of Fourier transform analysis. Nevertheless, in the time domain, the operator (2.63) is equivalent to the integro-differential equation

$$u_{,x} + \frac{1}{c}u_{,t} + \frac{1}{\kappa c}\int_0^t \frac{J_1(\kappa c(t-y))}{t-y}u(y)dy = 0. \tag{2.64}$$

It describes the propagation of waves travelling only in a positive direction. It is a so-called forward one-way wave equation. Hence, pseudo-differential operators may be used to analyze the solutions of one-way wave equations.

The technique of Engquist and Majda involves constructing Padé approximations of the DtN-map kernel $ik(\omega)$ described by (2.62). The point of approximation is $\omega \to \infty$. Padé approximations are rational functions. The coefficients of their Taylor series agree up to a defined degree with those of the Taylor series expansion of the function being approximated (see [BO78] or [BG96] for further details). Before applying the Taylor series expansion technique we rewrite (2.62) in the form

$$ik = \frac{i\omega}{c}\sqrt{1 + \left(\frac{\kappa c}{i\omega}\right)^2} = \frac{i\omega}{c}\sqrt{1 + (\kappa z)^2}\,, \qquad (2.65)$$

where $z = c/i\omega$. Now we can compute the Taylor series expansion of the root in (2.65) at $z = 0$. This is equivalent to an asymptotic expansion at $\omega \to \infty$.

The first absorbing boundary condition is obtained keeping only the first term of the Taylor series expansion of (2.65) which is of order $O(1)$. This yields

$$u_{,x} + i\frac{\omega}{c}u = 0. \qquad (2.66)$$

Using the relations $ik \leftrightarrow \partial/(\partial x)$ and $i\omega \leftrightarrow \partial/(\partial t)$ we obtain the associated time-domain boundary condition

$$u_{,x} + \frac{1}{c}u_{,t} = 0. \qquad (2.67)$$

The last equation is the nondispersive one-way wave equation. Interpreting (2.67) as a boundary condition we multiply it with Young's modulus $E = \rho c^2$ and obtain:

$$Eu_{,x} = \sigma = -\rho c u_{,t}. \qquad (2.68)$$

The physical interpretation of this boundary condition is a viscous damper. That is, the stress induced by the boundary condition is proportional to the time derivative of the deflection at the boundary. (2.68) is called a viscous boundary condition and was first introduced by Lysmer and Kuhlemeyer [LK69].

The next two absorbing boundary conditions are defined by the [1,0] and [1,2] Padé approximants of the DtN-map kernel (2.65), where the symbol [m,n] means that the Padé approximants has a numerator polynomial of degree m and a denominator polynomial of degree n. This yields the equations

$$i\omega u_{,x} + \frac{(i\omega)^2}{c}u + \frac{c}{2}\kappa^2 u = 0, \qquad (2.69.\text{a})$$

$$(i\omega)^2 u_{,x} + \frac{(\kappa c)^2}{4}u_{,x} + \frac{(i\omega)^3}{c}u + \frac{3}{4}(c\kappa)^2 i\omega u = 0. \qquad (2.69.\text{b})$$

The relations $ik \leftrightarrow \partial/\partial x$ and $i\omega \leftrightarrow \partial/\partial t$ allow us to transform of (2.69.a) and (2.69.b) in the time domain

$$u_{,xt} + \frac{1}{c}u_{,tt} + \frac{c}{2}\kappa^2 u = 0, \qquad (2.70.\text{a})$$

$$u_{,xtt} + \frac{\kappa^2 c^2}{4}u_{,x} + \frac{1}{c}u_{,ttt} + \frac{3}{4}c\kappa^2 u_{,t} = 0. \qquad (2.70.\text{b})$$

We will call these boundary conditions asymptotic boundary conditions of first and second order.

Figure 2.7 shows how these boundary conditions approximate the DtN-map. For better illustration, the function $f(\omega) = g(\omega) - i\omega$ has been used instead of kernel function $g(\omega)$. The viscous boundary condition has been omitted since $f(\omega) = g(\omega) - i\omega = i\omega - i\omega = 0$. The approximations are particularly good for large frequencies. This is because the approximation scheme is based on the asymptotic expansion for $\omega \to \infty$. In contrast, the approximations completely neglect the non-zero real part of the DtN-map kernel for low frequencies. This is due to

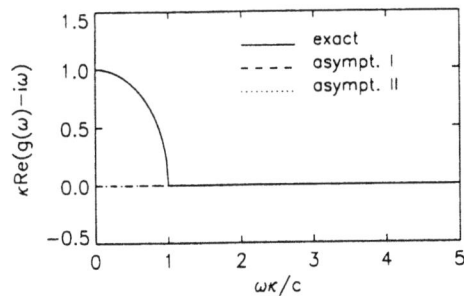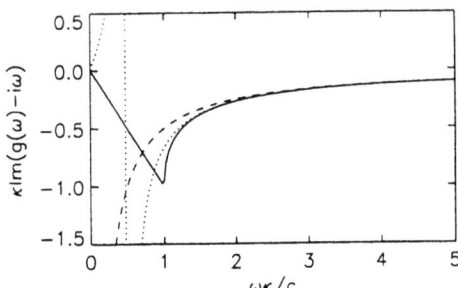

Figure 2.7: Real and imaginary part of the kernels of the asymptotic boundary conditions

the fact, that the convergence radius of the asymptotic expansion is limited by the singularity of the DtN-map kernel at the cut-off frequency. Moreover, the asymptotic absorbing boundary conditions of first and second order exhibit singularities below the cut-off frequency.

Combining the equation of motion (2.3) with (2.70.a) and (2.70.b) we obtain a more concise form of these boundary conditions

$$\left(\frac{\partial}{\partial x}+\frac{1}{c}\frac{\partial}{\partial t}\right)u = 0 \qquad \left(\frac{\partial}{\partial x}+\frac{1}{c}\frac{\partial}{\partial t}\right)^2 u = 0 \qquad \left(\frac{\partial}{\partial x}+\frac{1}{c}\frac{\partial}{\partial t}\right)^3 u = 0 . \qquad (2.71)$$

This property turns out to be general so that an asymptotic boundary condition of order N is generated by applying N times the differential operator:

$$\left(\frac{\partial}{\partial x}+\frac{1}{c}\frac{\partial}{\partial t}\right). \qquad (2.72)$$

The operator (2.72) has an interesting property. The Laplace transform of the Green's function of the equation of motion (2.3) describing a forward propagating wave is

$$u(x, s) = \frac{1}{k(s)}e^{-k(s)x+st} \quad \text{with} \quad k(s) = \sqrt{(s/c)^2+\kappa^2}. \qquad (2.73)$$

The dominant term of the asymptotic expansion of the amplitude for $s \to \infty$ is of the order $O(1/s)$. The application of the operator (2.72) to (2.73) deletes this dominant term. Moreover, after N applications of the operator (2.72) to (2.73) the dominant term is of the order $O(1/s^{N+1})$. Therefore, the operator (2.72) acts as a high-frequency annihilator. Because of the initial value theorem for Laplace transforms we conclude that asymptotic boundary conditions will accurately describe the short-time solution but fails to capture the long-time one.

Remarks:

• Observe that it is because of the rational approximation that we can use the relations $ik \leftrightarrow \partial/(\partial x)$ and $i\omega \leftrightarrow \partial/(\partial t)$ to define the absorbing boundary conditions in the time domain.

• The relations (2.71) were generalized by Higdon [Hig86b] defining the operator

$$\prod_{k=1}^{N}\left(\frac{\partial}{\partial x}+\frac{\alpha_k}{c}\frac{\partial}{\partial t}\right) = 0 . \qquad (2.74)$$

These absorbing boundary conditions are called multidirectional boundary conditions.

• The asymptotic boundary conditions don't allow such a nice intuitive interpretation like the viscous boundary condition.

- In engineering numerical analysis, the asymptotic boundary conditions are sometimes referred to as paraxial boundary conditions. This label stems from Claerbout [Cla70] who developed a theory to derive one-way wave equations in the context of seismic wave propagation problems. The central idea is to approximate a second order differential operator by a product of two operators, one describing a forward and the other a backward one-way wave equation.

2.2.2 Doubly-asymptotic boundary conditions

The doubly-asymptotic boundary condition is a generalization of the viscous boundary condition. It was proposed by Geers [Gee78] for fluids. Later, it was extended to elastic wave propagation problems [UG80]. Originally, it was developed to increase the accuracy of wave propagation problems with evanescent waves solutions. The doubly-asymptotic boundary condition allows to capture exact constant solutions as $t \to \infty$.

The classical doubly-asymptotic boundary condition is obtained by adding the dominant terms of the asymptotic expansions of a DtN-map at low ($\omega \to 0$) and high frequencies ($\omega \to \infty$). With (2.65) this approach yields the boundary condition

$$u_{,x} + \left(i\frac{\omega}{c} + \kappa \right) u = 0 \,. \tag{2.75}$$

Transformed to the time domain this yields

$$u_{,x} + \frac{1}{c} u_{,t} + \kappa u = 0. \tag{2.76}$$

The same result is obtained using the technique of Claerbout [Cla70] briefly discussed before. Assume we try to approximate the differential operator with a product of differential operators defining a forward and a backward one-way wave equation, respectively

$$\left(\frac{\partial}{\partial x} + \frac{1}{c}\frac{\partial}{\partial t} + \alpha \right)\left(\frac{\partial}{\partial x} - \frac{1}{c}\frac{\partial}{\partial t} + \beta \right) u = 0 \,, \tag{2.77}$$

where the coefficients α and β have to be defined accordingly. Subtracting the last equation from the equation of motion (2.3) gives

$$-(\kappa^2 + \alpha\beta)u - \left(2\frac{\varepsilon}{c} - \frac{\alpha - \beta}{c} \right)u_{,t} - (\alpha + \beta)u_{,x} = 0 \,. \tag{2.78}$$

Now we may get rid of the terms containing $u_{,x}$ and u imposing $\alpha + \beta = 0$ and $\kappa^2 + \alpha\beta = 0$. This yields the solution $\alpha = -\beta = \kappa$ and the forward wave equation is thus exactly (2.76). It is easy to see that getting rid of the terms containing $u_{,t}$ and u results in complex coefficients α and β provided $\varepsilon < \kappa$. Therefore, this alternative has to be discarded.

Similar to the viscous boundary condition, the doubly-asymptotic boundary condition has a nice physical interpretation. Expressed as a stress equilibrium condition it is given by

$$\sigma + \rho c u_{,t} + \rho c^2 \kappa u = 0 \,. \tag{2.79}$$

The second term is the dashpot of the viscous boundary condition. The third term corresponds to a linear spring. According to the analysis of the energy flux given in section 2.1.5 the linear spring doesn't contribute to the dissipation of energy at the boundary. Notice that for $\varepsilon = \kappa$ the term with $u_{,t}$ vanishes and (2.77) becomes an exact factorization. Hence, the doubly-asymptotic boundary condition is an exact boundary condition of the dissipative restrained rod with $\varepsilon = \kappa$.

An improved doubly-asymptotic boundary condition has been proposed by Engquist and Halpern [EH88]. It is given by

$$u_{,xt} + \frac{1}{c} u_{,tt} + c\kappa u_{,x} + \kappa u_{,t} + c\kappa^2 u = 0. \tag{2.80}$$

They obtain it in analogy to the operator product description of the asymptotic boundary conditions given in (2.71)

$$\left(\frac{\partial}{\partial x} + \frac{1}{c} \frac{\partial}{\partial t} \right) \left(\frac{\partial}{\partial x} + \kappa \right) u = 0. \tag{2.81}$$

The first operator is a high frequency annihilator while second is a low frequency annihilator. Compared to the doubly-asymptotic boundary condition, (2.81) contains an additional rational term which balances the term κu for $\omega \to \infty$.

$$u_{,x} + \left(i\frac{\omega}{c} + \kappa \right) u - \frac{\kappa i \omega}{i\omega + c\kappa} u = 0. \tag{2.82}$$

Further analysis will show, that this additional term has important consequences on the accuracy of the absorbing boundary condition.

Figure 2.8 shows how the doubly asymptotic and the Engquist-Halpern boundary conditions approximate the DtN-map. As in Figure 2.7 we have plotted the function $f(\omega) = g(\omega) - i\omega$ instead of the kernel $g(\omega)$. We note that the kernels have a nonvanishing real part. Although still very rough, the best approximation of the real part is provided by the Engquist-Halpern boundary condition. In addition, it has no poles on the imaginary axis. Observe, that both boundary conditions are exact at $\omega = 0$. The imaginary part of the DtN-map kernel is approximated less accurately than with the viscous and the asymptotic boundary conditions.

2.2.3   The reflection coefficient

The study of the reflection coefficient of the steady state solution of a finite model is the common method to analyze the efficiency of absorbing boundary conditions. The reflection coefficient is defined to be the ratio of the amplitude of the reflected to the amplitude of the incident wave. Consider the steady state solution

$$u(t, x) = A e^{-ik(\omega)x + i\omega t} + B e^{ik(\omega)x + i\omega t}, \tag{2.83}$$

then the reflection coefficient is defined as

$$r(\omega) = \frac{B}{A}. \tag{2.84}$$

Generally, the reflection coefficient is complex. However, when studying absorbing boundary conditions the main interest is focused on the magnitude of the reflected wave and not on its phase. Hence, we shall consider the absolute value or the modulus of the reflection coefficient

$$|r(\omega)| = \frac{|B|}{|A|}. \tag{2.85}$$

In this work, when we speak of the reflection coefficient we always refer to its absolute value.

Let us assume that the absorbing boundary condition is given by $u_{,x} + g(\omega)u = 0$. The reflection coefficient is obtained inserting the steady state solution in the boundary condition

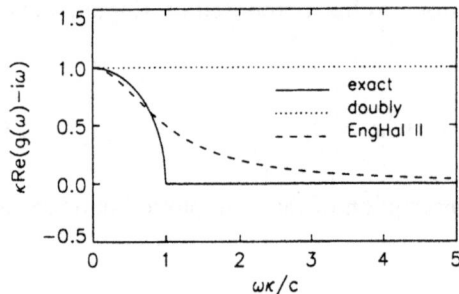 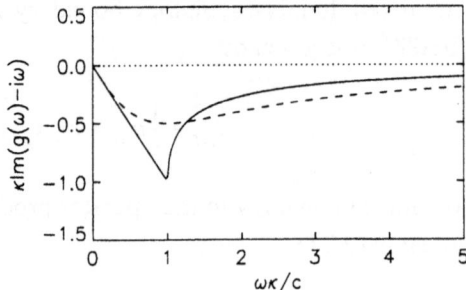

Figure 2.8: Real and imaginary part of the kernels of the doubly-asymptotic and Engquist-Halpern boundary conditions

$$|r(\omega)| = \frac{|ik(\omega) - g(\omega)|}{|ik(\omega) + g(\omega)|}. \tag{2.86}$$

The values of the reflection coefficient range from $r(\omega) = 0$ (total absorption) to $|r(\omega)| = 1$ (total reflection). From its very definition, an absorbing boundary condition is efficient if it is close to complete absorption in a large frequency interval.

Figure 2.9 (left) shows the reflection coefficient $|r(\omega)|$ of the viscous and of the asymptotic boundary conditions discussed in section 2.2.1. We note immediately that all absorbing boundary conditions are totally reflecting below the cut-off frequency. Beyond the cut-off frequency, the efficiency improves dramatically. The reflection coefficient goes quickly from complete reflection at the cut-off frequency to complete absorption. This general behavior is not really surprising since the associated kernels $g(\omega)$ approximate the DtN-map kernel with increasing accuracy as the frequency increases (see Figure 2.7). This extremely poor performance below the cut-off frequency is because these boundary conditions have zero real part so that they capture mainly the transient portion of the wave propagation process. The nonpropagating wave mode at low frequencies is completely neglected.

A much better efficiency below the cut-off frequency shows the doubly-asymptotic boundary condition as is shown in Figure 2.9 (right). This is due to the "spring element" κu. In fact, this boundary condition is by definition exact at $\omega = 0$ (static case). However, its effectiveness decreases rapidly with increasing frequency. Above the cut-off frequency, the doubly-asymptotic boundary condition shows the worst performance of all absorbing boundary conditions discussed so far. Apparently, the spring element, responsible for the improvement of the accuracy below the cut-off frequency, reduces accuracy above the cut-off frequency. The reason of this bad performance will become evident when we discuss the approximation of the dissipative rod. The Engquist-Halpern boundary condition (2.80) is much more effective below as well as above the cut-off frequency, where it behaves like the viscous boundary condition. This improvement is due to the better approximation of the Engquist-Halpern boundary condition as can be seen in Figure 2.8.

Remarks:

- The singularities of the asymptotic boundary conditions don't emerge in the reflection coefficient. This is because the zeroes and poles of the kernels are mapped to $r(\omega) = 1$.

- All absorbing boundary conditions are totally reflecting at the cut-off frequency. This fact is easily deduced from (2.86) since for all boundary conditions we have $ik(\omega_c) = 0$. Therefore the reflection coefficient becomes $r(\omega_c) = g(\omega_c)/g(\omega_c) = 1$.

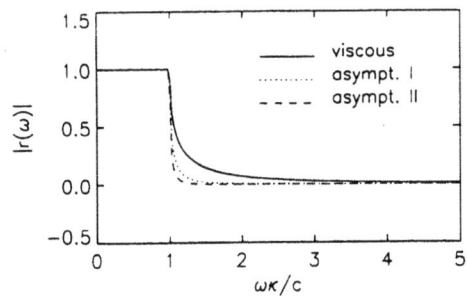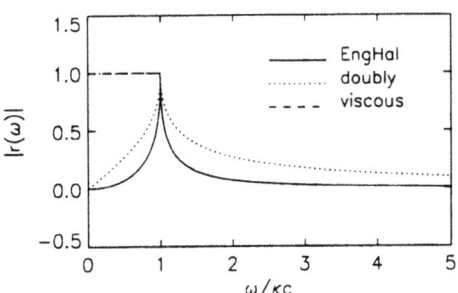

Figure 2.9: Reflection coefficients for the viscous and the asymptotic boundary conditions

2.2.4 Increasing the size of the near field

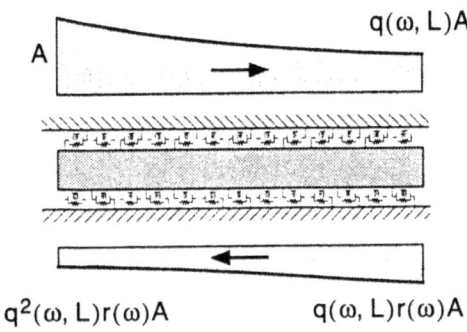

Figure 2.10: Reflection of a non-propagating wave at the right boundary

The extremely poor absorption performance of the viscous and the asymptotic boundary conditions below the cut-off frequency is definitely unsatisfactory. Apparently, there is no simple method to increase their efficiency in this frequency range without losing the accuracy above the cut-off frequency, as is demonstrated by the doubly-asymptotic and the Engquist-Halpern boundary conditions. But there is still another method, which is frequently used in practice, to obtain a significant improvement below the cut-off frequency.

Consider a finite restrained rod of length L (see Figure 2.10). According to the analysis performed in section 2.1.2, the rod shows a decaying non-propagating mode in the frequency range below the cut-off frequency. Thus, the amplitude of the deflection measured at the left boundary is reduced at the right boundary according to a specific decay law to qA, where A is the amplitude of the harmonic wave at the left boundary and q stands for the decay factor $q(\omega, L)$. This decay factor is positive and smaller or equal to unity and is a function of the circular frequency and of the length L of the rod. Hence, at the right boundary only the amplitude qA is partially reflected according to the absolute value of the reflection coefficient, that is $q|r(\omega)|A$. On its way back to the left boundary, the amplitude of the reflected wave decays by exactly the same decay factor q. Therefore, the amplitude of the reflected wave is reduced to $q^2|r(\omega)|A$ when it hits the left boundary again. Hence, measuring the reflection coefficient at the left end of the rod we obtain the generalized reflection coefficient

$$r_L(\omega) = q^2(\omega, L)r(\omega). \qquad (2.87)$$

The generalized reflection coefficient doesn't depend of the frequency alone but also of the length of the rod. The section of the rod at the left boundary is called the reference section. Observe, that by this definition the portion of the rod on the right hand side of the reference section is a part of the absorbing boundary condition. In this work, we shall call these types of boundary conditions generalized absorbing boundary conditions. From their very definition we obtain $q(\omega, 0) = 1$ and thus $r_L(\omega, 0) = r(\omega)$. Inspecting (2.27) we can identify $q(\omega, L)$ with the absolute value of the exponential term $\exp(-ikL)$. It is unity at $L = 0$ and decreases continuously with increasing distance of the reference section from the boundary. Above the cut-off frequency the absolute value of $\exp(ikL)$ is unity because its argument is purely imaginary so that the generalized reflection coefficient is identical to the reflection coefficient. By this approach, we obtain a consistent improvement of the absorbing boundary condition below the cut-off frequency.

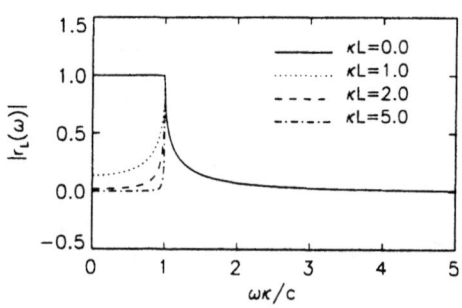

Figure 2.11: Generalized reflection coefficient of the viscous boundary condition

The generalized reflection coefficient of the viscous boundary condition is shown in Figure 2.11. A comparison with Figure 2.9 shows a drastic increase of the absorption below the cut-off frequency. The reflection coefficient can be made arbitrarily small by increasing the distance between the reference section and the boundary. Observe, that the absorption characteristics remain unaltered above the cut-off frequency. The part of the DtN-map below the cut-off frequency is approximated by the finite portion of the rod. Above the cut-off frequency, the approximation is performed by the original absorbing boundary conditions.

We study further the approximation properties of the generalized boundary conditions and define the DtN-map kernel of the finite model with absorbing boundary condition with respect to the reference section. According to (2.26) the displacement of the finite model is given by

$$u(x, \omega) = A(\omega)(e^{-ikx} + q(\omega)e^{ikx}) \tag{2.88}$$

with $q(\omega)$ defined by (2.27). Because the DtN-map kernel $ik(\omega)$ is associated with the displacement $v(x, \omega)$ by $v_{,x}(x, \omega) = -ik(\omega)v(x, \omega)$, we define the DtN-map kernel of the finite model $g_L(\omega)$ as

$$g_L(x, \omega) = -\frac{u_{,x}(x, t)}{u(x, t)} , \tag{2.89}$$

where $u(x, \omega)$ is the displacement of the finite rod given by (2.88). Evaluating (2.89) at the reference section ($x = 0$) we obtain

$$g_L(\omega) = ik(\omega)\frac{1 - q(\omega)}{1 + q(\omega)} . \tag{2.90}$$

Obviously the kernel $g_L(x, \omega)$ is a good approximation of the DtN-map kernel $ik(\omega)$ if the reflection coefficient $q(\omega)$ is small compared to unity.

Figure 2.12 shows the DtN-map kernels $g_L(x, \omega)$ of the finite model with the viscous boundary condition. We note that increasing the length L of the model, the kernel $g_L(\omega)$ gets closer and closer to the DtN-map kernel below the cut-off frequency. Above the cut-off frequency, $g_L(\omega)$ oscillates around the DtN-map kernel with periods depending on the length L of the model. The amplitude of the oscillation decays with increasing frequency. Hence, above the cut-off frequency the accuracy doesn't increase because it is determined by the properties of viscous boundary condition.

The analogous plot with the asymptotic boundary condition of second order is given in Figure 2.13. An excellent approximation is realized with a dimensionless length of five units. The approximation above the cut-off frequency is remarkably good. Observe, that the influence of the singularity at $\omega\kappa/c = 1/2$ is clearly visible for a dimensionless length of $\kappa L = 1$. The perturbation decays rapidly when the length of the finite model is increased.

The less efficient performance of the doubly asymptotic boundary condition above the cut-off frequency is shown in Figure 2.14. The amplitude of the oscillations is much larger than that of the finite model with viscous boundary condition and doesn't decay with increasing frequency. A detailed asymptotic analysis shows that the amplitude decay of the finite model with a viscous boundary condition is proportional to $1/\omega$ whereas the doubly-asymptotic boundary condition

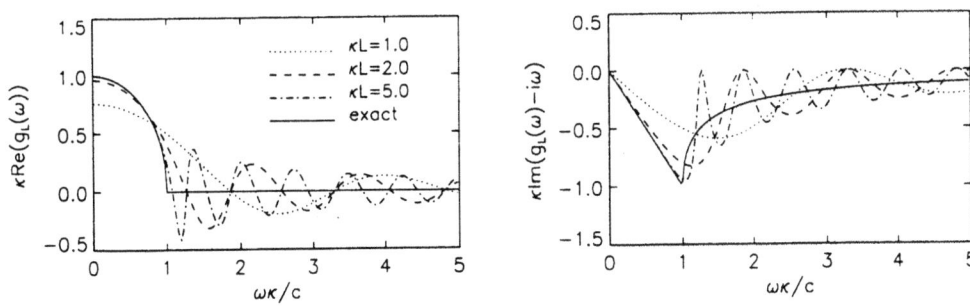

Figure 2.12: DtN-map kernels of the finite model with the viscous boundary condition

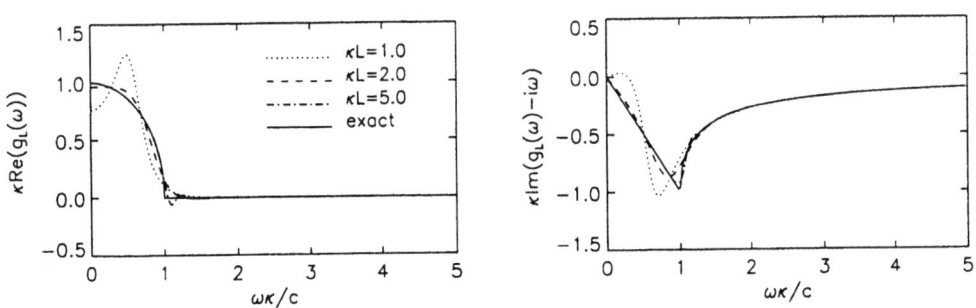

Figure 2.13: DtN-map kernels of the finite model with the asymptotic boundary condition of second order

exhibits constant amplitude κ. As we shall see, this behavior is related to the "spring" term proportional to κ.

2.2.5 Mean energy flux and dissipativity

Next, we shall analyze in detail the mean energy flux in the finite rod with different absorbing boundary conditions. In section 2.1.5 we found a simple relation for the mean energy flux of harmonic waves

$$S_m \sim Im(u^* u_{,x}). \qquad (2.91)$$

When $u_{,x}$ is replaced by the absorbing boundary condition we obtain

$$S_m \sim Im(g(\omega))|u(L, \omega)|^2. \qquad (2.92)$$

We call an absorbing boundary condition with $Im(g(\omega)) > 0$ for every $\omega > 0$ dissipative. As is shown in Figure 2.15, the viscous and the doubly-asymptotic boundary conditions are dissipative because both have $Im(g(\omega)) = i\omega$. The dissipativity of these boundary conditions can be shown directly in the time domain. Indeed, combining (2.37) with (2.67) we obtain for the viscous boundary condition the instantaneous energy flux S

$$S = EAu_{,t}u_{,x} = \rho cA(u_{,t})^2 \geq 0. \qquad (2.93)$$

Hence, the viscous boundary condition is dissipative irrespective of the time history of the displacement. To show the dissipativity of the doubly-asymptotic boundary condition is somewhat more intricate because the linear spring element allows to store energy. When this is released the instantaneous energy flux S may thus become negative. However, the integral of the instantaneous energy flux over a finite time interval shows that the boundary condition is dissipative

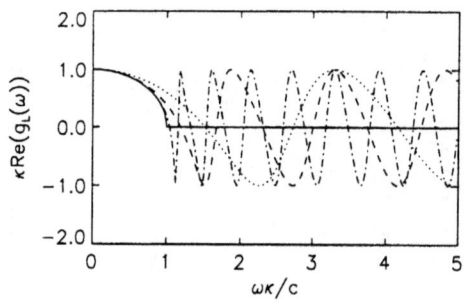 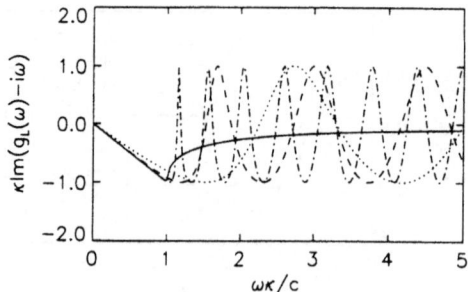

Figure 2.14: DtN-map kernels of the finite model with the doubly-asymptotic boundary condition

$$\int S dt = A \int \left(\rho c (u_{,t})^2 + \frac{\rho c^2 \kappa}{2} (u^2)_{,t} \right) dt = A \int \rho c (u_{,t})^2 dt + \frac{\rho c^2 \kappa}{2} A u^2 \ge 0. \qquad (2.94)$$

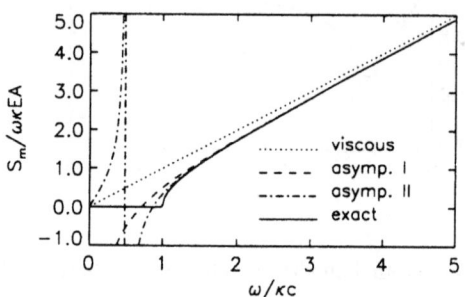

Figure 2.15: Mean energy fluxes of several absorbing boundary conditions

In contrast, the asymptotic boundary conditions of first and second order aren't dissipative (see Figure 2.15). In fact, below the cut-off frequency, the mean energy flux can become negative. That is, at these frequencies the energy flows from right to left. Hence, the asymptotic boundary conditions produce energy instead of absorbing it. However, as we will see in a forthcoming section, a negative mean energy flux does not imply that a correspondingly initial boundary value problem is necessarily ill-posed. Above the cut-off frequency, the mean energy fluxes associated with the asymptotic absorbing boundary conditions are positive and approach quickly the exact mean energy flux of the DtN-map.

Generalized absorbing boundary conditions with viscous and doubly-asymptotic boundary conditions are dissipative too (see Figure 2.16). The mean energy fluxes with respect to the reference section have been determined assuming a solution in the form of an incoming and a reflected harmonic wave:

$$u(x, t) = u_0 (e^{-ik(\omega)x} + q(\omega) e^{ik(\omega)x}) e^{i\omega t}, \qquad (2.95)$$

where $|u_0| = 1$. The mean energy flux computed with a finite model of length $\kappa L = 1$ is shown in Figure 2.16 (left). Comparing these mean energy fluxes with those plotted in Figure 2.15 we observe that the amplitude of the dissipation below the cut-off frequency is substantially reduced. The frequency range with a negative mean energy flux for the asymptotic boundary conditions is still there. However, increasing the length of the rod L improves the mean energy fluxes as is confirmed by Figure 2.16 right. With $\kappa L = 5$ the agreement between the mean energy fluxes of the generalized absorbing boundary conditions and the exact mean energy flux is excellent.

Remarks:

• Because of the steady state solution and because the rod is nondissipative every section of the rod has equal energy flux.

• The energy flux of finite models with asymptotic boundary conditions of order one and two are zero at the point of singularity of this boundary conditions. At these points, the only possible solution at $x = L$ is $u(L, \omega) = 0$. Otherwise, an infinitely strong force would act on the rod at $x = L$.

- Observe, that for a finite model with viscous boundary condition the mean energy flux above the cut-off frequency may be smaller than the exact mean energy flux.

2.2.6 The dissipative restrained rod

Similar to the nondissipative rod, the starting point for constructing absorbing boundary conditions of dissipative rods is the asymptotic expansion of the DtN-map at $s \to \infty$. This is given up to the fourth term by

$$\sqrt{(s/c)^2 + 2(\varepsilon/c)s + \kappa^2} \approx \frac{s}{c} + \varepsilon + \frac{c(\kappa^2 - \varepsilon^2)}{2} \frac{1}{s} - \frac{\varepsilon c(\kappa^2 - \varepsilon^2)}{2} \frac{1}{s^2} + O\!\left(\frac{1}{s^3}\right). \tag{2.96}$$

The major difference to the asymptotic expansion of the nondissipative rod is that terms of even order appear in (2.96). They are proportional to the normalized dissipation parameter ε. In addition, the odd order terms are modified by powers of ε.

We just start constructing the family of asymptotic boundary conditions of the dissipative rod. Since they are assumed to be a generalization of those of the nondissipative rod, we give them the same label provided they are of the same order in s and we obtain those derived in section 2.2.1 by setting $\varepsilon = 0$. The asymptotic boundary condition of first order is identical to the viscous boundary condition. The next higher absorbing boundary condition is constructed including the first three terms of the asymptotic expansion:

$$u_{,x} + \left(\frac{s}{c} + \varepsilon + \frac{c(\kappa^2 - \varepsilon^2)}{2}\frac{1}{s}\right)u = 0 \quad . \tag{2.97}$$

Observe, that the term εu appears in (2.97). Similar to the doubly-asymptotic boundary condition, it can be interpreted as a linear "spring element". However, the spring parameter ε is related to the dissipation of the rod and not to its properties under static loads as is suggested by the parameter κ. Hence, at high frequencies, the spring term of the doubly-asymptotic boundary condition is interpreted as originated by a dissipation term $\varepsilon = \kappa/c$ in the original equation of motion of the rod. Obviously, in a nondissipative rod this term doesn't exist. That's why the doubly-asymptotic boundary condition performs badly above the cut-off frequency. Hence, the doubly-asymptotic boundary condition is asymptotically inconsistent of order $O(1)$ for the limit $s \to \infty$. Observe, that the doubly-asymptotic boundary condition is a DtN-map of a dissipative restrained rod with the dissipation parameter $\varepsilon = \kappa/c$.

The reflection coefficients of the asymptotic boundary condition (2.97) are plotted in Figure 2.16. The qualitative behavior of the reflection coefficients is the same as that of the nondissipative rod discussed in section 2.2.5: poor efficiency below and good performance above the cut-off frequency. However, increasing the dissipation improves the efficiency below the cut-off frequency as is shown in the right plot in Figure 2.17 for a finite model of length $\kappa L = 2$. Clearly, increasing the length L, the dissipation in the rod reduces the amplitude of the reflected wave.

The behavior of the mean energy fluxes are qualitatively similar to those of the nondissipative rod as is confirmed by Figure 2.18. The mean energy flux of the viscous boundary conditions is positive and therefore dissipative. It oscillates around the exact mean energy flux. The second asymptotic boundary condition (2.97) show a negative mean energy flux in a considerable frequency range below the cut-off frequency. However, if the dimensionless length is increased to $\kappa L = 2$, the material damping in the rod is already large enough to compensate the energy creation of the boundary condition.

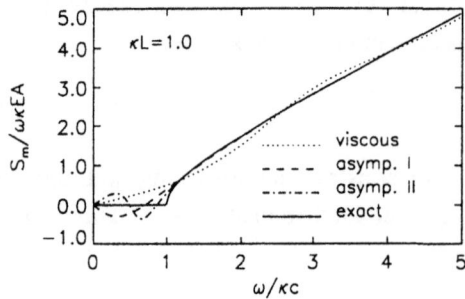 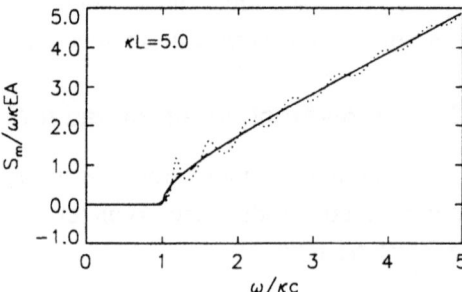

Figure 2.16: Mean energy fluxes of several generalized absorbing boundary conditions

2.2.7 Well-posedness of initial boundary value problems

Consider a finite model of an infinite restrained rod with absorbing boundary conditions. Assume that we wish to find a solution for given initial conditions and driving forces. If such a solution exists, then the initial boundary value problem is said to be well-posed. This means that

    a) a solution exists

    b) the solution is unique

    c) the solution is stable for small perturbations of the set of input data.

The first two conditions are self-explanatory. Of major concern is the third condition. It is closely related to the concept of stability.

The problem of well-posedness of IBVP is a classical problem of the theory of PDE's and has been studied in depth. The usual method to test for well-posedness is to construct an estimate of the solution of the IBVP in terms of the initial conditions and forcing functions with respect to an adequate norm. This implies that the solution of the IBVP depends continuously on the input data. On the other hand, Hersh ([Her63], [Her64]) developed an algebraic method for linear mixed initial boundary value problems governed by linear PDE. Consider the multidimensional domain $t > 0$, $x > 0$ and $-\infty < y_j < \infty$ (the hyperplane built up by the coordinates y_j is perpendicular to x). According to this theory, IBVP are well-posed if all solutions of the IBVP of the form

$$u(t, x) = F(x)e^{-kx + st + i(\eta \cdot y)} \tag{2.98}$$

satisfying the boundary condition

$$B(\partial_x, \partial_t, \partial_{y_j})u(t, x, y_j)\big|_{x=0} = 0. \tag{2.99}$$

are bounded. That is, we have no solution with $Re(k) > 0$ and $Re(s) \geq 0$ The function $F(x)$ grows at most like a polynomial in x for $x \to \infty$. PDE with constant coefficients have $F(x) = c$.

A similar criterion was derived by Kreiss [Kre70] for IBVP of hyperbolic systems with smoothly varying variable coefficients. The the so-called uniform Kreiss condition states that IBVP are well-posed if

$$|B(k, s, \eta_j)| \geq \delta > 0, \text{ for } Re(s) \geq 0 \tag{2.100}$$

holds, where $B(k, s, \eta_j)$ is obtained combining the phase function of (2.98) with the boundary condition (2.99). The well-posedness criterion can be summarized as follows: A one boundary problem with homogeneous boundary conditions is well-posed if and only if it doesn't admit solutions consisting entirely of waves travelling inward from the boundary. A multi-boundary problem, being the rule in applications, is well-posed if and only if each boundary condition satisfies the uniform Kreiss condition. This condition may be relaxed a little requiring only $Re(s) > 0$ in (2.100). In this case we say that the IBVP is weakly well-posed. The stronger con-

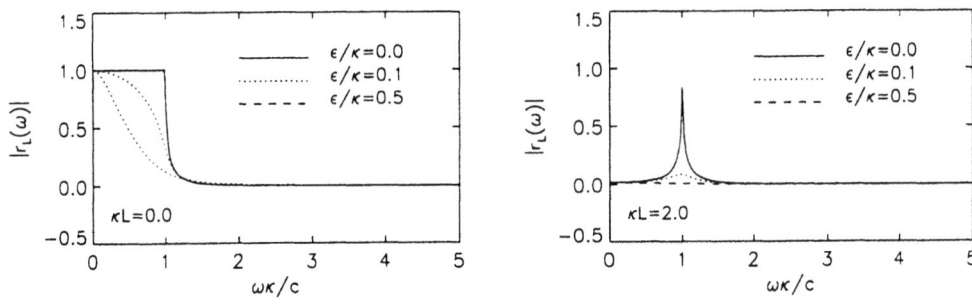

Figure 2.17: Generalized reflection coefficients of the asymptotic boundary condition (2.97)

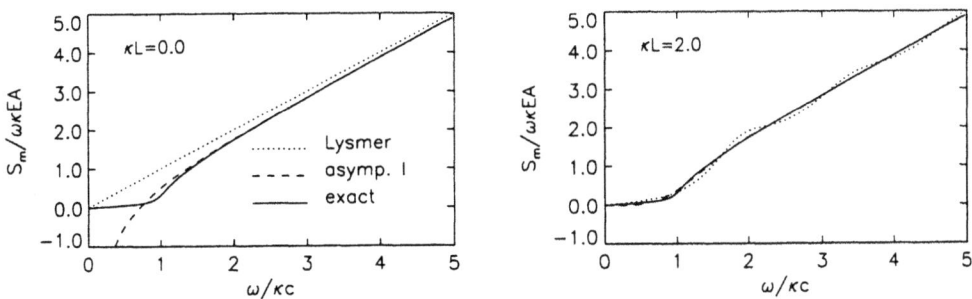

Figure 2.18: Average energy fluxes at the reference section

dition $Re(s) \geq 0$ leads to a strongly well-posed IBVP. A physical interpretation of this algebraic criterion in terms of group velocity was given by Higdon [Hig86b].

In a single space dimension, the uniform Kreiss condition simplifies to

$$|B(k, s)| \geq \delta > 0, \text{ for } Re(s) \geq 0 \tag{2.101}$$

A brief consideration gives a simple interpretation of the this condition. Assume that an IBVP is given by a linear PDE with constant coefficients and a nonhomogeneous boundary condition

$$B(\partial_x, \partial_t)u(t, x)\big|_{x=0} = h(t), \tag{2.102}$$

Suppose that $h(t)$ is an arbitrary function of time whose Laplace transform is analytic in the complex half-plane $Re(s) \geq 0$. We perform the Laplace transform of the PDE and the boundary condition and look for a solution which is bounded in the domain $x \geq 0$. It is given by $u(t, x) = a(s)e^{-k(s)x + st}$ with $Re(k(s)) > 0$. Inserting the last equation into (2.102) yields

$$a(s) = h(s)/B(s) . \tag{2.103}$$

Hence, the uniform Kreiss condition (2.101) implies that $a(s)$ is bounded in the half-plane $Re(s) \geq 0$. Since $h(s)$ is analytic in the complex half-plane $Re(s) \geq 0$ the well-posedness condition for the IBVP is equivalent to the requirement that $a(s)$ must be an analytic function in the half-plane $Re(s) \geq 0$. Weakly well-posedness means that $a(s)$ may have a singularity on the imaginary axis. For example a pole at $s = i\omega$. In this case the IBVP has a solution which is oscillatory for $t \to \infty$: $u(t, x) \sim a(i\omega)e^{-k(i\omega)x + i\omega t}$. Observe that for a pole at $\omega = 0$ we obtain for $t \to \infty$ the constant solution $u(t, x) \sim a(0)e^{-k(0)x}$. If $a(s)$ has a pole at s^* with $Re(s^*) > 0$ then the solution of the IBVP increases exponentially as time advances: $u(t, x) \sim a(s^*)e^{Re(s^*)t}$.

Remarks:

• Generally, it easier to show ill-posedness than well-posedness. Therefore, we look for solutions of $B(s) = 0$. If there exists a solution $s*$ with $Re(s*) > 0$ then according to the uniform Kreiss condition the IBVP is ill-posed.

- If the boundary conditions are defined by rational functions, the problem is reduced to that of finding the zeroes of a polynomial. If the polynomial has no zeroes $s*$ with $Re(s*) > 0$ the IBVP is well-posed.

Now let us apply this theory to the dissipative restrained rod. Consider a semi-infinite rod defined for $x \leq 0$. The absorbing boundary condition is applied at the right end $x = 0$. The "inward" wave solution is given by

$$u(t, x) = a(s)e^{\sqrt{(s/c)^2 + 2\varepsilon s/c + \kappa^2}x + st}.$$ (2.104)

Combining the last equation with the boundary condition yields

$$u_{,x}(x) + g(s)u(x)\big|_{x=0} = \sqrt{(s/c)^2 + 2\varepsilon s/c + \kappa^2} + g(s) = 0 .$$ (2.105)

Since $g(s) = p(s)/q(s)$ is assumed to be a rational function it is convenient to bring the last equation in the more tractable form

$$((s/c)^2 + 2\varepsilon s/c + \kappa^2) \cdot q^2(s) - p^2(s) = 0 .$$ (2.106)

A short reflection shows that the solutions of (2.105) is a subset of the solutions of (2.106). The solutions of (2.105) are found by inspection inserting those of (2.106) in (2.105).

Inserting the viscous boundary conditions in (2.106) yields

$$2\varepsilon s/c + \kappa^2 = 0.$$ (2.107)

Its solution has a negative real part and hence the viscous boundary condition generates a well-posed IBVP. Similarly, we can show that both the doubly-asymptotic as well as the Engquist and Halpern boundary condition generate well-posed IBVP's. The last result has been proved with classical methods in [EH88].

To prove the well-posedness of an IBVP with the asymptotic boundary conditions derived in this chapter is much more laborious because of the polynomials of increasing degree involved in (2.106). The Hurwitz criterion [Leh66] is very helpful to test if a polynomial has roots on the half-plane $Re(s) > 0$. After some lengthy calculations we obtain that the asymptotic boundary condition of first order is well-posed provided that $\kappa > \varepsilon$. If $\kappa = \varepsilon$ it is weakly well-posed. The asymptotic boundary condition of second order is well-posed except if $\kappa = \varepsilon$. In this case, the IBVP is weakly well-posed. For $\kappa = \varepsilon$ they transform to the doubly asymptotic boundary condition. Hence, at least in the parameter range $\kappa > \varepsilon$ all absorbing boundary conditions treated so far lead to a well-posed IBVP. For $\kappa = 0$ and $\varepsilon = 0$ all absorbing boundary conditions transform to the viscous boundary condition. In this case the IBVP is weakly well-posed because $|B(s = 0)| = 0$.

Another criterion to test for well-posedness of IBVP's was proposed by Barry, Bielak and McCamy [BBM88]. It is based on a dissipativity concept. In section 2.1.5 we have already introduced the concept of dissipativity when analyzing the average energy flux. There we found that an absorbing boundary condition is dissipative if:

$$Im(g(\omega)) > 0 \quad \text{for} \quad \omega > 0.$$ (2.108)

Barry, Bielak and MacCamy have slightly generalized this concept. An absorbing boundary condition with kernel $g(s)$ is called dissipative if for every $s = \xi + i\eta$ in the right complex half-plane the following conditions hold:

$$Im(g(\xi + i\eta)) > 0 \quad \text{for} \quad \eta > 0, \xi \geq 0, \qquad (2.109.\text{a})$$

$$Im(g(\xi + i\eta)) < 0 \quad \text{for} \quad \eta < 0, \xi \geq 0, \qquad (2.109.\text{b})$$

$$Re(g(\xi)) > 0 \quad \text{for} \quad \eta = 0, \xi \geq 0. \qquad (2.109.\text{c})$$

With this definition, we can prove that dissipative absorbing boundary conditions lead always to well-posed IBVP's. As we will see, the dissipativity condition is much more restrictive than the Kreiss condition. However, in many cases it has the advantage of being much simpler to apply than the Kreiss condition.

The idea of the proof is very simple: We try to show that the unique solution of a homogeneous equation of motion of a finite rod with a dissipative absorbing boundary condition and vanishing initial values is the trivial solution. Suppose that $u(s, x)$ is a solution of the Laplace transformed equation of motion (2.3). Multiplying the transformed equation of motion with the conjugate complex of the solution and integrating from 0 to L yields

$$\frac{s^2}{c^2}\|u\|^2 + 2\frac{\varepsilon}{c}s\|u\|^2 + \kappa^2\|u\|^2 + \|u_{,x}\|^2 = u_{,x}^L \bar{u}^L, \qquad (2.110)$$

where $\|u\|$ is the Euclidian norm. Including the absorbing boundary condition $u_{,x}^L + g(s)u^L = 0$ into (2.110) and considering the real and imaginary part separately we obtain

$$\frac{\xi^2 - \eta^2}{c^2}\|u\|^2 + 2\frac{\varepsilon}{c}\xi\|u\|^2 + \kappa^2\|u\|^2 + \|u_{,x}\|^2 + Re(g(\xi + i\eta))|\bar{u}^L|^2 = 0, \qquad (2.111.\text{a})$$

$$\frac{2\xi\eta}{c^2}\|u\|^2 + 2\frac{\varepsilon}{c}\eta\|u\|^2 + Im(g(\xi + i\eta))|\bar{u}^L|^2 = 0. \qquad (2.111.\text{b})$$

Using the dissipativity conditions (2.109.a) and (2.109.b) in the last equations yields the trivial solution $u(s, x) \equiv 0$. Hence, the IBVP is well-posed if the kernel of the absorbing boundary condition is dissipative. This result agrees very well with our intuition because a boundary condition which absorbs energy at each frequency impedes a unbounded growth of the solution of an IBVP. If the kernel $g(s)$ of the absorbing boundary condition is analytic in any bounded subset of the half-plane $Re(s) > 0$ and continuous on the imaginary and if $g(s) \to cs$ for $s \to \infty$, where $c \geq 0$, then (2.109.a) and (2.109.b) can be replaced by the condition (2.108). This follows directly from the fact that the imaginary part of an analytic function is a harmonic function which attains its maximum and minimum on the boundary.

Observe, that the DtN-map of the restrained rod, the viscous as well as the doubly-asymptotic and the Engquist-Halpern boundary conditions are dissipative. In contrast, the asymptotic boundary conditions of first and second order aren't dissipative. But we know from the uniform Kreiss criterion (2.98) that these absorbing boundary conditions leads to well-posed IBVP's too.

If $\kappa = \varepsilon = 0$ the Kreiss as well as the dissipativity condition of Barry, Bielak and McCamy predict weakly well-posedness of the IBVP for the viscous boundary condition. This result is not intuitive because the physical image of the viscous boundary condition, the dashpot, is obviously dissipative. We will study this point and observe that for $s = 0$ $g(0) = 0$ such that (2.111.a) yields $u_{,x} \equiv 0$. This implies an x-independent solution of the IBVP $u(t, x) = c(t)$. Inserting the solution into the equation of motion we obtain $c(t) = c_0 + c_1 t + c_2 t^2$. This function must fulfill the viscous boundary condition as well. This is the case only if $c_1 = c_2 = 0$. Hence we obtain the constant solution $u(t, x) = c_0$. This result suggests that the IBVP may have a non-unique solution. Now consider the following general two-boundary IBVP with nonvanishing initial data $u(0, x) = g(x)$ and $u_{,t}(0, x) = h(x)$:

$$u_{,xx} - \frac{1}{c^2} u_{,tt} = f(t, x) \quad \text{in } \Omega = \{-L < x < L\}, \tag{2.112.a}$$

$$\left(u_{,x} - \frac{1}{c^2} u_{,t} \right)\bigg|_{-L} = b_{-L}(t) \quad \text{and} \quad \left(u_{,x} + \frac{1}{c^2} u_{,t} \right)\bigg|_{L} = b_L(t). \tag{2.112.b}$$

Assume that we have a solution of the IBVP given by $u(t, x)$. Then $w(t, x) = c_0 + u(t, x)$ fulfills both the equation of motion (2.112.a) and the boundary conditions (2.112.b). But $w(t, x)$ doesn't fulfill the initial data unless $c_0 = 0$. Hence, the IBVP problem has a unique solution and is therefore well-posed. In this case both the uniform Kreiss as well as the dissipativity condition predict weakly well-posedness because they don't include the initial data in their analysis. Hence, weakly well-posedness doesn't imply that necessarily something will go wrong. It may go wrong. However, this depends on the particular IBVP and needs a more detailed analysis.

Chapter 3
DtN-maps for fluids and solids

In this chapter, we develop DtN-maps for fluids and solids. The far fields are two-dimensional semi-infinite strata. The wave dynamics of these strata is governed by linear scalar and elastic wave equations. The DtN-maps are derived using the so-called semi-analytical technique developed by Lysmer and Waas [LW72]. The harmonic wave solution in the frequency domain is factorized into two functions. The first is a phase function and depends only on the space variable whose axis shows in the direction of the infinite part of the domain. The second function describes the solution perpendicular to this direction and is defined by an eigenvalue problem. Its solution is obtained using the method of finite elements. To be successful, the semi-analytical technique needs a fundamental prerequisite: both the governing differential equations and the boundary conditions must factorize in the sense described above. This furnishes two well-defined problems which can be solved separately. Finally, these solutions are coupled together. First, we derive DtN-maps for two-dimensional strata governed by the acoustic wave equation. Then, we consider the DtN-map of a half-space with a semi-circular hole. Next, we develop DtN-maps of two-dimensional strata governed by the elastic wave equations and finally we consider DtN-maps of coupled fluid and solid strata.

3.1 The homogeneous fluid stratum

3.1.1 The governing equations

Consider the fluid stratum sketched in Figure 3.1. The stratum has an infinite extension in the x-direction and a finite one in z-direction. Furthermore, we assume that the state of the fluid is invariant with respect to translations in the y-direction. We suppose that the fluid is homogeneous and the displacements and velocities of the body are small enough to be described by the linearized theory (small perturbation around a static equilibrium). In addition, we assume that the fluid is inviscid and that its motion is non-rotational. These assumptions allow us to apply the theory of acoustic waves. The conservation of linear momentum and mass yields

$$\rho v_{,t} + \nabla p = \nabla f \quad \text{and} \quad p_{,t} + \rho c^2 \nabla v_{,t} = 0, \tag{3.1}$$

41

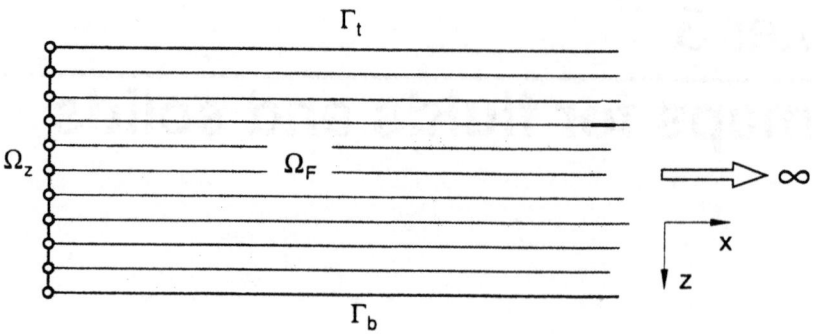

Figure 3.1: Semi-infinite fluid stratum with the finite element discretization of Ω_z

where $p(t, x, z)$ is the pressure field, $v(t, x, z)$ is the velocity field, ρ is the static mass density and c is the speed of sound in the fluid. The term ∇f considers driving forces. For a detailed derivation of these equations we refer to textbooks of fluid mechanics like [Méh76, Pat83].

In order to obtain a symmetric system of equations when modelling with finite elements, we will introduce the velocity potential $\phi(t, x, z)$ as the field variable. The velocity of the fluid $v(t, x, z)$ is described by the gradient of the velocity potential

$$v = \nabla \phi. \tag{3.2}$$

Combining (3.2) with (3.1) yields the equation of motion

$$\nabla^2 \phi = \frac{\ddot{\phi}}{c^2} + f. \tag{3.3}$$

The force function on the right hand side of (3.3) is assumed to be independent of the space variable x: $f(t, z)$. Therefore any particular solution of (3.3) is independent of the variable x. The relation between the pressure and the velocity potential is

$$p = -\rho \dot{\phi}. \tag{3.4}$$

The energy flux S across an arbitrary boundary is easily derived using the stress-energy tensor and is found to be

$$S = -\frac{\partial \phi}{\partial n} \dot{\phi} = (v \cdot n) \frac{p}{\rho}. \tag{3.5}$$

n is the unit normal vector at the boundary pointing outside of the domain Ω_F. At the bottom and at the top of the stratum denoted by Γ_b and Γ_t we formulate essential or natural boundary conditions. A brief discussion of the effects of boundary conditions on the general solution and the DtN-map follows in section 3.1.4.

3.1.2   The harmonic wave solution

To construct the DtN-map we need a solution of the homogeneous equation of motion (3.3). We will search for solutions of (3.3) in the form of harmonic waves

$$\phi(t, x) = \Phi(z) e^{-ik(\omega)x + i\omega t}. \tag{3.6}$$

The parameter $k(\omega)$ can be interpreted as a wave number. Although (3.6) is similar in its structure to a plane wave solution, the variation of the function $\Phi(z)$ with z allows for an arbitrary wave motion. Inserting (3.6) into the homogeneous equation of motion (3.3) yields

$$\Phi_{,zz} + \left(\frac{\omega^2}{c^2} - k^2\right)\Phi = 0 \quad \text{in } \Omega_z. \tag{3.7}$$

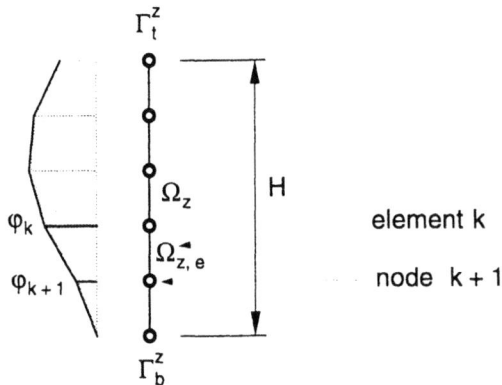

Note that (3.7) is formulated in the one-dimensional interval $\Omega_z = (0, H)$ (see Figure 3.2). Consequently, the boundary conditions are formulated at the points $\Gamma_t^z = 0$ and $\Gamma_b^z = H$. The equation (3.7) together with the boundary conditions, which have to be specified separately, define an eigenvalue problem. For homogeneous boundary conditions the eigenvalue problem is easily solved with standard analytical techniques. However, in view of more complicated boundary conditions, we shall proceed to the construction of approximated solutions with the method of finite elements. We first construct a weak formulation of the problem and multiply (3.7) with a trial function

Figure 3.2: Discretization of the one-dimensional domain Ω_z with finite elements

$\Xi(z)$. It satisfies the essential boundary conditions $\Xi(\Gamma) = 0$, if there are any. Next, we integrate the resulting expression in the interval Ω_z

$$\int_{\Omega_z} \Phi_{,zz}\Xi d\Omega_z + \int_{\Omega_z}\left(\frac{\omega^2}{c^2} - k^2\right)\Phi\Xi d\Omega_z = 0. \tag{3.8}$$

The partial integration of the first term yields

$$-\int_{\Omega_z} \Phi_{,z}\Xi_{,z} d\Omega_z + \int_{\Omega_z}\left(\frac{\omega^2}{c^2} - k^2\right)\Phi\Xi d\Omega_z + \Phi_{,z}\Xi\big|_{\Gamma_t^z} - \Phi_{,z}\Xi\big|_{\Gamma_b^z} = 0. \tag{3.9}$$

The domain Ω_z is now subdivided into an arbitrary number of isoparametric finite elements. Using standard procedures equation (3.9) transforms to the matrix equation

$$(k^2 A + G - \omega^2 M)\varphi = 0 \tag{3.10}$$

φ is the vector of the nodal velocity potentials. The symmetric global matrices A, G, C and M are assembled using the following element matrices

$$A_e = \int_{\Omega_z^e} N_z^T N_z d\Omega_z^e, \tag{3.11.a}$$

$$G_e = \int_{\Omega_z^e} N_{z,z}^T N_{z,z} d\Omega_z^e, \tag{3.11.b}$$

$$M_e = \frac{1}{c^2}\int_{\Omega_z^e} N_z^T N_z d\Omega_z^e = \frac{1}{c^2}A_e. \tag{3.11.c}$$

Here we list the element matrices for the two node isoparametric finite elements. Those for the three node isoparametric finite elements are easily obtained

$$A_e = \frac{h}{6}\begin{bmatrix} 2 & 1 \\ 1 & 2 \end{bmatrix} \qquad G_e = \frac{1}{h}\begin{bmatrix} 1 & -1 \\ -1 & 1 \end{bmatrix} \qquad M_e = \frac{h}{6c^2}\begin{bmatrix} 2 & 1 \\ 1 & 2 \end{bmatrix} \tag{3.12}$$

Material damping is defined via the Rayleigh damping term $C = \alpha_R M + \beta_R G$. The matrix eigenvalue problem (3.10) becomes

$$(k^2 A + G + i\omega C - \omega^2 M)\varphi = 0. \tag{3.13}$$

The matrices A and M are positive definite. The matrix G is positive semi-definite and becomes positive definite if an essential boundary condition is applied. The matrix C is positive definite if $\alpha_R > 0$. Otherwise it exhibits the same definiteness of G.

3.1.3 Analysis of the eigenvalue problem

The homogeneous matrix equation (3.13) has the formal structure of an eigenvalue problem. It defines the relationship between a set of n wave numbers $k_j(\omega)$ (n is the dimension of (3.13)) and the circular frequency ω. Hence, solving (3.13) for a fixed circular frequency ω yields a set of wave numbers $k_j(\omega)$ with the associated eigenvectors $\varphi_j(\omega)$. Because the wave number k is squared in (3.10) we obtain two solutions $k_j(\omega)$ and $-k_j(\omega)$. The first one is related to a wave travelling in the positive x-direction and the second to one travelling in the negative x-direction. The eigenvector associated with the eigenvalue $-k_j(\omega)$ is identical to that associated with the eigenvalue $k_j(\omega)$. For $C = 0$ (no damping) $(k_j(\omega), \varphi_j(\omega))$ and $(\bar{k}_j(\omega), \bar{\varphi}_j(\omega))$ are eigenvalue-eigenvector pairs. This is because the wave-numbers are either real or imaginary. Because the matrix A is positive definite the eigenvalues $k_j(\omega)$ are bounded and continuous functions of the frequency ω except at $\omega \rightarrow \infty$.

The eigenvalue problem (3.13) is very simple if all boundary conditions are homogeneous or frequency-independent. Consider the reduced eigenvalue problem

$$(\lambda^2 A - G)x = 0, \tag{3.14}$$

where A is positive definite and G is positive semi-definite or positive definite. The congruence transformation with the set of independent real eigenvectors X of (3.14) diagonalize simultaneously both matrices. We assume that the eigenvectors are scaled such that $X^T A X = I$. The congruence transformation maps G to $G_D = X^T G X = \Lambda_G^2$. The real diagonal matrix Λ_G^2 contains the eigenvalues of (3.14). Since the matrices M and C are proportional to A and G the eigenvalue problem (3.13) is diagonalized by the congruence transformation with X. It becomes

$$K^2 + \Lambda_G^2 + i\omega\Lambda_C^2 - \left(\frac{\omega}{c}\right)^2 I = 0. \tag{3.15}$$

Obviously, K^2 contains the eigenvalues $k_j(\omega)$ of (3.13). Hence, for homogeneous boundary conditions, the eigenvectors of (3.14) and (3.13) are frequency-independent. In this case, the eigenvalue problem (3.13) needs to be solved only once, e.g. for $\omega = 0$, resulting in the eigenvalues λ and the real eigenvectors φ_λ. Then, to obtain the wave numbers $k_j(\omega)$ for an arbitrary circular frequency, we only need to apply the congruence transformation with X. This gives

$$k_j^2 = -\lambda_j^2 - i\omega\left(\alpha_R\lambda_j^2 + \frac{\beta_R}{c^2}\right) + \omega^2/c^2. \tag{3.16}$$

Observe, that $k \rightarrow \pm\omega/c$ for $\omega \rightarrow \infty$. Hence, all wave numbers $k_j(\omega)$ are equal for $\omega \rightarrow \infty$. This result holds even for more complex frequency-dependent boundary conditions as long as the matrices A and M remain proportional. In fact, for $\omega \rightarrow \infty$, the eigenvalue problem reduces to

$$(k^2 - (\omega^2/c^2))A\varphi = 0, \tag{3.17}$$

because the contributions of the matrices G and C are negligible for $\omega \to \infty$. Since A is positive definite, that is $A\varphi \neq 0$ for $\|\varphi\| \neq 0$, we obtain $k^2 - (\omega^2/c^2) = 0$. In addition, every set of independent vectors spanning the full vector space is a set of eigenvectors of (3.17).

If the eigenvalue problem can be reduced to the form (3.15) the dynamics of the layer is completely determined by a set of independent propagation modes. Each of these modes is described by the equation of motion of a dissipative restrained rod with wave speed c, restrain parameter $\kappa = \lambda_j$ and damping parameter $\varepsilon = (c/2)(\alpha_R \lambda_j^2 + \beta_R/c^2)$. This simplicity is lost if the boundary conditions are frequency-dependent. Then, the propagation modes are coupled. The strength of this coupling is frequency-dependent. As shown before, for high frequencies, this coupling becomes very weak. This is also the case at very low frequencies. Hence, we expect the strongest coupling at intermediate frequencies.

3.1.4 The effects of boundary conditions

Let us first consider homogeneous boundary conditions. A familiar homogeneous boundary conditions is the free fluid surface condition. There the pressure vanishes ($p_t(t) = 0$) and according to (3.4) the boundary condition becomes

$$\phi = 0. \tag{3.18}$$

A second frequently used homogeneous boundary condition is the rigid wall condition

$$\frac{\partial \phi}{\partial n} = 0. \tag{3.19}$$

The fluid is only allowed to move perpendicular to the wall. A generalization of the last boundary condition is the flexible wall condition

$$\frac{\partial \phi}{\partial n} + \kappa \phi = 0. \tag{3.20}$$

These boundary conditions produce a set of frequency-independent eigenvectors because only the matrix G is modified. Therefore, the eigenvalue problem is reduced to (3.15).

This property is lost if the stratum has frequency-dependent boundary conditions. Consider the frequently used boundary condition

$$\frac{\partial \phi}{\partial n} + q\frac{\dot{\phi}}{c} = 0. \tag{3.21}$$

It provides a simplified interaction model of a fluid with another fluid or solid. The parameter q of the damper-like term is defined as

$$q = \frac{\rho c}{\rho_s c_s}. \tag{3.22}$$

ρ_s and c_s are the mass density and the wave velocity of the adjacent fluid or solid. The parameter q is related to the reflection coefficient α by

$$\alpha = \frac{1-q}{1+q}. \tag{3.23}$$

Observe that (3.21) provides for an energy dissipation mechanism because the energy flux across the boundary is

$$S = q\frac{\dot{\phi}^2}{c}. \tag{3.24}$$

This boundary condition modifies the matrix C in (3.13) so that it is no longer proportional to A and G. As a consequence, we cannot reduce the eigenvalue problem to (3.16) with one set of eigenvectors being independent of the frequency.

If the effects of surface waves are considered, the boundary condition on the fluid surface is

$$\frac{\partial \phi}{\partial n} + \frac{\ddot{\phi}}{g} = 0, \tag{3.25}$$

where g is the acceleration due to gravity. This boundary condition modifies the mass matrix M. Therefore, the matrix A is no longer proportional to M. Hence, the simultaneous diagonalization of the eigenvalue problem is impossible. As we shall see, this has an important effect on the structure of DtN-map kernels for high frequencies.

3.1.5   The DtN-map

We derive the DtN-map using the eigenvectors of (3.10) and express the vector of nodal velocity potentials by

$$\varphi(x) = \Psi_\omega \gamma(x). \tag{3.26}$$

The columns of the matrix Ψ_ω contain the eigenvectors φ_j and the vector γ contains the functions $\gamma_j(x) = a_j e^{-ik_j(\omega)x}$. Inside an element, the velocity potential is interpolated according to the shape functions of the finite element

$$\varphi_e(x, y) = N_e(z)\Psi_\omega \gamma(x). \tag{3.27}$$

The velocity in the x-direction, found by differentiating the last equation with respect to x, is

$$\varphi_{e,x}(x, y) = N_e(z)\Psi_\omega \gamma_{,x}(x) = -iN_e(z)\Psi_\omega K\gamma(x). \tag{3.28}$$

K is the diagonal matrix containing the wave numbers $K_{jj} = k_j(\omega)$. Premultiplication with the transposed matrix of the shape functions and integration over the domain of the element Ω_z^e yields the vector of nodal fluxes

$$f_e = -i\int_{\Omega_y^e} N_e^T(z)N_e(z)\Psi_\omega K\gamma(x)d\Omega_y^e = -iA_e\Psi_\omega K\gamma(x). \tag{3.29}$$

The global vector of nodal fluxes is obtained by assembling together the contributions of all elements and replacing the vector $\gamma(x)$ with $\varphi(x)$ using (3.26):

$$f = -H_\omega \varphi = -iA\Psi_\omega K\Psi_\omega^{-1}\varphi. \tag{3.30}$$

Hence, the DtN-map kernel is given by

$$H_\omega = iA\Psi_\omega K\Psi_\omega^{-1}. \tag{3.31}$$

Because of (3.17) the DtN-map kernel is singular at $\omega \to \infty$. Setting $\Psi_\omega = I$ in (3.31), this is allowed because all eigenvalues of (3.17) are $k = \omega/c$ and A is nonsingular, we obtain

$$H_\omega^\infty = i\frac{\omega}{c}A. \tag{3.32}$$

This corresponds exactly to the kernel of the viscous boundary condition.

Before closing this section we compute the mean energy flux S_m across the boundary Γ_{NF}. The mean energy flux through an element is

$$S_{m,e} = -\frac{\omega}{2}\mathrm{Im}(\varphi^*\varphi_{,x}) = -\frac{\omega}{2}\mathrm{Im}(-i\varphi N_y^T N_y \Psi_\omega K \Psi_\omega^{-1}\varphi). \tag{3.33}$$

Integration of (3.33) over the height of the stratum yields

$$S_m = -\frac{\omega}{2}\mathrm{Im}(-i\varphi^H A\Psi_\omega K\Psi_\omega^{-1}\varphi) = \frac{\omega}{2}\mathrm{Im}(\varphi^H H_\omega \varphi) = \frac{\omega}{2}\varphi^H\mathrm{Im}(H_\omega)\varphi. \tag{3.34}$$

Hence, the mean energy flux is essentially defined by the imaginary part of the DtN-map kernel. In addition, the DtN-map is dissipative if $\mathrm{Im}(H_\omega)$ is positive definite for any frequency $\omega \geq 0$. If the mean energy flux is formulated with respect to the eigenvectors Ψ_ω it is

$$S_m = \frac{\omega}{2}\mathrm{Re}(\varphi^H A\Psi_\omega K\Psi_\omega^{-1}\varphi) = \frac{\omega}{2}\mathrm{Re}(\gamma^H \Psi_\omega^H A\Psi_\omega K\gamma), \tag{3.35}$$

where γ are the participation factors. If the eigenvectors are frequency-independent ($\Psi_\omega = \Psi$), we may scale the eigenvectors Ψ such that $\Psi^H A\Psi = I$ and obtain

$$S_m = \frac{\omega}{2}\mathrm{Re}(\gamma^H K\gamma) = \sum S_{m,j}, \tag{3.36}$$

where $S_{m,j}$ is the mean energy flux of the j-th mode. Hence, with respect to the basis system formed by the eigenvectors Ψ, the mean energy flux is the sum of the mean energy fluxes of every propagation mode.

3.1.6 Changing the basis of DtN-maps

The DtN-map kernel given in (3.43) is suitable for applications in frequency-domain. However, this form is not very suitable for the goal of approximation. This is because each component of the matrix valued DtN-map kernel is found by a superposition of distinct wave numbers and eigenvectors. Let us analyze this point in detail. Suppose we wish to construct the DtN-map of a stratum with homogeneous boundary conditions. We formulate the DtN-map with respect to a basis defined by the set of eigenvectors of the reduced eigenvalue problem (3.14). Defining the vector of the modal velocity potential φ_ψ by $\varphi = \Psi_\omega \varphi_\psi$ and the vector of the modal fluxes f_ψ by $f_\psi = \Psi_\omega^T f$, respectively, (3.43) becomes

$$f_\psi = -iK\varphi_\psi. \tag{3.37}$$

The DtN-map kernel is reduced to a diagonal matrix. Obviously, (3.37) is much easier to handle than the full matrix (3.43).

The crux with this formulation is that for general boundary conditions the eigenvectors are complex and frequency-dependent. This impedes a diagonalization of the DtN-map kernel in the complete frequency range. However, if the frequency dependence of the eigenvectors is not too strong, we may find a basis with respect to which the DtN-map kernel is diagonally dominant. In addition, because the variables are real, in time domain it is preferable to formulate the DtN-map with respect to a set of real basis vectors.

Assume that the columns of the real matrix Θ contain the new set of basis vectors. With respect to this new basis the vector φ is defined by $\varphi = \Theta\varphi_\theta$. Then, the DtN-map is given by

$$f_\theta = -iA\Theta^{-1}\Psi_\omega K\Psi^{-1}\Theta\varphi_\theta = -iA_\Theta YKY^{-1}\varphi_\theta \tag{3.38}$$

with $A_\Theta = \Theta^T A\Theta$ and $Y = \Theta^{-1}\Psi_\omega$ or $\Psi_\omega = \Theta Y$, since Θ is nonsingular. The last equation expresses the eigenvectors of (3.10) with respect to the new basis. The DtN-map kernel is then

$$H_\Psi = iA_\Theta YKY^{-1} \tag{3.39}$$

To find the matrix Y we insert $\Psi_\omega = \Theta Y$ into the eigenvalue problem (3.44)

$$A\Theta YK^2 + B\Theta Y = 0 \tag{3.40}$$

Multiplying the last equation from the left with Θ^T yields

$$\Theta^T A\Theta YK^2 + \Theta^T B\Theta Y = A_\Theta YK^2 + B_\Theta Y = 0. \tag{3.41}$$

The matrices A_Θ and B_Θ result from the congruence transformation of A and B with Θ. Hence, Y are the eigenvectors of the transformed eigenvalue problem (3.41). The DtN-map kernel in (3.38) is symmetric since a congruence transformation preserves symmetry and definiteness of matrices. The matrix Y is of course frequency-dependent and describes the deviation of the eigenvectors Ψ_ω from the frequency-independent basis Θ. If the eigenvectors are real and frequency-independent we may choose $\Theta = \Psi_\omega$ as a basis system. This yields $Y = I$ and $H_\omega = iA_\Theta K$. Furthermore, if the new set of basis vectors is normalized such that $\Theta^T A\Theta = I$, we obtain $H_\Psi = iK$ and the DtN-map kernel becomes a diagonal matrix. The criterion for choosing an appropriate modal basis is obvious: It should provide a diagonally dominant DtN-map kernel.

In addition, the basis transformation gives us a way to approximate the DtN-map by discarding the components related to the modes associated with the higher eigenvalues. We can do this, transforming (3.10) to (3.41) by a congruence transformation with a rectangular matrix Θ_r, containing only $r < n$ basis vectors (n is the number of degrees of freedom of (3.13)). Clearly, the reduced matrix K_r of the wave numbers, as well as the reduced eigenvectors contained in Y_r are approximations of the associated full scale problem because the eigenvalue problem is projected to a smaller subspace. This approach reduces the size of the eigenvalue problem (3.41) and allows to consider only those propagation modes which are physically relevant. We will call the basis associated with Θ the *modal basis*. The former basis which is related to (3.10) is called the *natural basis*.

Let us consider the modal basis defined by the eigenvectors of the eigenvalue problem

$$(G - \omega^2 M)\theta = 0. \tag{3.42}$$

The eigenvalues ω_j are the cut-off frequencies of the stratum. Observe, that due to (3.42) $\Theta^T G\Theta$ is a diagonal matrix. If A is proportional to M, this basis system is identical to that generated by (3.14). The advantage of the formulation (3.42) becomes more evident in section 3.1.8 when the particular solution is integrated into the DtN-map and eventually when we will construct the DtN-maps of solid strata.

3.1.7 The symmetry of the DtN-map kernel

As we will see, the symmetry of DtN-map kernels becomes important when the equations of motion of the near field are coupled to those which define the far field by way of an absorbing boundary condition. Hence, we dedicate a short section to the analysis of this aspect of DtN-map kernels. We have seen before that if the eigenvectors of (3.13) are frequency-independent the DtN-map kernel formulated with respect to the basis composed by the eigenvectors is diagonal and hence symmetric. Considering that the vector of velocity potential is defined by $\varphi = \Psi \varphi_\Psi$, the vector of modal fluxes by $f_\Psi = \Psi^T f$ and that we may normalize the eigenvectors such that $\Psi^T A \Psi = I$, the DtN-map kernel formulated with respect to the natural basis is

$$H_\omega = iA\Psi K \Psi^T A. \tag{3.43}$$

This expression is evidently symmetric. However, (3.43) doesn't cover the general case with frequency-dependent eigenvectors. To prove the symmetry of the DtN-map kernel for this case we proceed as follows. Consider the eigenvalue problem (3.13) rewritten as

$$A\Psi K^2 + B\Psi = 0, \tag{3.44}$$

where the matrix B is equal to $B = G + i\omega C - \omega^2 M$. B is obviously symmetric. Combining (3.31) with (3.44) we obtain

$$A\Psi K \Psi^{-1} \Psi K + B\Psi = iH\Psi K + B\Psi = -HA^{-1}H + B = 0 \tag{3.45}$$

Because the matrix A is symmetric and positive definite ($A = U^T \Lambda U$, with U unitary, that is, $U^T U = I$), A^{-1} is symmetric and positive definite: $A^{-1} = U\Lambda^{-1}U^T$). Because of the symmetry of B we obtain

$$HA^{-1}H = H^T A^{-1} H^T. \tag{3.46}$$

But the last equation holds only if $H = H^T$. Hence the DtN-map kernel is symmetric. This results holds irrespective of the basis system used in the formulation of the DtN-map kernel.

3.1.8 Including the particular solution in the DtN-map

A particular solution of the equation of motion emerges from the force term on the right hand side of (3.3). It will be sufficient to derive the particular solution for an arbitrary force term $f(t, z)$. The particular solution associated with the boundary conditions (3.53) may be treated by the same formalism.

We assume that the force term in (3.3) as well as the velocity term in (3.53) is independent of the space variable x so that the particular solution is also independent of x. Hence, the differential equation (3.3) reduces to the one dimensional wave equation

$$\phi_{p,zz} = \frac{\ddot{\phi}_p}{c^2} + f \quad \text{in } \Omega_z. \tag{3.47}$$

Modeling (3.47) with finite elements yields

$$M\ddot{\varphi}_p + C\dot{\varphi}_p + G\varphi_p = f. \tag{3.48}$$

The matrices M, C and G are identical to those in (3.10). The right hand side vector f is assembled using the element vectors

$$f^e = \int_{\Omega_z^e} N^T f d\Omega_z^e. \tag{3.49}$$

To include the particular solution into the DtN-map we use the fact that the velocity potential is the sum of the homogeneous solution and of the particular solution:

$$\varphi = \varphi_h + \varphi_p. \tag{3.50}$$

The DtN-map is defined as a map $H_\omega: \varphi \to f_\omega$. But since the particular solution φ_p is independent of the space variable x the map $\varphi_p \to f_\omega$ must be identical to the zero map. Hence, only the homogeneous solution φ_h gives a contribution to the DtN-map. Therefore, it is given by

$$f_\omega = H_\omega \varphi_h = H_\omega(\varphi - \varphi_p). \tag{3.51}$$

Obviously, this relation is valid with respect to any basis system.

Equation (3.48) can be solved directly in the time domain without going through a frequency-domain formulation. If the boundary conditions are not frequency-dependent, (3.48) can be simplified using the modal basis defined in (3.42):

$$I\ddot{\varphi}_{p\theta} + c^2 C_\theta \dot{\varphi}_{p\theta} + \Omega \varphi_{p\theta} = c^2 f_\theta. \tag{3.52}$$

The diagonal matrix $\Omega = \Theta^T G \Theta$ contains the squares of the cut-off frequencies (eigenfrequencies of (3.42)). The matrix $C_\theta = \Theta^T C \Theta$ is the congruence-transformed damping matrix. It is diagonal too, if Rayleigh damping has been considered. The right hand side vector becomes $f_\theta = \Theta^T f$. (3.52) is a set of independent ODE's of second order which can be solved more efficiently than the coupled ODE system (3.48).

Earthquake ground motion can be considered by way of the boundary condition

$$\frac{\partial \phi}{\partial n} = v_s \cdot n \tag{3.53}$$

at the bottom of the fluid stratum. It represents a boundary exited by the velocity v_s. The particular solution associated with this boundary condition is obtained by defining the vector f in (3.48) accordingly: f is zero everywhere except for the degree of freedom f_{j_b} at the bottom of the stratum. There we have $f_{j_b} = v_s \cdot n$.

3.2   Propagation of SH-waves in semi-infinite solid strata

The formalism developed in section 3.1 may be successfully applied to construct DtN-maps for solids subjected to out of plane shear-waves (SH-waves) motion. Consider the semi-infinite plane solid stratum shown in Figure 3.1. We assume that the motion of the solid does not depend of the y-coordinate. The material of the solid is assumed to behave linear-elastically. Furthermore, we assume that the displacements and velocities of the body are small enough so that the motion can be described by the linearized theory of elasticity. In order to avoid the mere repetition of the equations of section 3.1 we shall analyze the SH-wave motion of a transversely isotropic solid. The plane of isotropy are the (x, y)-planes parallel to the surface.

3.2.1   The governing equations

Assume that the variable $u(t, x, z)$ refers to the deformations of the solid perpendicular to the (x, z)-plane. We deduce the equation of motion using the Lagrangian formalism. The kinetic energy density is equal to

$$T = \frac{1}{2}\rho(u_{,t})^2, \tag{3.54}$$

where ρ is the mass density of the solid. The potential energy density is the elastic energy density

$$V = (\tau_{yx}\varepsilon_{yx} + \tau_{zy}\varepsilon_{zy}) = 2(G_x\varepsilon_{xy}^2 + G_z\varepsilon_{zy}^2) = \frac{1}{2}(G_xu_{,x}^2 + G_zu_{,z}^2), \tag{3.55}$$

where G_x and G_z are the shear moduli in x- and z-direction. The Lagrangian density is

$$L = T - V = \frac{1}{2}\rho(u_{,t})^2 - \frac{1}{2}(G_xu_{,x}^2 + G_zu_{,z}^2). \tag{3.56}$$

The Euler equation of (3.56) yields the equation of motion of the solid

$$\rho u_{,tt} - G_xu_{,xx} - G_yu_{,zz} = 0. \tag{3.57}$$

This equation corresponds to the a generalized scalar wave equation. For isotropic solids we have $G_x = G_z = G$ and the last equation becomes the scalar wave equation

$$\Delta u = \frac{1}{c^2}u_{,tt}, \tag{3.58}$$

where the constant c is the shear wave speed $c = \sqrt{G/\rho}$.

3.2.2 The DtN-map

We consider a solution of the scalar wave equation in the form of a product of harmonic waves in the x-direction (phase function) and an amplitude function depending only on the variable z.

$$u(t, x, z) = U(z)e^{-ikx + i\omega t} \tag{3.59}$$

Substitution into the scalar wave equation yields the ODE

$$G_zU_{,zz} - k^2G_xU + \omega^2U = 0. \tag{3.60}$$

If (3.60) together with the boundary conditions is modeled with isoparametric finite elements we obtain the matrix eigenvalue problem

$$(k^2A + G - \omega^2M)\varphi = 0 \tag{3.61}$$

The element matrices of two-node isoparametric elements are

$$A_e = \frac{hG_x}{6}\begin{bmatrix} 2 & 1 \\ 1 & 2 \end{bmatrix} \qquad G_e = \frac{G_z}{h}\begin{bmatrix} 1 & -1 \\ -1 & 1 \end{bmatrix} \qquad M_e = \frac{h\rho}{6}\begin{bmatrix} 2 & 1 \\ 1 & 2 \end{bmatrix} \tag{3.62}$$

In virtue of the formal agreement of the matrix eigenvalue problem (3.61) with that defined by (3.10) we refer to section 3.1.3 for the analysis of the eigenvalue problem.

We construct the DtN-map following the method shown in section 3.1.5. The stress-strain relation of solids subjected to SH-waves is

$$\tau_{yx} = 2G_x\varepsilon_{xy} = G_xu_{,x}. \tag{3.63}$$

If the displacements are represented with respect to the basis system Ψ_ω, $u(x) = \Psi_\omega\gamma(x)$ we find that

$$\tau_{yx}(x, z) = G_xN_e(z)\Psi_\omega\gamma_{,x}(x) = -iG_xN_e(z)\Psi_\omega K\gamma(x). \tag{3.64}$$

K is the diagonal matrix containing the wave numbers $K_{jj} = k_j(\omega)$ defined by the matrix eigen-value problem (3.61). The matrix Ψ_ω contains the associated eigenvectors. $\gamma(x)$ is the vector of the participation factors defined according to $\gamma_j(x) = a_j e^{-ik_j(\omega)x}$. Premultiplication with the transposed matrix of the shape functions and integration over the domain of the element Ω_e^e yields the vector of nodal forces

$$f_e = -iG_x \int_{\Omega_e^e} N_e^T(y)N_e(y)\Psi_\omega K\gamma(x)d\Omega_y^e = -iG_x A_e \Psi_\omega K\gamma(x). \qquad (3.65)$$

Assembling together the contributions of all elements and replacing the vector $\gamma(x)$ with $u(x)$ yields

$$f = -H_\omega \varphi = -iG_x A\Psi_\omega K\Psi_\omega^{-1} u. \qquad (3.66)$$

Hence, the DtN-map kernel is

$$H_\omega = iG_x A\Psi_\omega K\Psi_\omega^{-1} \qquad (3.67)$$

Observe, that the form of the DtN-map kernel (3.67) is identical to the form of the kernel (3.31). Therefore, the DtN-map kernel (3.67) is symmetric. The singular part of the DtN-map kernel for $\omega \to \infty$ is diagonal and is given by

$$H_\omega^\infty = i\omega\rho c_x A, \qquad (3.68)$$

where $c_x = \sqrt{G_x/\rho}$.

3.3   The half space with a semi-circular hole

The conceptual framework derived in section 3.1 may also be successfully applied to problems formulated in other coordinate systems than the orthogonal Cartesian coordinate system. Consider the half-space with an infinite cylindrical hole with semi-circular section as shown in Figure 3.3. We shall describe the motion with respect to a cylindrical coordinate system with its y-axis parallel to the cylindrical hole. We will assume that the motion of the fluid is independent of the y-coordinate.

3.3.1   The harmonic wave solution

Assuming an harmonic wave solution, the equation of motion (3.3) formulated in cylindrical coordinates becomes

$$\frac{1}{r}(ru_{,r})_{,r} + \frac{1}{r^2}u_{,\varphi\varphi} + \frac{\omega^2}{c^2}u = 0. \qquad (3.69)$$

On the surface of the half-space we impose the boundary conditions

$$\phi(r, -\pi/2) = \phi(r, \pi/2) = 0 \qquad r \geq R. \qquad (3.70)$$

We try to find a solution in form of a product of two separable functions

$$u(r, \varphi) = U(r)V(\varphi), \qquad (3.71)$$

where $U(r)$ describes the component of the solution in radial direction and $V(\varphi)$ describes that in circumferential direction. The partial differential equation (3.69) is split into two independent ordinary differential equations for $U(r)$ and $V(\varphi)$, respectively

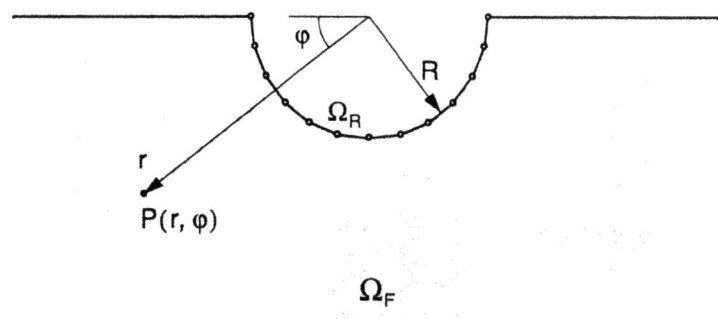

Figure 3.3: Half space with a semi-circular hole and the finite element discretization of Ω_R

$$\frac{d^2V}{d\varphi^2} + m^2 V = 0, \tag{3.72.a}$$

$$\frac{d^2U}{dr^2} + \frac{1}{r}\frac{dU}{dr} + \left(k^2 - \frac{m^2}{r^2}\right)U = 0, \tag{3.72.b}$$

where m is a constant which needs to be specified and k is defined as $k = \omega/c$. The boundary conditions (3.70) lead to the two boundary conditions

$$V(-\pi/2) = V(\pi/2) = 0. \tag{3.73}$$

These boundary conditions together with the ODE (3.72.a) define an eigenvalue problem. Observe that the boundary condition (3.73) is sufficient to fulfil the boundary condition (3.70) irrespective of the solution of the ODE (3.72.b).

3.3.2 The solution in the radial direction

The eigenvalues m_j induce a set of associated radial solutions of the ODE (3.72.b) given by:

$$U_j(r) = A_j H_j^{(1)}(kr) + B_j H_j^{(2)}(kr) \tag{3.74}$$

where the functions $H_j^{(1)}(kr)$ and $H_j^{(2)}(kr)$ are the j-th Hankel-functions of first and second kind, respectively. Hence, the general solution of (3.69) is

$$u(r, \varphi) = \sum_{j=1}^{\infty} A_j H_j^{(1)}(kr)v_j(\varphi) + \sum_{j=1}^{\infty} B_j H_j^{(2)}(kr)v_j(\varphi) \tag{3.75}$$

where $v_j(\varphi)$ are the eigenvalues. In order to discern the outgoing from the incoming waves we apply to (3.75) the Sommerfeld radiation condition:

$$\lim_{r \to \infty} r^{1/2}\left(\frac{\partial u}{\partial r} - iku\right) = 0. \tag{3.76}$$

The asymptotic expressions of Hankel functions for large arguments are [AS84]

$$H_m^{(1)}(z) \sim \sqrt{\frac{2}{\pi z}}e^{i\left(z - \frac{1}{2}m\pi - \frac{1}{4}\pi\right)}, \tag{3.77.a}$$

$$H_m^{(2)}(z) \sim \sqrt{\frac{2}{\pi z}} e^{-i\left(z - \frac{1}{2}m\pi - \frac{1}{4}\pi\right)}, \tag{3.77.b}$$

where $z = kr \rightarrow \infty$. Applied to (3.75) we obtain the asymptotic expression

$$U(t, r, \varphi) \sim \sum_{j=0}^{\infty} \sqrt{\frac{2}{\pi k r}} \cdot \left(A_j e^{i\omega\left(\frac{r}{c} + t\right)} + B_j e^{i\omega\left(-\frac{r}{c} + t\right)} \right) \tag{3.78}$$

Inserting this expression into the Sommerfeld radiation condition (3.76) and considering the orthogonality of the modes, we obtain

$$i\frac{\omega}{c} A_j e^{i\omega\left(\frac{r}{c} + t\right)} = 0 \quad \text{for all } j. \tag{3.79}$$

The only solution of (3.79) being valid for all ω is $A_j \equiv 0$. Hence, the outgoing wavetrain solution of (3.74) is

$$u_j(r) = B_j H_j^{(2)}(kr). \tag{3.80}$$

3.3.3   The solution in the circumferential direction

The general solution of (3.72.a) is easy to find and is given by

$$V(\varphi) = a_1 \cos(m\varphi) + a_2 \sin(m\varphi) \ . \tag{3.81}$$

The eigenvalues m_j and the associated eigenfunctions $v_j(\varphi)$ are found combining the general solution with the boundary conditions (3.73). The solutions are two distinct set of eigenvalues and orthonormal eigenfunctions:

$$m_j = 2j - 1, \quad v_j(\varphi) = \sqrt{\frac{2}{\pi}} \sin(2j-1)\varphi \, ; \text{for } j = 1, 2, 3, \dots . \tag{3.82}$$

$$m_j = 2j, \quad v_j(\varphi) = \sqrt{\frac{2}{\pi}} \cos(2j)\varphi \, ; \text{for } j = 1, 2, 3, \dots . \tag{3.83}$$

The eigenvalues are integers and define the order of the Hankel functions. (3.82) defines a set of skew symmetric propagation modes and (3.83) a set of symmetric propagation modes.

Although the analytical solution of the eigenvalue problem (3.72.a) and (3.73) is a simple task, we may solve the problem with the scheme developed in section 3.1. The modeling of (3.72.a) and (3.73) with finite elements leads to the matrix eigenvalue problem

$$(m^2 A - G)v = 0 . \tag{3.84}$$

The global matrices A and G are built up as before and share the properties of the equivalent matrices discussed in section 3.1. Notice, that the shape functions are functions of the variable φ. Since A and G are symmetric and positive definite we obtain a well-defined set of real eigenvalues M with associated real eigenvectors. The eigenvectors are collected in the matrix Ψ and normalized such that $\Psi^T A \Psi = I$. The eigenvalues and eigenvectors are approximations of the exact solution of the eigenvalue problem (3.72.a) and (3.73). But the set of eigenvalues M computed with (3.84) will consist of real numbers which may differ from an integer. Therefore, in practical computations we have to round off the eigenvalues M to integer values.

3.3.4   The DtN-map

We shall first formulate the DtN-map with respect to the basis system defined by the eigenvalue problem (3.72.a) and (3.73). In this basis system the j-th modal component of the radial velocity is

$$u_{j,r}(r) = B_j H^{(2)}_{j,r}(kr).$$ (3.85)

Combining (3.70) with (3.85) we may replace the constant B_j and obtain the DtN-map of the j-th modal component

$$u_{j,r}(r) = B_j H^{(2)}_{j,r}(kr) = \frac{H^{(2)}_{j,r}(kr)}{H^{(2)}_j(kr)} u_j(r).$$ (3.86)

Therefore, in virtue of its very definition, the DtN-map kernel at $r = R$ is given by

$$H_\Psi(\omega) = \frac{1}{R} \text{diag}\left(kR \frac{H^{(2)\prime}_j(kR)}{H^{(2)}_j(kR)} \right).$$ (3.87)

The prime in $H^{(2)\prime}_j(kr)$ symbolizes the derivative with respect to the argument kr. Observe that the DtN-map components depend on the variable R so that at different distances from the origin we have different DtN-map kernels. At the frequency $\omega = 0$ the real part takes the value $Re(H_{jj}) = j/R$. For large frequencies the real part of every modal component goes asymptotically to $Re(H_{jj}) = 1/2R$. For $\omega \to \infty$ all DtN-map kernel components approach the value $i\omega/c$. Furthermore, increasing the distance R of the artificial boundary, every component of the DtN-map kernel converges to $Re(H_{jj}) = 0$ and to $Im(H_{jj}) = i\omega/c$.

In Figure 3.4 the first three components of the DtN-map $H_\Psi(\omega)$ are plotted. The real as well as the imaginary parts of the components of the DtN-map are positive in the complete frequency range. The real and imaginary parts of the higher modes have a certain affinity with those of the dissipative restrained rod (Figure 3.4). For high frequencies, the DtN-map kernels approach the function ω/c. For low frequencies, the DtN-map kernels of the higher modes have a dominant real part and a small imaginary part.

To formulate the DtN-map kernel on Ω_R with respect to the system of natural coordinates we use the fact that in the circumferential direction we have the relation $\phi = \Psi \phi_\Psi$. In addition, we define the vector of modal forces as $f_\Psi = \Psi^T f$. With the normalization condition $\Psi^T A \Psi = I$ we obtain the DtN-map kernel

$$H_\omega = R A \Psi H_\Psi \Psi^T A.$$ (3.88)

The radius R appears in (3.88) because the integration path along the semi-circular boundary is magnified by the factor R with respect to the integration interval $[-\pi/2, \pi/2]$.

3.4   The homogeneous and isotropic solid stratum

Consider the semi-infinite solid stratum sketched in Figure 3.7. We assume that the stratum is infinite in the x-direction and has a finite height in the z-direction. Furthermore, we assume that the state of the solid is invariant with respect to translations in the y-direction which is perpendicular to the x- and z-directions. We consider only linear elastic solids. In addition, in order to apply the linearized theory, we assume that the displacements and velocities of the body are small. λ and μ are the Lamé constants and ρ_s is the mass density.

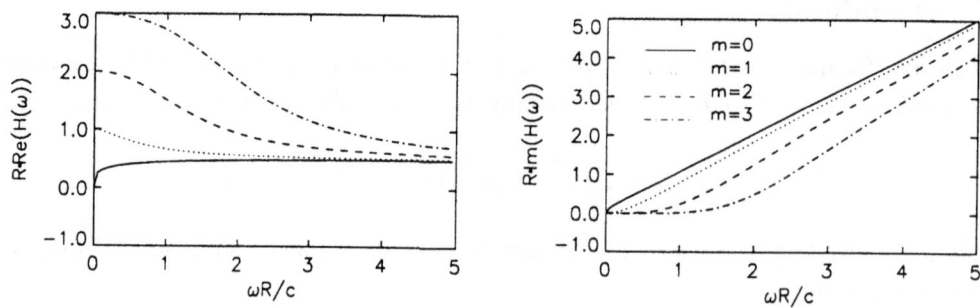

Figure 3.4: DtN-map kernels of the first four modes of the half-space with a semi-circular hole

3.4.1 Equations for the state of plane strain and plane stress

Before starting with the construction of DtN-maps for solid strata we briefly collect all the relevant equations describing the plane motion of a solid. For a detailed derivation of these equations we refer to textbooks of solid mechanics and wave propagation like [Fun65], [BD94], [Gra75]. Consider the equations of a homogeneous solid in a state of plane strain parallel to the (x, z)-plane. This is achieved requiring that the displacements as well as the external forces are independent of the y-direction ($v_{,y} \equiv 0 \quad f_{,y} \equiv 0$). As a consequence, we obtain the following strain components $\varepsilon_{yy} = \varepsilon_{yx} = \varepsilon_{yz} = 0$. The nonzero components of the strain tensor are given by

$$\varepsilon_{xx} = u_{,x} \qquad \varepsilon_{zz} = w_{,z} \qquad \varepsilon_{zx} = \varepsilon_{xz} = (u_{,z} + w_{,x}) . \tag{3.89}$$

Let us define the strain vector $\varepsilon^T = [\varepsilon_{xx}, \varepsilon_{zz}, \varepsilon_{xz}]$ as well as the stress vector $\sigma^T = [\sigma_{xx}, \sigma_{zz}, \sigma_{xz}]$. Then, the stress-strain relation becomes

$$\sigma = \begin{bmatrix} \sigma_{xx} \\ \sigma_{zz} \\ \sigma_{xz} \end{bmatrix} = E\varepsilon = \begin{bmatrix} \lambda + 2\mu & \lambda & 0 \\ \lambda & \lambda + 2\mu & 0 \\ 0 & 0 & \mu \end{bmatrix} \begin{bmatrix} \varepsilon_{xx} \\ \varepsilon_{zz} \\ \varepsilon_{xz} \end{bmatrix} \tag{3.90}$$

Due to the constraints, the stress component σ_{yy} is given by $\sigma_{yy} = \lambda(\varepsilon_{xx} + \varepsilon_{zz})$. The equations of motion of a homogeneous solid in plane strain are

$$\begin{aligned} (\lambda + 2\mu)u_{,xx} + \mu u_{,zz} + (\lambda + \mu)w_{,xz} + f_x &= \rho u_{,tt} \\ (\lambda + 2\mu)w_{,zz} + \mu w_{,xx} + (\lambda + \mu)u_{,zx} + f_z &= \rho w_{,tt} \end{aligned} \tag{3.91}$$

A much more compact notation is obtained introducing a displacement vector U defined by $U^T = [u, w]$. Then, we can rewrite (3.91) as

$$A_{xx}U_{,xx} + A_{xz}U_{,xz} + A_{zz}U_{,zz} + F = A_{tt}U_{,tt}, \tag{3.92}$$

where $A_{xx}, A_{xz}, A_{zz}, A_{tt}, F$ are given by

$$A_{xx} = \begin{bmatrix} \lambda + 2\mu & 0 \\ 0 & \mu \end{bmatrix} \qquad A_{xz} = \begin{bmatrix} 0 & \lambda + \mu \\ \lambda + \mu & 0 \end{bmatrix} \qquad A_{zz} = \begin{bmatrix} \mu & 0 \\ 0 & \lambda + 2\mu \end{bmatrix}$$

$$A_{tt} = \begin{bmatrix} \rho & 0 \\ 0 & \rho \end{bmatrix} \qquad F = \begin{bmatrix} f_x \\ f_z \end{bmatrix} \tag{3.93}$$

If the stress components $\sigma_{yy}, \sigma_{yx}, \sigma_{yz}$ vanish everywhere in the solid we obtain a state of plane

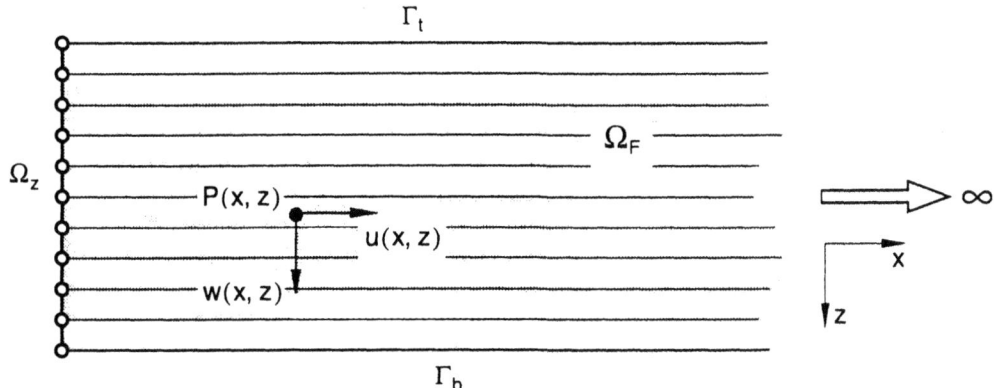

Figure 3.5: The isotropic and homogeneous elastic stratum with the finite element discretization of Ω_z

stress parallel to the (x, z)-plane. It is well-known that the equation of motion of a solid in plane stress can be cast into the form (3.91) if the Lamé constant λ is replaced by

$$\bar{\lambda} = \frac{2\mu\lambda}{\lambda + 2\mu}.$$ (3.94)

Using the parameters μ and $\bar{\lambda}$, the components of the strain tensor are given by

$$\varepsilon_{xx} = u_{,x}, \ \varepsilon_{yy} = v_{,y} = -\frac{\bar{\lambda}}{2\mu}(\varepsilon_{xx} + \varepsilon_{zz}), \ \varepsilon_{zz} = w_{,z}$$

$$\varepsilon_{xy} = \varepsilon_{yx} = 0, \ \varepsilon_{xz} = \varepsilon_{zx} = \frac{1}{2}(u_{,z} + w_{,x}), \ \varepsilon_{yz} = \varepsilon_{zy} = 0$$ (3.95)

With the strain vector $\varepsilon^T = [\varepsilon_{xx}, \varepsilon_{zz}, \varepsilon_{xz}]$ and the stress vector $\sigma^T = [\sigma_{xx}, \sigma_{zz}, \sigma_{xz}]$, the stress-strain relation matrix becomes

$$\sigma = \begin{bmatrix} \sigma_{xx} \\ \sigma_{zz} \\ \sigma_{xz} \end{bmatrix} = E\varepsilon = \begin{bmatrix} \bar{\lambda} + 2\mu & \bar{\lambda} & 0 \\ \bar{\lambda} & \bar{\lambda} + 2\mu & 0 \\ 0 & 0 & 2\mu \end{bmatrix} \begin{bmatrix} \varepsilon_{xx} \\ \varepsilon_{zz} \\ \varepsilon_{xz} \end{bmatrix}.$$ (3.96)

The equations of motion are given by the two differential equations

$$(\bar{\lambda} + 2\mu)u_{,xx} + \mu u_{,zz} + (\bar{\lambda} + \mu)w_{,xz} + f_x = \rho u_{,tt}$$

$$(\bar{\lambda} + 2\mu)w_{,zz} + \mu w_{,xx} + (\bar{\lambda} + \mu)u_{,zx} + f_z = \rho w_{,tt}$$ (3.97)

Because (3.96) and (3.97) have the same structure of (3.90) and (3.91), in the following we don't make a distinction between plane strain and plane stress.

3.4.2 The harmonic waves solution

We first derive the DtN-map of a free vibrating system so that we set $F = 0$ in (3.92) and search for a solution in the form of a harmonic wave

$$U(x, z) = U_z e^{ikx - i\omega t},$$ (3.98)

where the amplitude function U_z depends only of the coordinate z. Inserting (3.98) into (3.92) yields

$$k^2 A_{xx} U_z - ik A_{xz} U_{z,z} - A_{zz} U_{z,zz} - \omega^2 A_{tt} U_z = 0. \tag{3.99}$$

To this equations we add the boundary conditions at the bottom and at the top of the stratum. The resulting system of equations represents an eigenvalue problem in the domain $\Omega_z = (0, H)$. It defines for each circular frequency ω an infinite set of wave-numbers $k_j(\omega)$ with associated eigenvectors. Analytical solutions of the eigenvalue problem (3.99) are available only for some mixed boundary conditions [Gra91] so that (3.99) must generally be solved with numerical methods. We use the finite element method to find a solution of the eigenvalue problem (3.99). Therefore, analogously to the fluid stratum, the vertical domain $\Omega_z = (0, H)$ is subdivided into finite elements. Applying standard techniques of isoparametric finite elements we obtain the matrix eigenvalue problem

$$(k^2 A + ki B + G - \omega^2 M) U = 0. \tag{3.100}$$

The system matrices A, B, G and M are constructed assembling all the corresponding element matrices. How these are constructed is the objective of the next two sections.

3.4.3 The element matrices: a first method

The first method starts from the eigenvalue problem (3.99) and constructs a weak formulation of the problem. According to this procedure, we multiply the equation (3.99) from the left with the transpose of the trial functions vector V_z and integrate the resulting equation over the element domain Ω_e. This yields

$$k^2 \int_{\Omega_e} V_z^T A_{xx} U_z - ik \int_{\Omega_e} V_z^T A_{xz} U_{z,z} - \int_{\Omega_e} V_z^T A_{zz} U_{z,zz} - \omega^2 \int_{\Omega_e} V_z^T A_{tt} U_z = 0. \tag{3.101}$$

The partial integration of the third term transforms the last equation to

$$k^2 \int_{\Omega_e} V_z^T A_{xx} U_z d\Omega_e - ik \int_{\Omega_e} V_z^T A_{xz} U_{z,z} d\Omega_e + \int_{\Omega_e} V_{z,z}^T A_{zz} U_{z,z} d\Omega_e$$
$$- \omega^2 \int_{\Omega_e} V_z^T A_{tt} U_z d\Omega_e - V_z^T A_{zz} U_{z,z} \Big|_{\Gamma_e^1}^{\Gamma_e^2} = 0 \tag{3.102}$$

In order to assure the continuity of the stresses at the element interfaces (see Figure 3.6), we have to replace the expression $A_{zz} U_{z,z}$ in (3.102) with an expression containing the stresses. Combining (3.90) with (3.98) yields

$$\sigma_{zz} = (\lambda + 2\mu) w_{,z} + ik\lambda u$$
$$\sigma_{xz} = \sigma_{zx} = \mu(ikw + u_{,z}) \tag{3.103}$$

Defining the vector S by

$$S = \begin{bmatrix} \sigma_{xz} \\ \sigma_{zz} \end{bmatrix}, \tag{3.104}$$

we rewrite (3.103) as

$$S = A_{zz} U_{z,z} + ki B_{xz} U_z, \tag{3.105}$$

where B_{xz} is given by

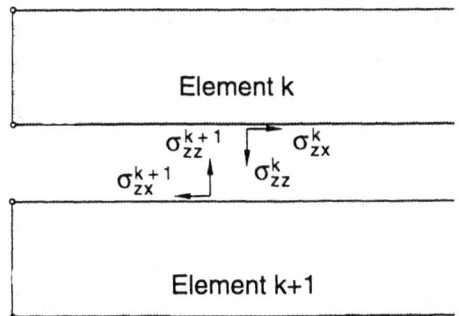

Figure 3.6: Continuity of the stresses at the interface between two elements

$$B_{xz} = \begin{bmatrix} 0 & \mu \\ \lambda & 0 \end{bmatrix}. \tag{3.106}$$

Rearranging (3.105) we obtain

$$A_{zz}U_{z,z} = S - kiB_{xz}U_z. \tag{3.107}$$

Combining (3.107) with (3.102) yields

$$k^2 \int_{\Omega_e} V_z^T A_{xx} U_z d\Omega_e - ik \int_{\Omega_e} V_z^T A_{xz} U_{z,z} d\Omega_e + \int_{\Omega_e} V_{z,z}^T A_{zz} U_{z,z} d\Omega_e$$

$$-\omega^2 \int_{\Omega_e} V_z^T A_{tt} U_z d\Omega_e + k V_z^T B_{xz} U_z \Big|_{-H}^{H} = V_z^T S \Big|_{\Gamma_e^1}^{\Gamma_e^2} \tag{3.108}$$

Now we are ready to compute the element matrices. We perform a detailed derivation of these matrices only for two-node isoparametric finite elements. Generalization to three or higher node isoparametric elements is straightforward. The displacement vector U_z is approximated by

$$U_z = H(\xi)U_e, \tag{3.109}$$

where U_e contains the displacements u and w of the first and second node

$$U_e^T = [u_1, w_1, u_2, w_2] \tag{3.110}$$

$H(\xi)$ contains the shape functions $h_1 = (1/2)(1-\xi)$, $h_2 = (1/2)(1+\xi)$ and is defined as

$$H(\xi) = \begin{bmatrix} h_1 & 0 & h_2 & 0 \\ 0 & h_1 & 0 & h_2 \end{bmatrix}. \tag{3.111}$$

The values of ξ range between $(-1, 1)$. The expression (3.109) holds also for the vector of trial functions V_z. The element matrices A_e, G_e and M_e are computed applying standard methods of isoparametric finite elements. These are given by

$$A_e = \int_{\Omega_e} H^T A_{xx} H d\Omega_e = \frac{h}{6} \begin{bmatrix} 2(\lambda + 2\mu) & 0 & \lambda + 2\mu & 0 \\ 0 & 2\mu & 0 & \mu \\ \lambda + 2\mu & 0 & 2(\lambda + 2\mu) & 0 \\ 0 & \mu & 0 & 2\mu \end{bmatrix} \tag{3.112}$$

$$G_e = \int_{\Omega_e} H_{,x}^T A_{zz} H_{,x} d\Omega_e = \frac{1}{h} \begin{bmatrix} \mu & 0 & -\mu & 0 \\ 0 & \lambda+2\mu & 0 & -(\lambda+2\mu) \\ -\mu & 0 & \mu & 0 \\ 0 & -(\lambda+2\mu) & 0 & \lambda+2\mu \end{bmatrix} \qquad (3.113)$$

$$M_e = \int_{\Omega_e} H^T A_\rho H d\Omega_e = \frac{h\rho}{6} \begin{bmatrix} 2 & 0 & 1 & 0 \\ 0 & 2 & 0 & 1 \\ 1 & 0 & 2 & 0 \\ 0 & 1 & 0 & 2 \end{bmatrix} \qquad (3.114)$$

The element matrix B_e is composed by the second and last term in the right-hand side of equation (3.108):

$$B_e = \int_{\Omega_e} H_{,x}^T A_{xz} H d\Omega_e - \begin{bmatrix} -B_{xz} & 0 \\ 0 & B_{xz} \end{bmatrix} = \frac{1}{2} \begin{bmatrix} 0 & -\lambda+\mu & 0 & \lambda+\mu \\ \lambda-\mu & 0 & \lambda+\mu & 0 \\ 0 & -(\lambda+\mu) & 0 & \lambda-\mu \\ -(\lambda+\mu) & 0 & -\lambda+\mu & 0 \end{bmatrix} . \qquad (3.115)$$

Observe that A_e and M_e are positive definite. The element matrix G_e is positive semi-definite and has rank two. The element matrix B_e is a full rank skew-symmetric matrix: $B_e^T = -B_e$.

3.4.4 The element matrices: a second method

The second method is also based on a weak formulation of the problem. However, instead of the equation of motion it uses the linear momentum equations

$$\sigma_{ij,j} + f_i = \rho \ddot{u}_i. \qquad (3.116)$$

Multiplying these with the trial functions v_i and integrating over a domain Ω we obtain

$$\int_\Omega v_i \sigma_{ij,j} + \int_\Omega v_i f_i = \int_\Omega \rho v_i \ddot{u}_i. \qquad (3.117)$$

The partial integration of the first term on the right-hand side yields the usual expression of the weak formulation

$$-\int_\Omega v_{i,j} \sigma_{ij} d\Omega + \int_{\Gamma_\sigma} v_i \sigma_{ij} d\Gamma + \int_\Omega v_i f_i d\Omega = \int_\Omega \rho v_i \ddot{u}_i d\Omega. \qquad (3.118)$$

Since the boundary integral (second term on the right-hand side) contains the stress tensor the continuity of the stresses are implicitly enforced in this formulation. This is an advantage when compared with the first method where the continuity of the stresses must be enforced explicitly.

Now let us compute the first term. Defining the strain vector $\varepsilon(u)$ by

$$\varepsilon(u) = \begin{bmatrix} \varepsilon_{xx} \\ \varepsilon_{zz} \\ 2\varepsilon_{xz} \end{bmatrix} = \begin{bmatrix} u_{,x} \\ w_{,z} \\ u_{,z}+w_{,x} \end{bmatrix} \qquad (3.119)$$

the first term in (3.118) can be written as

$$\int_\Omega v_{i,j}\sigma_{ij}d\Omega = \int_\Omega \varepsilon^T(v)E\varepsilon(u)d\Omega \;, \tag{3.120}$$

where $\varepsilon(v)$ are the strains associated with the trial functions v_i and the matrix D is

$$E = \begin{bmatrix} \lambda+2\mu & \lambda & 0 \\ \lambda & \lambda+2\mu & 0 \\ 0 & 0 & \mu \end{bmatrix}. \tag{3.121}$$

Assuming a solution in form of an harmonic wave travelling in x-direction we obtain for $\varepsilon(u)$:

$$\varepsilon(u) = \left(ik\begin{bmatrix} u_z \\ 0 \\ w_z \end{bmatrix} + \begin{bmatrix} 0 \\ w_{z,z} \\ u_{z,z} \end{bmatrix} \right)e^{ik-i\omega} \tag{3.122}$$

The trial functions are defined to be in the form $v_j = v_{jz}e^{-ik+i\omega}$. Then, the associated strain vector is

$$\varepsilon(v) = \left(-ik\begin{bmatrix} \tilde{u}_z \\ 0 \\ \tilde{w}_z \end{bmatrix} + \begin{bmatrix} 0 \\ \tilde{w}_{z,z} \\ \tilde{u}_{z,z} \end{bmatrix} \right)e^{-ik+i\omega}\;. \tag{3.123}$$

Observe that the trial functions have been chosen in such a way that the product of the exponential terms is cancelled. Combining (3.120) with (3.122) and (3.123) yields

$$k^2 V_z^T A_{xx} U_z + k V_{z,z}^T B_{xz} U_z - k V_z^T B_{xz}^T U_{z,z} + V_{z,z}^T A_{zz} U_{z,z} = 0, \tag{3.124}$$

where A_{xx} and A_{zz} are defined according to (3.93) and B_{xz} is given by (3.106). With (3.124) and (3.118) the weak formulation becomes

$$k^2 \int_{\Omega_z} V_z^T A_{xx} U_z d\Omega_z - ki \int_{\Omega_z} (V_z^T B_{xz}^T U_{z,z} - V_{z,z}^T B_{xz} U_z)d\Omega_z + \int_{\Omega_z} V_{z,z}^T A_{zz} U_{z,z} d\Omega_z$$
$$-\omega^2 \int_{\Omega_z} V_z^T A_{tt} U_z d\Omega_z = V_z^T S\Big|_{\Gamma_*}^{\Gamma_*} \tag{3.125}$$

The discretization of the first, third and fourth term in (3.125) yields the element matrices A_e, G_e and M_e which are given in (3.112), (3.113) and (3.114). Defining the matrix

$$D = \int_{\Omega_z} H^T B_{xz}^T H_{,z} d\Omega_e = \frac{1}{2}\begin{bmatrix} 0 & -\lambda & 0 & \lambda \\ -\mu & 0 & \mu & 0 \\ 0 & -\lambda & 0 & \lambda \\ -\mu & 0 & \mu & 0 \end{bmatrix} \tag{3.126}$$

the second term yields the matrix $B_e = D - D^T$ which is exactly the matrix B_e given in (3.115).

3.4.5   Analysis and solution of the eigenvalue problem

The matrix eigenvalue problem determining the wave-numbers $k(\omega)$ of a solid stratum is

$$(k^2 A + kiB + G - \omega^2 M)U = 0 \;. \tag{3.127}$$

It is quadratic in $k(\omega)$. According to the element matrices derived in the foregoing section, the global system matrices A and M are positive definite whereas the matrix B is skew symmetric so that iB is Hermitian. The matrix G is positive definite or positive semi-definite. Its definiteness depends on the boundary conditions. In order to include an energy dissipation mechanism, we generalize the matrix eigenvalue problem to

$$(k^2 A + kiB + G + i\omega C - \omega^2 M)U = 0 \,, \tag{3.128}$$

where the symmetric matrix C may be positive definite or positive semi-definite.

If the vector of the nodal displacements is rearranged in the partitioned form

$$U = \begin{bmatrix} u_z \\ w_z \end{bmatrix} \tag{3.129}$$

where the subvector u_z refers to the horizontal displacements and w_z to the vertical displacements, then the eigenvalue problem (3.128) can be written in the partitioned form

$$\left(k^2 \begin{bmatrix} A_x & 0 \\ 0 & A_z \end{bmatrix} + k \begin{bmatrix} 0 & iB \\ -iB^T & 0 \end{bmatrix} + \begin{bmatrix} G_x & 0 \\ 0 & G_z \end{bmatrix} + i\omega \begin{bmatrix} C_x & 0 \\ 0 & C_z \end{bmatrix} - \omega^2 \begin{bmatrix} M_x & 0 \\ 0 & M_z \end{bmatrix} \right) \begin{bmatrix} u_z \\ w_z \end{bmatrix} = 0 \,. \tag{3.130}$$

This structure follows directly from the structure of the element matrices A_e, B_e, G_e and M_e. We assume that the damping matrix C has the partitioned form given in (3.130). In addition, we assume that the boundary conditions preserve the structure of (3.130). From (3.130) we see that the matrix B couples the subvector u_z of the horizontal displacements to the subvector w_z of the vertical displacements. Hence, the eigenvectors of (3.130) generally contain horizontal as well as vertical displacement components.

Observe that if $k(\omega) = 0$, the horizontal displacements uncouples from the vertical one. The circular frequencies for which $k(\omega) = 0$ are called the cut-off frequencies of the stratum. The cut-off frequencies of the stratum are defined by the reduced eigenvalue problem

$$\left(\begin{bmatrix} G_x & 0 \\ 0 & G_z \end{bmatrix} - \omega^2 \begin{bmatrix} M_x & 0 \\ 0 & M_z \end{bmatrix} \right) \begin{bmatrix} u_z \\ w_z \end{bmatrix} = 0 \,. \tag{3.131}$$

Since the submatrices G_x and G_z are positive semi-definite and M_x as well as M_z are positive definite, the cut-off frequencies are real. Moreover, these properties of the submatrices allows to diagonalize the matrices G and M with respect to the eigenvectors of (3.131).

For some special mixed boundary conditions the eigenvectors can be uncoupled in the complete frequency range. In these cases, the eigenvectors u_j are real and the eigenvalues $k_j(\omega)$ are

$$k_j^2(\omega) = \frac{\omega^2 m - i\omega c - g}{a} \,, \tag{3.132}$$

where the numbers $m = u_j^T M u_j$, $c = u_j^T C u_j$, $g = u_j^T G u_j$ and $a = u_j^T A u_j$ are real and larger or equal to zero. If $c = 0$ the wave-numbers $k_j(\omega)$ are either real or purely imaginary.

Now assume that $k(\omega)$ is an eigenvalue of (3.130) and that U_k is the associated eigenvector given by

$$U_k^T = \begin{bmatrix} u_z^{(k)} & w_z^{(k)} \end{bmatrix} \,. \tag{3.133}$$

Then it follows from (3.130) that

$$\xi(\omega) = -k(\omega) \quad \text{with} \quad U_\xi^T = \begin{bmatrix} -u_z^{(k)} & w_z^{(k)} \end{bmatrix} \tag{3.134}$$

is an eigenvalue-eigenvector pair of (3.130). If the pair (k, U_k) is assumed to be a forward travelling harmonic wave then the eigenvector of the backward travelling harmonic wave (ξ, U_ξ) differs only in the sign of the horizontal displacement components. Furthermore if $C = 0$ and (k, U_k) is an eigenvalue-eigenvector pair of (3.130) then (k^*, U_ξ^*) is also an eigenvalue-eigenvector pair. As a consequence, (ξ^*, U_k^*) is an eigenvalue-eigenvector pair of (3.130) too. Because the matrix A is positive definite the eigenvalues $k(\omega)$ are bounded and continuous functions of the frequency except for $\omega \to \infty$.

The asymptotic analysis of the eigenvalue system (3.130) shows that for $\omega \to \infty$ the horizontal displacements uncouple from the vertical one because the influence of the coupling term due to the matrix B becomes infinitesimally small. Because the global matrix A is proportional to M the wave-numbers $k(\omega)$ associated with the horizontal displacements are all equal to $k_p = \pm\omega/c_p$, where $c_p = \sqrt{(\lambda + 2\mu)/\rho}$ is the velocity of longitudinal waves. Those associated with the vertical displacements are all equal to $k_s = \pm\omega/c_s$, where $c_s = \sqrt{\mu/\rho}$ is the velocity of transverse (shear) waves. Because all eigenvalues are equal in these subspaces, the eigenvectors can be chosen arbitrarily provided they are linearly independent and span these subspaces.

Assume that the columns of the matrix Ψ collect the eigenvectors of (3.131). Then, an arbitrary displacement vector is described by $U_s = \Psi X$. Inserting the last equation in (3.128) and multiplying the resulting equation by Ψ^T from the left yields the eigenvalue problem

$$(k^2 A_\Psi + k B_\Psi + G_\Psi + i\omega C_\Psi - \omega^2 M_\Psi)X = 0 . \tag{3.135}$$

As stated before, the matrices G_Ψ and M_Ψ are diagonal. Since the matrix A is proportional to M A_Ψ is diagonal too. If C is a Rayleigh-type damping matrix C_Ψ is also diagonal. B_Ψ is nondiagonal and Hermitian and is responsible for the coupling between the new basis vectors. Furthermore, if we define a matrix Ψ_m which contains only $m < n$ columns of the matrix Ψ so that the columns of Ψ_m spans only a subspace S_m of the displacement vector space S, we can reduce the size of the eigenvalue problem (3.135) from n by n to m by m. This allows a very efficient computation of the DtN-map associated with the subspace S_m. Of course, the eigenvalues and eigenvectors of (3.135) are approximations of the eigenvalues and eigenvectors of (3.128).

In order to be handled by standard numerical eigenvalue analysis procedures, the eigenvalue problem (3.135) must be cast in the standard form. This is achieved defining the new vector

$$Y^T = [X, kX] . \tag{3.136}$$

Then, (3.135) transforms to

$$\left(k \begin{bmatrix} iB_\Psi & A_\Psi \\ A_\Psi & 0 \end{bmatrix} + \begin{bmatrix} G_\Psi + i\omega C_\Psi - \omega^2 M_\Psi & 0 \\ 0 & -A_\Psi \end{bmatrix} \right) \begin{bmatrix} X \\ kX \end{bmatrix} = 0 . \tag{3.137}$$

The first matrix is Hermitian and has full rank and the second is complex symmetric.

3.4.6 The DtN-map

The DtN-map of a solid stratum relates the nodal forces with the nodal displacements. This relationship is constructed starting with the two stress-strain relations

$$\sigma_{xx} = (\lambda + 2\mu)\varepsilon_{xx} + \lambda\varepsilon_{zz}$$
$$\sigma_{zx} = 2\mu\varepsilon_{xz}$$

(3.138)

Combining the last equation with the harmonic wave solution (3.98) we obtain

$$\begin{bmatrix} \sigma_{xx} \\ \sigma_{xz} \end{bmatrix} = \begin{bmatrix} \lambda + 2\mu & 0 \\ 0 & \mu \end{bmatrix} \begin{bmatrix} u_z \\ w_z \end{bmatrix} + \begin{bmatrix} 0 & \lambda \\ \mu & 0 \end{bmatrix} \begin{bmatrix} u_{z,z} \\ w_{z,z} \end{bmatrix}.$$

(3.139)

Defining the stress vector in the right-hand side of the last equation by

$$S_x = \begin{bmatrix} \sigma_{xx} \\ \sigma_{xz} \end{bmatrix}$$

(3.140)

the last equation becomes

$$S_x = ikA_{xx}U_z + B_{xz}^T U_{z,z}$$

(3.141)

Multiplying (3.141) from the left with V_z^T and integrating the resulting expression over the domain of an element, we obtain the desired relationship between nodal forces F_e and the nodal displacements U_e

$$F_e = (ikA_e + D_e)U_e .$$

(3.142)

Assembling the contributions of all elements yields the global force-displacement relationship which is given by

$$F = (ikA + D)U .$$

(3.143)

If the displacement vector U is expressed with respect to a basis defined with the eigenvectors of (3.128), say $U = V\Gamma$, the forces of the stratum become

$$F = (iAVK + DV)\Gamma ,$$

(3.144)

where K is the diagonal matrix containing those eigenvalues of (3.128) which are related to harmonic waves travelling in the positive x-direction, V is the matrix of the associated eigenvectors and Γ is the vector of the participation factors. Replacing Γ with the displacement vector U we obtain the relation (3.144) with respect to the natural basis

$$F = (iAVKV^{-1} + D)U = H(\omega)U .$$

(3.145)

Hence, with respect to this basis, the DtN-map $H(\omega)$ is given by

$$H(\omega) = iAVKV^{-1} + D .$$

(3.146)

Next we derive the DtN-map with respect to a modal basis defined by the columns of the matrix Ψ_s. Note that with respect to this modal basis the modal force vector is defined by $F_\Psi = \Psi^T F$. Then (3.144) becomes

$$F_\Psi = (iA_\Psi X K_\Psi + D_\Psi X)\Gamma_\Psi$$

(3.147)

where K_Ψ is the diagonal matrix of the eigenvalues of (3.135), X is the matrix of the associated eigenvectors, $D_\Psi = \Psi^T D\Psi$ and Γ_Ψ is the vector of the participation factors. Replacing Γ_Ψ in (3.147) and taking into account the relation $\Gamma_\Psi = X^{-1}U_\Psi$ we obtain

$$F_\Psi = (iA_\Psi X K_\Psi X^{-1} + D_\Psi) U_\Psi = H_\Psi(\omega) U_\Psi \ . \tag{3.148}$$

Therefore, with respect to the basis vectors Ψ, the DtN-map is given by

$$H_\Psi(\omega) = iA_\Psi X K_\Psi X^{-1} + D_\Psi \ . \tag{3.149}$$

If the basis vectors Ψ are scaled such that $\Psi^T A \Psi = I$, observe that this scaling is always possible because A is positive definite, the DtN-map simplifies further to

$$H_\Psi(\omega) = iX K_\Psi X^{-1} + D_\Psi \ . \tag{3.150}$$

The DtN-map $H_\Psi(\omega)$ is transformed to the natural basis using the relations $U = \Psi U_\Psi$, $F_\Psi = \Psi^T F$ and $\Psi^T A \Psi = I$. Combining these equations we obtain

$$U_\Psi = \Psi^{-1} U = \Psi^T A U \quad \text{and} \quad F = \Psi^{-T} F_\Psi = A \Psi F_\Psi, \tag{3.151}$$

where Ψ^{-1} as well as Ψ^{-T} have to be interpreted as pseudo-inverses if Ψ is a rectangular matrix. Combining (3.151) with (3.148) we obtain the DtN-map with respect to the natural basis

$$H_n(\omega) = A \Psi H_\Psi(\omega) \Psi^T A \ . \tag{3.152}$$

Obviously, $H_n(\omega) \neq H(\omega)$, where $H(\omega)$ is given by (3.145), if the columns of Ψ spans only a subspace of the displacements.

In section 3.4.5, we noted that for $\omega \to \infty$ the eigenvalue problem (3.128) reduces to

$$(k^2 A_\Psi - \omega^2 M_\Psi) u = 0 \tag{3.153}$$

and observed that there are only two distinct eigenvalues $k(\omega)$. The first is associated with the subspace of the purely horizontal displacements and the second to that of the purely vertical displacements. We found that the eigenvalues of (3.128) are proportional to ω. As a consequence, the DtN-map is singular for $\omega \to \infty$. Because the constant term D_Ψ is negligible for $\omega \to \infty$ the DtN-map simplifies to

$$H_{\Psi, \infty} = iA_\Psi X K X^{-1} \ , \tag{3.154}$$

where X and K are defined by the eigenvalue problem (3.153). Because the horizontal displacements are uncoupled from the vertical displacements and A is proportional to M the singular part of the DtN-map kernel is diagonal

$$H_{\Psi, \infty} = i\omega\rho \begin{bmatrix} c_p I_h & 0 \\ 0 & c_s I_v \end{bmatrix} \ . \tag{3.155}$$

The singular part of the DtN-map kernel corresponds to the viscous boundary condition.

3.4.7 The symmetry of DtN-maps of solid strata

The DtN-maps defined by the relations (3.146) and (3.150) don't reveal if they are symmetric. Hence, we must prove the symmetry of these DtN-maps explicitly. That is, we have to show that

$$\Delta_H = H(\omega) - H^T(\omega) = iAVKV^{-1} - iV^{-T} K V^T A + D - D^T = 0 \ . \tag{3.156}$$

Waas [Waa72] proved that the DtN-map of the undamped homogeneous solid stratum is symmetric. Here we show that the DtN-maps of any solid stratum with a symmetric damping matrix is symmetric. Moreover, this result is independent of the basis with respect to which the DtN-map is formulated.

Considering that $B = D - D^T$ (3.156) is rewritten as

$$\Delta_H = V^{-T}(iV^T A V K - iK V^T A V + V^T B V) V^{-1} = 0 .$$ (3.157)

Since V has full rank (3.157) is equivalent to

$$W = V^T A V K - K V^T A V - i V^T B V = 0 .$$ (3.158)

Now consider an arbitrary component W_{mn}. It is given by

$$W_{mn} = (k_n - k_m) V_m^T A V_n - i V_m^T B V_n$$ (3.159)

where V_i represents the i-th column of V and is thus the eigenvector associated with the i-th eigenvector k_i. Observe that if $m = n$ we obtain $W_{mm} = -i V_m^T B V_m$. However, since B is skew-symmetric we have also $W_{mm} = i V_m^T B V_m = -W_{mm}$. Hence $W_{mm} = 0$ so that the diagonal components of W vanish.

To show that the nondiagonal terms vanish we proceed as follows. Because of the eigenvalue problem (3.128) we obtain the following equation

$$A V_n k_n^2 - i B V_n k_n + F V_n = 0 ,$$ (3.160)

where $F = F^T = G + i\omega C - \omega^2 M$ is a complex symmetric matrix. Multiplying (3.160) from the left with V_m^T we obtain

$$V_m^T A V_n k_n^2 - i V_m^T B V_n k_n + V_m^T F V_n = 0$$ (3.161)

Similarly, we found

$$V_n^T A V_m k_m^2 - i V_n^T B V_m k_m + V_n^T F V_m = 0 .$$ (3.162)

Transposing the last equation and subtracting it from (3.161) yields

$$V_m^T A V_n (k_n^2 - k_m^2) - i V_m^T B V_n (k_n + k_m) = 0 .$$ (3.163)

Since the wave-numbers k_m and k_n are associated with harmonic waves traveling in the same direction, $k_m + k_n \neq 0$ for all combinations of indices m, n and all frequencies ω except for the special case that k_m as well as k_n vanish simultaneously. However, if k_m and k_n vanish, the horizontal displacements uncouple from the vertical one so that we obtain $V_m^T B V_n = 0$. As a consequence, the right-hand side of (3.159) vanish and we obtain $W_{mn} = 0$. Hence, in (3.163) $(k_n + k_m)$ can be factored out so that (3.163) becomes

$$V_m^T A V_n (k_n - k_m) - i V_m^T B V_n = 0$$ (3.164)

The left-hand side of the last equation is identical to the component W_{mn} given in (3.159) so that we obtain $W_{mn} = 0$ and the symmetry property of the DtN-map $H(\omega)$ is proved.

3.4.8   The horizontally layered solid stratum

A generalization of homogeneous and isotropic strata which may exist frequently in practice is the horizontally layered stratum shown in Figure 3.7. The stratum is composed by two or more layers of homogeneous and isotropic solid materials. The procedure describing homogeneous and isotropic strata applies essentially unchanged also for the horizontally layered stratum. Furthermore, because the layers are homogeneous and isotropic, the global matrices are assembled using the element matrices A_e, B_e, G_e, M_e and D_e derived in the sections before.

However, several qualitative results stated for the homogeneous and isotropic stratum will change. The most important new fact is that in a horizontally layered stratum the matrix A is no longer proportional to the matrix M. Therefore, the arguments used in the section section 3.4.6 to prove that $H_{\Psi,\infty}$ is diagonal don't hold any more. Since for $\omega \to \infty$ the partitioned form (3.128) is valid also for a horizontally layered stratum, the horizontal displacements uncouple from the vertical one and the eigenvalue problem $(k^2 A - \omega^2 M)x = 0$ can be solved separately in each displacement subspace. Both matrices are positive definite so that there will be in general n different eigenvalues where n is the dimension of the associated displacement subspace. The eigenvalues will be proportional to $k = \omega/c$ with $c = u^T A u / u^T M u > 0$. c has the dimension of a velocity. Generally, c will neither be equal to the longitudinal nor to the transverse wave velocity of any solid layer.

Next we study the structure of the singular part of the DtN-map kernel $H_{\Psi,\infty}$. We assume that the set of basis vectors Ψ are a subset of the eigenvectors associated with the cut-off frequencies defined by the eigenvalue problem $(G - \lambda M)x = 0$. With respect to this basis the matrices M_Ψ and G_Ψ are diagonal. C_Ψ is diagonal if it is defined by a Rayleigh damping term. A_Ψ and B_Ψ are nondiagonal. Therefore, with respect to the basis Ψ, A_Ψ contributes to the coupling of the horizontal and vertical displacements.

Now consider the case where $\omega \to \infty$. The eigenvalue problem (3.135) reduces to the eigenvalue problem $(k^2 A_\Psi - \omega^2 M_\Psi)x = 0$. Assume that the solution is given by

$$A_\Psi X K^2 - \omega^2 M_\Psi X = 0 \tag{3.165}$$

where K collects the real eigenvalues and X the associated real eigenvectors. An explicit expression for K is found by multiplying (3.165) from the left with X^T. This yields

$$X^T A_\Psi X K^2 - \omega^2 X^T M_\Psi X = 0. \tag{3.166}$$

Now observe that $D_A = X^T A_\Psi X$ as well as $D_M = X^T M_\Psi X$ are positive definite diagonal matrices. Hence K is given by

$$K = \omega D_A^{-1/2} D_M^{1/2} = \omega D_M^{1/2} D_A^{-1/2} \tag{3.167}$$

The singular part of the DtN-map is

$$H_{\Psi,\infty} = i A_\Psi X K X^{-1} = i X^{-T} X^T A_\Psi X K X^{-1} = i X^{-T} D_A K X^{-1}. \tag{3.168}$$

Inserting (3.167) in the last equation yields

$$H_{\Psi,\infty} = i\omega X^{-T} D_A^{1/2} D_M^{1/2} X^{-1}. \tag{3.169}$$

As expected $H_{\Psi,\infty}$ is symmetric. Furthermore, since both diagonal matrices are positive definite $X^{-T} D_A^{1/2} D_M^{1/2} X^{-1}$ is positive definite too. Hence $X^{-T} D_A^{1/2} D_M^{1/2} X^{-1}$ has positive dominant terms

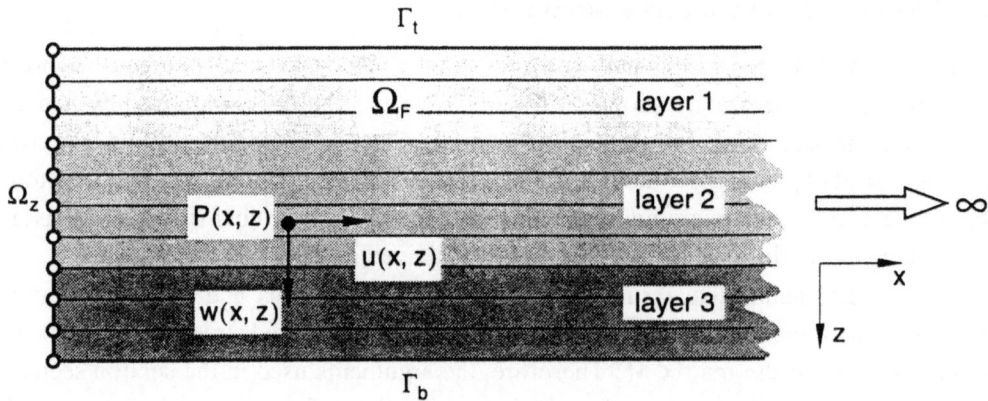

Figure 3.7: The horizontally layered solid stratum with the finite element discretization of Ω_z

on their diagonal. $H_{\Psi,\infty}$ can be simplified scaling X such that $D_A = X^T A_\Psi X = I$ or $D_M = X^T M_\Psi X = I$. The associated singular parts of the DtN-maps become

$$H_{\Psi,\infty} = i\omega X^{-T} D_M^{1/2} X^{-1} \quad \text{or} \tag{3.170.a}$$

$$H_{\Psi,\infty} = i\omega X^{-T} D_A^{1/2} X^{-1} \tag{3.170.b}$$

Now let us see if $H_{\Psi,\infty}$ is diagonal. We assume that the eigenvectors Ψ are scaled such that $M_\Psi = \Psi^T M \Psi = I$. As a consequence, the matrix X becomes unitary if X is scaled such that $D_M = X^T M_\Psi X = X^T X = I$. Therefore we have $D_A^{1/2} = X^T A_\Psi^{1/2} X$ and

$$H_{\Psi,\infty} = i\omega X^{-T} D_A^{1/2} X^{-1} = i\omega X^{-T} X^T A_\Psi^{1/2} X X^{-1} = i\omega A_\Psi^{1/2}. \tag{3.171}$$

Since A_Ψ is not diagonal it turns out that $H_{\Psi,\infty}$ is not diagonal with respect to the basis associated with the eigenvalue problem $(G - \lambda M)x = 0$.

Now let us assume that the basis vectors Ψ are the eigenvectors associated with the reduced eigenvalue problem $(G - \lambda A)x = 0$. Assume that the basis vectors are scaled such that $A_\Psi = \Psi^T A \Psi = I$. The solution of the eigenvector problem for $\omega \to \infty$ is given by

$$XK^2 - \omega^2 M_\Psi X = 0 \quad . \tag{3.172}$$

Scaling X such that $D_A = X^T X = I$ and using the same arguments as before we obtain

$$H_{\Psi,\infty} = i\omega X^{-T} D_M^{1/2} X^{-1} = i\omega X^{-T} X^T M_\Psi^{1/2} X X^{-1} = i\omega M_\Psi^{1/2}. \tag{3.173}$$

Since M_Ψ is nondiagonal $H_{\Psi,\infty}$ is also nondiagonal. This point has to be taken into account when the DtN-map is regularized for $\omega \to \infty$.

3.4.9 Including external forces in the DTN-map

If the external forces don't vanish in (3.91) the solution of the homogeneous problem (3.99) must be combined with the particular solution. We assume that the vector of the external forces $f^T = [f_x, f_z]$ is a function of the z-coordinate only. Therefore the particular solution U_p is independent of the x-coordinate and is defined by the equation

$$(G + i\omega C - \omega^2 M)U_p = f . \tag{3.174}$$

If $C \neq 0$ there exists a well-defined solution for every frequency ω :

$$U_p = (G + i\omega C - \omega^2 M)^{-1} f \; . \tag{3.175}$$

For $C = 0$ the matrix pencil $G - \omega^2 M$ is singular at the cut-off frequencies. At these frequencies, (3.174) has a solution only if $f(\omega)$ vanish there.

Since the solution is independent of the x-coordinate the vector of internal nodal forces (3.143) is

$$F_p = DU_p \; . \tag{3.176}$$

Combining the solution of the homogeneous problem with the particular solution, the vectors of the total nodal displacements U_T and internal forces F_T are

$$U_T = U_0 + U_p \, , \tag{3.177.a}$$

$$F_T = F_0 + F_p \; . \tag{3.177.b}$$

Using (3.145) we may rewrite the last equation to

$$F_T = H(\omega) U_0 + F_p \; . \tag{3.178}$$

Using (3.177.a) we replacing U_0 in the last equation and obtain the expression of the DtN-map which includes the effects of external forces

$$F_T = H(\omega)(U_T - U_p) + DU_p \; . \tag{3.179}$$

Observe that the sum of the internal forces depends of the difference $U_T - U_p$. These equations are valid also with respect to any modal basis. If the matrices G, C, M are simultaneously diagonalizable with respect to the modal basis, the computation of the particular solution is reduced to a set of scalar equations.

Earthquake loadings are treated in the same way. We assume that the ground motion is composed of vertically propagating SV- and P-waves. Furthermore, we assume that the ground motion of the free field is specified on the surface of the stratum by the horizontal and vertical components $U_g^T = [U_{gx}(t), U_{gy}(t)]$. Hence, to obtain the motion at the bottom of the stratum we have to compute the displacements U_b solving the system

$$(G + i\omega C - \omega^2 M) U_p = 0 \tag{3.180}$$

with the condition that the displacement on the surface is U_g. Observe that because of this additional condition the boundary conditions of (3.180) are different from those in (3.174) so that the system (3.180) differs from that of (3.174).

3.5   The layered solid-fluid stratum

So far, we have developed DtN-maps of plane semi-infinite strata being composed by fluids or solids. With the same method, we can construct DtN-maps of semi-infinite strata consisting of layers of solids and fluids. In these strata we have to impose special conditions at the interface between solids and fluids so that the displacement or the velocity and the stresses are continuous. In this section, we analyze a plane semi-infinite stratum Ω_F which is composed by a solid layer Ω_s coupled with a fluid layer Ω_f as shown in Figure 3.8. We assume that the solution is independent of the coordinate y which is orthogonal to the (x,z)-plane. For the sake of simplicity, we assume that the fluid as well as the solid are homogeneous and isotropic. The motion of the fluid is described by the velocity potential. The governing equations for the fluid and the solid are found in the sections 3.1.1 and 3.2.1. Boundary conditions are formulated at the boundaries Γ_b and Γ_t of the stratum. Except for several details, this approach corresponds to that developed by Lotfi et al. [LTF86].

3.5.1   The harmonic wave solution

To construct the DtN-map we proceed as in chapters 4 and 5 with fluids and solids strata. The solutions of the homogeneous equations of motion in the form of harmonic waves are

$$\varphi(t, x,z) = \varphi_z(z)e^{-ik_{\omega}x + i\omega t} \quad , \tag{3.181.a}$$

$$U(x, z) = U_z(z)e^{-ik_{\omega}x + i\omega t} \quad . \tag{3.181.b}$$

Inserting the last equations into the equations of motions and modelling the resulting equations defined on Ω_z with finite elements we obtain the two matrix equations

$$(k^2A_f + G_f - \omega^2 M_f)\Phi = f_{\Phi}^i \quad , \tag{3.182.a}$$

$$(k^2A_s + kiB_s + G_s - \omega^2 M_s)U = f_U^i \quad . \tag{3.182.b}$$

The vectors f_{Φ}^i, f_U^i consider the coupling conditions between the solid and the fluid layers at the solid-fluid interface Γ_i. The first condition is a kinematic condition and establishes the continuity of the velocity perpendicular to the interface

$$v_s \cdot n_{sf} = v_f \quad . \tag{3.183}$$

Expressing the velocity of the fluid by the velocity potential and considering that the interface is parallel to the x-axis, we obtain

$$i\omega U_z = \varphi_{z,z} \quad . \tag{3.184}$$

The second condition enforces the continuity of the stresses at the interface Γ_i

$$\sigma \cdot n_{sf} = -p \quad , \tag{3.185}$$

where the stress vector is given by $\sigma^T = [\sigma_{zx}, \sigma_{zz}]$ and p is the pressure in the fluid. Because the interface is parallel to the x-axis (3.185) transforms to

$$\sigma_{zx} = 0 \qquad \sigma_{zz} = -i\omega\rho_f\varphi_z \quad . \tag{3.186}$$

In order to obtain symmetric matrices we multiply (3.100) as well as (3.184) with $-\rho_f$. Combining the resulting equations with (3.182.b) and (3.186) yields

$$\left(k^2\begin{bmatrix} -\rho A_f & 0 \\ 0 & A_s \end{bmatrix} + k\begin{bmatrix} 0 & 0 \\ 0 & iB_s \end{bmatrix} + \begin{bmatrix} -\rho G_f & 0 \\ 0 & G_s \end{bmatrix} + i\omega\begin{bmatrix} 0 & L \\ L^T & 0 \end{bmatrix} - \omega^2\begin{bmatrix} -\rho_f M_f & 0 \\ 0 & M_s \end{bmatrix}\right)\begin{bmatrix} \varphi \\ u \end{bmatrix} = 0 \quad . \tag{3.187}$$

The only nonzero component of the matrix L is $L_{mn} = \rho$, where m is the number of degrees of freedom of the fluid and n the number of degrees of freedom of the vertical displacement of the solid at the interface. Introducing in (3.187) a Rayleigh damping term $C_j = \alpha_R M_j + \beta_R G_j$ for

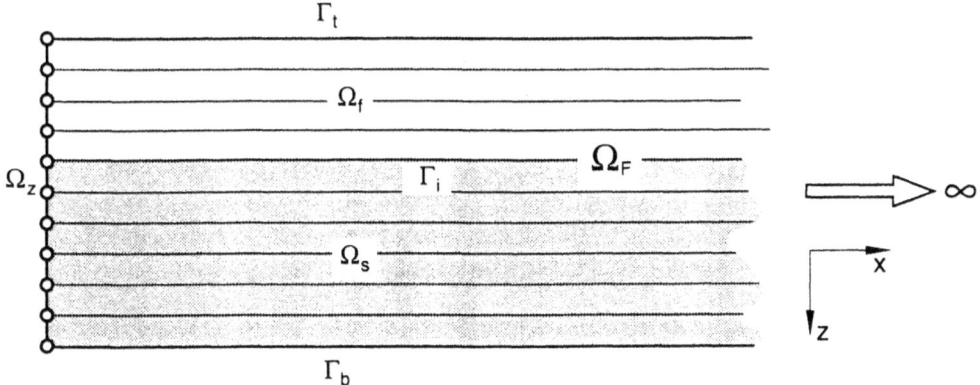

Figure 3.8: Coupled semi-infinite solid-fluid stratum with the finite element discretization of Ω_z

the fluid as well as for the solid we obtain

$$(k^2 A + kiB + G + i\omega L + i\omega C - \omega^2 M)U = 0 \quad . \tag{3.188}$$

where C is

$$C = i\omega \begin{bmatrix} -\rho C_f & 0 \\ 0 & C_s \end{bmatrix} . \tag{3.189}$$

Observe that (3.188) has the same formal structure of (3.128). However, because of the minus sign in front of the block matrices associated with the degrees of freedom of the fluid, the matrices A, G, C and M of (3.188) are still symmetric but indefinite.

3.5.2 Analysis of the eigenvalue problem

If the vector of the nodal degrees of freedom is rearranged in the partitioned form

$$U^T = [\varphi^T, u_z^T, w_z^T] , \tag{3.190}$$

where the vector φ contains the velocity potentials, u_z the horizontal displacements and w_z the vertical displacements, then the eigenvalue problem (3.188) can be written as

$$\left(k^2 \begin{bmatrix} -\rho A_f & 0 & 0 \\ 0 & A_{sx} & 0 \\ 0 & 0 & A_{sz} \end{bmatrix} + k \begin{bmatrix} 0 & 0 & 0 \\ 0 & 0 & iB_{xz} \\ 0 & -iB_{xz}^T & 0 \end{bmatrix} + \begin{bmatrix} -\rho G_f & 0 & 0 \\ 0 & G_{sx} & 0 \\ 0 & 0 & G_{sz} \end{bmatrix} + \right.$$

$$\left. i\omega \begin{bmatrix} -\rho C_f & 0 & L \\ 0 & C_{sx} & 0 \\ L^T & 0 & C_{sz} \end{bmatrix} - \omega^2 \begin{bmatrix} -\rho M_f & 0 & 0 \\ 0 & M_{sz} & 0 \\ 0 & 0 & M_{sz} \end{bmatrix} \right) \begin{bmatrix} \varphi \\ u_z \\ w_z \end{bmatrix} = 0 \tag{3.191}$$

This structure follows directly from the structure of the element matrices A_e, B_e, G_e and M_e of the solid and the fluid. In addition, we assume that the damping matrix C has the partitioned form given in (3.191) and that the boundary condition preserves the block structure of (3.191). Observe that the matrix B couples the subvector u_z of the horizontal displacements with the subvector w_z of the vertical displacements. In addition, the matrix L couples the subvector of the velocity potentials with that of the vertical displacements w_z. Hence, in general, the eigenvectors of (3.191) will contain nonzero components of each subvector.

Now assume $k(\omega)$ is an eigenvalue of (3.130) and U_k is the associated eigenvector

$$U_{(k)}^T = [\varphi^T{}_{(k)}, u_{z(k)}^T, w_{z(k)}^T] \tag{3.192}$$

Then it follows from (3.130) that $\xi(\omega) = -k(\omega)$ and

$$U_\xi^T = [\varphi^T{}_{(k)}, -u_{z(k)}^T, w_{z(k)}^T] \tag{3.193}$$

is also an eigenvalue-eigenvector pair of (3.130). If the pair (k, U_k) is assumed to be a harmonic wave travelling in the positive x-direction then the pair (ξ, U_ξ) is a harmonic wave travelling in the negative x-direction. Observe that the backward travelling harmonic wave differs only in the sign of the subvectors of the horizontal displacements. Because the matrix A has full rank, the eigenvalues $k(\omega)$ are bounded whenever the frequency ω is bounded. In addition, the eigenvalues $k(\omega)$ are continuous functions of ω.

An asymptotic analysis of (3.130) for $\omega \to \infty$ shows that the eigenvalue problem may be decomposed into three uncoupled eigenvalue problems formulated in the subspaces of the subvectors φ, u_z, w_z

$$\left(k^2 \begin{bmatrix} -\rho A_f & 0 & 0 \\ 0 & A_{sx} & 0 \\ 0 & 0 & A_{sz} \end{bmatrix} - \omega^2 \begin{bmatrix} -\rho M_f & 0 & 0 \\ 0 & M_{sz} & 0 \\ 0 & 0 & M_{sz} \end{bmatrix} \right) \begin{bmatrix} \varphi \\ u_z \\ w_z \end{bmatrix} = 0 \tag{3.194}$$

Because the block matrices A_j are proportional to M_j the wave-numbers $k(\omega)$ (eigenvalues) associated with the horizontal displacements are all equal to

$$k_f = \pm\omega c_f, \; k_x = \pm\omega c_p \; \text{and} \; k_z = \pm\omega c_s, \tag{3.195}$$

where c_f is the wave speed of the fluid and

$$c_p = \sqrt{\frac{(\lambda + 2\mu)}{\rho}} \qquad c_s = \sqrt{\frac{\mu}{\rho}} \tag{3.196}$$

are the speeds of the longitudinal and transverse waves. Because the eigenvalues are equal in every subspace, every set of independent vectors is a set of eigenvectors.

The cut-off frequencies of the stratum are defined by the reduced eigenvalue problem resulting from the additional condition $k = 0$ and $C = 0$

$$\left(\begin{bmatrix} -\rho G_f & 0 & 0 \\ 0 & G_{sx} & 0 \\ 0 & 0 & G_{sz} \end{bmatrix} + i\omega \begin{bmatrix} 0 & 0 & L \\ 0 & 0 & 0 \\ L^T & 0 & 0 \end{bmatrix} - \omega^2 \begin{bmatrix} -\rho M_f & 0 & 0 \\ 0 & M_{sx} & 0 \\ 0 & 0 & M_{sz} \end{bmatrix} \right) \begin{bmatrix} \varphi \\ u_z \\ w_z \end{bmatrix} = 0 \; . \tag{3.197}$$

Observe that the matrix L still couples the subvector of the velocity potentials with that of the vertical displacements. Since the subvector of the horizontal displacements is uncoupled from the other two subvectors the set of eigenvectors associated with the cut-off frequencies are composed of a set of eigenvectors with pure horizontal displacements and a set of eigenvectors which spans the vector space $V_\varphi \oplus V_{w_z}$. It can be shown that all the eigenfrequencies $\omega_{(k)}$ are real and the eigenvectors of the subspace $V_\varphi \oplus V_{w_z}$ are given in the form

$$v_{(k)} = \begin{bmatrix} i\varphi_{(k)} \\ w_{z(k)} \end{bmatrix}, \qquad (3.198)$$

where $\varphi^{(k)}$, $w_z^{(k)}$ are real subvectors.

The choice of an adequate modal basis for solid-fluid strata is not as straightforward as for pure fluid or solid strata. This is because the coupling between solid and fluid layers is much more complicated. In many cases, we may choose the set of eigenvectors Ψ_f and Ψ_s defined by the two reduced eigenvalue problems

$$(G_f - \omega^2 M_f)\psi_f = 0 \quad \text{and} \quad (G_s - \omega^2 M_s)\psi_s = 0 \qquad (3.199)$$

The boundary conditions for the node on the interface are assumed to be

$$\varphi_{z,z} = 0 \qquad \sigma_{zz} = 0 \qquad (3.200)$$

If the solid layer is assumed to be homogeneous and isotropic, this basis system has the advantage that it diagonalizes the singular part of the DtN-map kernel for $\omega \to \infty$. An alternative basis system may be defined with the eigenvalue problem (3.197). But it has the disadvantage that the singular part of the DtN-map kernel is not diagonalizable.

3.5.3 The DtN-map

The DtN-map is constructed using the results described in the sections 3.1.5 and 3.4.6. Assume that the displacements of the stratum are described with respect to the set of eigenvectors of (3.188)

$$U = \Psi_\omega \gamma . \qquad (3.201)$$

Then the force vector is given by

$$f = -(iA\Psi_\omega K + D\Psi_\omega)\gamma , \qquad (3.202)$$

where K is the matrix containing the eigenvalues of (3.188) and A corresponds to the matrix A in (3.188) and D is given by

$$D = \begin{bmatrix} 0 & 0 \\ 0 & D_s \end{bmatrix}, \qquad (3.203)$$

D_s is assembled with the element matrices D_e defined in (3.126). The minus sign in front of the block matrix A_f in (3.188) agrees with the finite element formulation of the fluid domain of the near field. Combining (3.201) with (3.202) we eliminate the participation factors γ and obtain

$$f = -(iA\Psi_\omega K\Psi_\omega^{-1} + D)U = -H_\omega U . \qquad (3.204)$$

Hence the DtN-map kernel is

$$H_\omega = iA\Psi_\omega K\Psi_\omega^{-1} + D \quad . \qquad (3.205)$$

This form is also valid with respect to every basis system. Applying the method of section 3.4.7 we can prove that the DtN-map kernel (3.205) is symmetric.

3.5.4   Including external forces in the DtN-map

As usual, we assume that the vector of the external forces $f^T = [-\rho f_\varphi, f_x, f_z]$ is a function of the z-coordinate only. Therefore the particular solution U_p is independent of the x-coordinate and will be defined by the equation

$$(G + i\omega L + i\omega C - \omega^2 M)U_p = f \quad . \tag{3.206}$$

For $C \neq 0$ there exists a well-defined solution

$$U_p = (G + i\omega L + i\omega C - \omega^2 M)^{-1} f \quad . \tag{3.207}$$

According to section 3.5.2 for $C = 0$ the matrix pencil $G + i\omega L - \omega^2 M$ is singular at the cut-off frequencies and (3.174) has a solution only if $f(\omega)$ vanishes there. If the matrices G, C, M are simultaneously diagonalizable with respect to the modal basis, the computation of the particular solution is reduced to the solution of a set of scalar equations.

Since the particular solution is independent of the x-coordinate the vector of the nodal forces simplifies to

$$F_p = DU_p , \tag{3.208}$$

where D is given by (3.203). Combining the solution of the homogeneous problem with the particular solution, the vectors of the total nodal displacements U_T and internal forces F_T are

$$U_T = U_0 + U_p , \tag{3.209.a}$$

$$F_T = F_0 + F_p . \tag{3.209.b}$$

Using (3.145), the last equation becomes

$$F_T = H_\omega U_0 + F_p \quad . \tag{3.210}$$

Replacing U_0 in (3.210) using (3.209.a) we obtain the DtN-map which includes the effects of external forces

$$F_T = H(\omega)(U_T - U_p) + DU_p \quad . \tag{3.211}$$

Observe that the total forces depends on the difference $U_T - U_p$. These equations are valid also with respect to any modal basis. Earthquake loads resulting from the ground motion of vertically propagating SV- and P-waves are included using the same procedure described in section 3.4.9.

Chapter 4

Approximation of scalar DtN-maps I

In this chapter, we analyse a class of linear time-invariant systems which are defined by convolution integrals. We shall investigate how the kernel of these convolution integrals have to be so that the underlying systems are causal and stable. Both these properties can be related to a class of analytic functions in the complex right half-plane. We shall show that DtN-map kernels generally belongs to this class of functions. Later, we shall develop two methods to uniformly approximate DtN-maps on the imaginary axis. These methods uses a Möbius transformation to simplify the approximation procedure. Finally, we shall establish a link between these approximations and transfer functions of systems of linear difference and differential equations.

4.1 Systems defined by convolution integrals

4.1.1 Normed function spaces

Throughout this work, we shall frequently work with so-called normed spaces. This is mainly because approximation problems obtain an exact meaning only with respect to these spaces. Therefore, we are obliged to introduce some very elementary concepts of linear normed spaces. For a detailed presentation of normed spaces which is readable even for engineers we refer to [Kre78, NS82] and, for Hilbert spaces, to [DM90].

A normed space is a pair $(U, \|\cdot\|)$, where U is a set (e. g. numbers, functions) and $\|\cdot\|$ is a norm defined on U. A norm is a real functional mapping every nonzero element u belonging to U to a real positive number. With the norm we can introduce a metric in a normed space $(U, \|\cdot\|)$. The metric allows to define a distance between two elements u and v of the normed space. The distance is defined by way of the norm as $d(u, v) = \|v - u\|$, where u and v are elements of the normed space.

In this work, the elements of normed spaces, that is of the set U, are always functions. These functions are defined in so-called domains of definition X. In this works, the domain of definitions are always subsets of the open complex plane. The range of the functions of U, that is the set of all values $y = f(x)$, where $f \in U$ and $x \in X$, are subsets of the complex plane too. Now let us list several such domains (Figure 4.1). In the following, we shall refer to the set of real

75

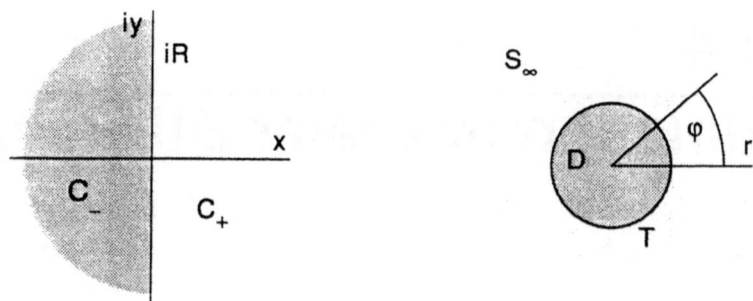

Figure 4.1: Domains in the complex plane

numbers by R, to the set of positive real numbers $x > 0$ by R_+ and to the set of negative real numbers $x < 0$ by R_-. The imaginary axis in the complex plane is referred by iR, the open complex half-plane $Re(s) > 0$ by C_+ and the open complex half-plane $Re(s) < 0$ by C_-. Furthermore, we refer with T to the unit circle $|z| = 1$ in the complex plane, with D to the open unit disk $|z| < 1$ and with S_∞ to the open domain $|z| > 1$.

The most important normed spaces are the spaces of Lebesgue measurable real or complex functions $u(x)$ denoted by $L_p(X)$. The norm of these spaces is defined by the integral

$$\|u\|_{L_p(X)} = \left(\int_X |u(x)|^p dx \right)^{\frac{1}{p}} , \tag{4.1}$$

where p is a real number with $1 \le p < \infty$. Obviously, the last integral have to exist, that is $\|u\|_{L_p(X)} < \infty$. Contrary, the normed space $L_\infty(X)$ of bounded real or complex functions is defined by

$$\|u\|_{L_\infty(X)} = \sup_{x \in X} |u(x)| . \tag{4.2}$$

Similarly, the Hardy spaces $H_p(X)$, with $1 \le p < \infty$, of analytic functions defined in the domain of definitions C_+ or C_- have a norm defined by

$$\|u\|_{H_p(D)} = \sup_{c \to 0} \left(\frac{1}{2\pi i} \int_{c - i\infty}^{c + i\infty} |u(s)|^p ds \right)^{\frac{1}{p}} , \tag{4.3}$$

The integral is computed along any vertical line $Re(s) = c$, where c is a real constant which belongs to the domain of definition. That is, $c > 0$ for C_+ and $c < 0$ for C_-. The Hardy spaces $H_\infty(C_+)$ and $H_\infty(C_-)$ of complex functions bounded in C_+ and in C_-, respectively, have the norm

$$\|u\|_{H_\infty(X)} = \sup_{s \in X} |u(s)| , \tag{4.4}$$

where $s \in C_+$ or $s \in C_-$. Next we consider the Hardy spaces $H_p(X)$, $1 \le p < \infty$, of analytic functions defined in the open disk D or in the open domain S_∞, respectively with norms defined by

$$\|u\|_{H_p(X)} = \sup_{|z| \to 1} \left(\frac{1}{2\pi i} \int_0^{2\pi} |u(z)|^p d\theta \right)^{\frac{1}{p}} , \tag{4.5}$$

where the integral is defined along any circle $z = re^{i\theta}$ with r real and positive. Where $r < 1$ for

$H_p(D)$ and $r > 1$ for $H_p(S_\infty)$. The Hardy spaces $H_\infty(X)$ of complex functions bounded in the open disk D and in the open domain S_∞, respectively, have the norm

$$\|u\|_{H_\infty(X)} = \sup_{z \in X} |u(z)|. \tag{4.6}$$

Remarks:
- The concept of Lebesgue measurable functions is very important for the study of many theoretical aspects of integration. However, DtN-map kernels are rather "regular" functions so that the integrals can be evaluated in the context of the Riemann integration theory.
- In this work, the domain of definitions X of the functions belonging to a Lebesgue space $L_p(X)$ are always one-dimensional. That is, dx in (4.1) is always a scalar.
- If the domain of definition X is infinite the functions belonging to $L_p(X)$ or $H_p(X)$, $1 \le p < \infty$, must vanish for $x \to \infty$. Contrary, the functions belonging to $L_\infty(X)$ or $H_\infty(X)$ needs only to be bounded.

In the context of normed spaces the so-called separable Hilbert spaces have particularly nice properties. In these spaces, we can introduce the concept of orthogonality. This makes them look very similar to the vector spaces of linear algebra. In fact, every separable Hilbert space has an countable orthonormal basis so that each element of the space can be defined with respect to this basis. That is, if $f(s)$ is an arbitrary element of a separable Hilbert space then it can be represented by the Fourier series expansion

$$f(s) = \sum_{k=-\infty}^{\infty} f_k \phi_k(s), \tag{4.7}$$

where the set $\{f_k, k \in N\}$ are the Fourier coefficients and the set of functions $\{\phi_k, k \in N\}$ is the orthonormal basis of the separable Hilbert space. The Fourier coefficients are defined by the inner product $f_k = \langle f_k, \phi_k \rangle$ which represents the projection of $f(s)$ to the basis $\phi_k(s)$. In Hilbert spaces, the norm is defined by an inner product which is essentially a generalization of the dot product of vector spaces.

The Lebesgue spaces $L_2(R)$ and its closed subspaces $L_2(R_+)$ and $L_2(R_-)$ are Hilbert spaces. Furthermore, $L_2(iR)$ is a Hilbert space. Its norm is induced by the inner product

$$\langle u, v \rangle = \int_{-\infty}^{\infty} u(x)\bar{v}(x)dx. \tag{4.8}$$

The orthonormal basis of some of these spaces will be introduced in section 4.2.2. The Hilbert space $H_2(C_+)$ with norm $\|u\|_{H_2(C_+)} = \langle u, u \rangle^{1/2}$ is induced by the inner product

$$\langle u, v \rangle = \frac{1}{2\pi i} \int_{c-i\infty}^{c+i\infty} u(s)\bar{v}(s)ds. \tag{4.9}$$

On the unit disk we may define the Hilbert space $H_2(D)$ whose norm $\|u\|_{H_2(D)} = \langle u, u \rangle^{1/2}$ is induced by the inner product

$$\langle u, v \rangle = \frac{1}{2\pi i} \int_0^{2\pi} u(z)\bar{v}(z)d\theta. \tag{4.10}$$

In this work, the Hilbert spaces $H_2(D)$, $H_2(S_\infty)$ and $L_2(T)$ will be of outstanding importance. These spaces have particularly simple orthonormal bases which are given by the sets of functions

$\{z^n, n > 0\}$, $\{z^{-n}, n > 0\}$ and $\{e^{in\theta}, \infty < n < \infty\}$, respectively. Observe, that the last set of functions is the classical orthonormal basis defined by trigonometric functions. Therefore, any complex function belonging to $L_2(T)$ can be expanded in a trigonometric Fourier series

$$f(\theta) = \sum_{k=-\infty}^{k=\infty} f_k e^{ik\theta} \tag{4.11}$$

with Fourier coefficients given by

$$f_k = \frac{1}{2\pi} \int_{-\pi}^{\pi} f(\theta) e^{-ik\theta} d\theta. \tag{4.12}$$

4.1.2 Causality and analyticity of systems defined by convolution integrals

Let us consider systems defined by the convolution integral

$$y(t) = \int_{-\infty}^{\infty} h(t-s)u(s)ds, \tag{4.13}$$

where $u(t)$ is an arbitrary bounded input function and $y(t)$ is the output function. The function $h(t)$ is the convolution kernel which is generally a distribution. It completely defines the input-output properties of the system. The system (4.13) is linear and time-invariant. If $h(t)$ is a regular function the output function $y(t)$ is continuous even when $u(t)$ is discontinuous.

The first concept we shall discuss is that of causality. Obviously, proper physical systems have to be causal. Avoiding mathematically correct but too technical definitions, we refer to causality in the following sense: a system is causal if and only if past output is independent of future input. It is well-known that systems described by convolution integrals like (4.13) are causal if the kernel function $h(t)$ defining the input-output properties of the system vanish for negative times [Zem65]. That is,

$$h(t) = 0 \text{ for } t < 0. \tag{4.14}$$

In this case the convolution integral simplifies to

$$y(t) = \int_{-\infty}^{t} h(t-s)u(s)ds. \tag{4.15}$$

Remarks:
- For initial value problems, we have $u(t) = 0$ for $t < 0$. The lower integration bound in (4.15) becomes zero. This corresponds to the usual definition of the convolution integral.
- The characterization of causality given in (4.14) is still valid if $h(t)$ is a distribution. In particular, $h(t)$ may be given by $h(t) = \tilde{h}(t) + a_0\delta(t) + a_1\dot{\delta}(t)$, where $\tilde{h}(t)$ is a regular function. In fact, we have

$$y(t) = \int_{0}^{t} \tilde{h}(t-s)u(s)ds + a_0 u(t) + a_1 \dot{u}(t) \tag{4.16}$$

which is obviously causal. The distributional terms in $h(t)$ describes the nonconvolutionary part of the system.
- DtN-map kernels are usually singular at $s \to \infty$ because of the distributional terms. These are

of the order $O(s)$ and $O(1)$. If these terms are subtracted from $h(s)$ the resulting kernel vanish for $s \to \infty$. In the following, we refer to DtN-map kernels which are regularized in this sense (if not stated otherwise).

The property (4.14) of the kernel $h(t)$ is easy to verify if it is given explicitly. However, usually we will handle with DtN-map kernels $h(t)$ which are defined by means of their Fourier transform. Now, what can be stated about causality if the kernel $h(t)$ is given by a complex function $h(\omega)$ defined on the imaginary axis iR? The Kramers-Kronig relations [Wal72] give a definite answer about the causality of $h(\omega)$. However, these relations are too complicated to be applicable in practice.

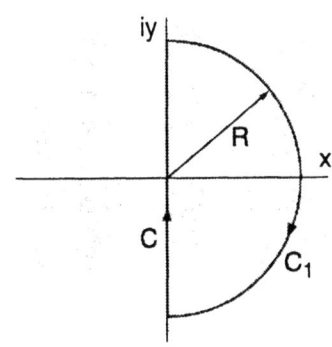

Causality is easy to establish for kernels $h(\omega)$ which are continuous and differentiable on iR. Let us assume that we are able to find a function $\hat{h}(s)$ defined on the complex plane which is analytic in $C_+ \cup iR$ and agree with the kernel $h(\omega)$ on the imaginary axis. That is, $\hat{h}(s = i\omega) = h(\omega)$. In order to recover the kernel $h(t)$ in the time domain, we must back-transform $h(\omega)$ with the inverse Fourier transform formula

$$h(t) = \int_{-\infty}^{\infty} h(\omega)e^{i\omega t}d\omega. \qquad (4.17)$$

Figure 4.2: Path for the contour integration with $t > 0$

This may be done by contour integration. Consider the path $C + C_1$ of Figure 4.2. The integration along the path C corresponds to (4.17) provided that $R \to \infty$. Now, because we assumed that $\hat{h}(s)$ is analytic in $C_+ \cup iR$ we have

$$\int_C \hat{h}(s)e^{st}ds + \int_{C_1} \hat{h}(s)e^{st}ds = 0 \qquad (4.18)$$

for any radius R. Assume that $t < 0$, then Jordan's lemma states that if $\hat{h}(s)$ is analytic in C_+ and $\lim_{s \to \infty} \hat{h}(s) \to 0$ for $Re(s) > 0$ then

$$\lim_{R \to \infty} \int_{C_1} \hat{h}(s)e^{st}ds = 0. \qquad (4.19)$$

Hence, (4.18) becomes

$$h(t) = \int_{-\infty}^{\infty} h(\omega)e^{i\omega t}d\omega = 0 \quad \text{for} \quad t < 0 \qquad (4.20)$$

Therefore, $h(\omega)$ is a causal function. Hence, analyticity in $C_+ \cup iR$ implies causality.

A generalization of this result is Titchmarsh's theorem: $h(\omega)$ is the Fourier transform of a causal function $h(t)$ if there exists a complex function $\tilde{h}(s) \in H_2(C_+)$ so that $\tilde{h}(s) \to h(\omega)$ for $Re(s) \to 0$. Note that $h(\omega)$ has not to be continuous nor continuously differentiable on iR in order to fulfill Titchmarsh's theorem.

Example: The regularized kernel of the DtN-map of the nondissipative restrained rod with $c = \kappa = 1$ is given by $h(\omega) = (\sqrt{1 - \omega^2} - i\omega)$. This function is continuous but not differentiable on iR. In fact, the differential doesn't exist at the cut-off frequency. The function $\tilde{h}(s) = (\sqrt{1 + s^2} - s)$ is equal to $h(\omega)$ on the imaginary axis and analytic in C_+. In addition, $\tilde{h}(s) \in H_2(C_+)$ because it is bounded in C_+ and it is of the order $O(1/s)$ for $s \to \infty$. Hence, because of Titchmarsh's theorem, $h(\omega)$ is the Fourier transform of a causal function. Moreover, the function $\bar{h}(s) = (\sqrt{1 + 2\varepsilon s + s^2} - \varepsilon - s)$ extends the regularized DtN-map kernel of the dissipative rod to the complex plane. It is analytic in $C_+ \cup iR$ because the singularity at the cut-off frequency is moved into C_-. Therefore, the causality of this kernel follows from the simple contour integral argument explained before.

Remarks:

- The extension of DtN-maps developed in chapter 4 to the complex plane is established by replacing $i\omega$ with the complex number s. These DtN-maps are bounded and continuous functions of s.

- Obviously, causality does not imply analyticity in C_+ because causal nonregularized DtN-map kernels may be unbounded when $s \to \infty$.

- The regularized DtN-map kernels $h(t)$ are real functions in R_+. Therefore, their Fourier transforms must obey the well-known relation $h(-\omega) = \bar{h}(\omega)$. Similarly, the Laplace transform of $h(t)$ obeys the relation $h(\bar{s}) = \bar{h}(s)$. This implies that $h(s)$ is a real function for real s. That is, the imaginary part of $h(s)$ vanish on the real axis R.

An alternative approach to causality allows the concept of passivity [Zem65]. A system defined by the convolution integral (4.13) is called to be passive if the functional $W(t)$ defined by

$$W(t) = \int_{-\infty}^{t} u(s)y(s)ds = \int_{-\infty}^{t} u(\tau)\left(\int_{-\infty}^{\infty} h(\tau - s)u(s)ds\right)d\tau \tag{4.21}$$

exists for every time t and is equal to or larger than zero. That is,

$$W(t) = \int_{-\infty}^{t} u(\tau) \int_{-\infty}^{\infty} h(\tau - s)u(s)ds d\tau \geq 0. \tag{4.22}$$

A remarkable result is that passive systems defined by convolution integrals are causal [Zem65]. Another important fact is that kernels of passive systems can be characterized by so-called *positive real functions* [New66]: Let $h(s)$ be the Laplace transform of the kernel $h(t)$. Then, the system (4.13) is passive if $h(s)$ is a positive real function. That is,

1) $h(s)$ is analytic in C_+
2) $h(\bar{s}) = \bar{h}(s)$ in C_+
3) The poles on iR are simple and the residues are positive
4) $\text{Re}(h(s)) \geq 0$ in C_+.

Example: Let us reconsider the regularized DtN-map kernel of the nondissipative restrained rod with $c = \kappa = 1$. It is given by $h(\omega) = (\sqrt{1 - \omega^2} - i\omega)$. This function is positive below the cut-off frequency and zero above. The function $\tilde{h}(s) = (\sqrt{1 + s^2} - s)$ is equal to $h(\omega)$ on the imaginary axis and is analytic in C_+. In addition, condition 3) is satisfied by definition because in the time domain the kernel $h(t)$ is a real function. Hence, $h(s)$ is real positive and therefore $h(\omega)$ is the Fourier transform of a causal function. The DtN-map kernel itself is passive because the function $g(s) = s$ is positive real in C_+ and the sum of two positive real functions is still positive real. Hence, the DtN-map of the nondissipative rod is a passive system and therefore causal. Analogously, we can show that the DtN-map of the dissipative restrained rod is passive and therefore causal.

Remarks:

- The functional (4.21) can be interpreted as an energy functional. We shall meet it again when we consider a stability problem in chapter 11.

- The linear systems defined by the viscous, the asymptotic and the doubly asymptotic boundary conditions discussed in section 3.1 are all passive and therefore causal.

4.1.3 Stability of systems defined by convolution integrals

The next topic we shall investigate concerns the stability of systems defined by convolution integrals defined by (4.15). We assume that $h(t)$ is a regularized kernel. In terms of an input-output description, a system is said to be bounded input/bounded output stable (BIBO-stable) if and only if given a bounded input $u(t)$ the output $y(t)$ is bounded too. That is, there exists a real K with $0 \leq K < \infty$ so that

$$\|y(t)\|_{L_\infty(R_+)} \leq K \|u(t)\|_{L_\infty(R_+)}, \tag{4.23}$$

A sufficient condition for stability is easily derived. Taking the absolute value of the output function $y(t)$ we obtain

$$|y(t)| = \left| \int_0^t h(t-s)u(s)ds \right| \leq \int_0^t |h(t-s)u(s)|ds < \|u(t)\|_{L_\infty(R_+)} \int_0^t |h(s)|ds \tag{4.24}$$

and therefore

$$\|y(t)\|_{L_\infty(R_+)} < \|h(t)\|_{L_1(R_+)} \|u(t)\|_{L_\infty(R_+)}. \tag{4.25}$$

With $K = \|h(t)\|_{L_1(R_+)}$ the convolution integral is BIBO-stable if $h(t)$ belongs to $L_1(R_+)$. This condition is even necessary [Che84].

Let us characterize the kernel functions $h(t) \in L_1(R_+)$ in terms of the input-output behaviour of the convolution integral. Without proof we list three results [Che84]:

1) Given an input function $u(t)$ of finite energy $u(t) \in L_2(R_+)$, the output function will be of finite energy: $y(t) \in L_2(R_+)$. In other words, there exists a constant $C < \infty$ so that $\|y(t)\|_{L_2(R_+)} < C \|u(t)\|_{L_2(R_+)}$.
2) Given an input function $u(t)$ which approaches to a constant u_∞ for $t \to \infty$, the output function $y(t)$ approaches to a constant y_∞ for $t \to \infty$.
3) Given a periodic input function $u(t)$ with a period T, say $u(t) = u(t + T)$, the output function $y(t)$ tends for $t \to \infty$ to a periodic function $y(t) = y(t + T)$ with the same period T, but not necessarily of the same shape as $u(t)$.

On physical grounds, these results reflect exactly what one would expect from proper DtN-map kernels. Item 1) is of great concern, because it implies that $|y(t)| \to 0$ for $t \to \infty$ whenever $|u(t)| \to 0$ for $t \to \infty$. That is, if a finite amount of energy is introduced into the system this energy will finally be completely radiated to infinity. In contrast, if the system would conserve the energy content provided by the input its response wouldn't vanish for $t \to \infty$. Item 2) implies that DtN-maps return a finite generalized force when subjected to static deformations. Item 3) finally implies that DtN-map kernels have no poles on the imaginary axis. In fact, poles on the imaginary axis would generate nondecaying harmonic oscillations so that the period of the output $y(t)$ would be different from that of the input $u(t)$. Therefore, kernel functions belonging to $L_1(R_+)$ characterize quite well the type of systems described by DtN-maps.

The Fourier transform $h(\omega)$ of functions belonging to $L_1(R_+)$ are continuous on iR and $\|h(\omega)\|_{L_\infty(iR)} < \|h(t)\|_{L_1(R_+)}$. That is, the Fourier transform of a function belonging to $L_1(R_+)$ is in $L_\infty(iR)$. Moreover, with the unilateral Laplace transform we can extend $h(\omega)$ to a function $h(s)$ analytic in C_+. As already pointed out, The DtN-maps developed in chapter 3 are bounded and continuous on iR. Hence, in the frequency domain, we may restrict the approximation problem to continuous functions $h(\omega)$ belonging to $L_\infty(iR)$.

Example: Let us reconsider the regularized kernel of the nondissipative constrained rod. In the time domain, the kernel is given by

$$h(t) = \frac{J_1(t)}{t} \quad \text{for} \quad t \geq 0, \tag{4.26}$$

where $J_1(t)$ is the Bessel function of the first kind and of the order one. The kernel is continuous and is of the order $O(t^{-3/2})$ for $t \to \infty$. Therefore, $h(t)$ belongs to $L_1(R_+)$. However, the function $f(t) = th(t) = J_1(t)$ doesn't belong to $L_1(R_+)$. As a consequence, $h(\omega)$ is continuous on iR but not differentiable everywhere. In fact, the first derivative doesn't exist at the cut-off frequency. On the other hand, the regularized kernel of the dissipative constrained rod is given by

$$h_\varepsilon(t) = e^{-\varepsilon t}\frac{J_1(t)}{t} \quad \text{for} \quad t > 0. \tag{4.27}$$

Obviously, the kernel as well as all its derivatives belong to $L_1(R_+)$. Therefore, its Fourier transform $h_\varepsilon(\omega)$ is continuous and has continuous derivatives of any order. As a consequence, it is analytic even on iR.

Remarks:

• The Fourier transform $h(\omega)$ of a kernel belonging to $L_1(R_+)$ need not have a continuous first derivative $dh(\omega)/d\omega$. The first derivative is continuous if $th(t)$ belongs to $L_1(R_+)$.

• Observe that not every function $h(\omega)$ belonging to $L_\infty(iR)$ is the Fourier transform of a function belonging to $L_1(R_+)$.

4.2 Uniform approximation of DtN-maps

4.2.1 The asymptotic series approach

The goal of this section is to demonstrate the limits of the approximation methods based on the asymptotic series approach. In chapter 2 we pointed out that the kernels of the asymptotic boundary conditions of the nondissipative restrained rod have poles on the imaginary axis. Both poles are located below the cut-off frequency and are mainly responsible for the poor accuracy of these boundary conditions in this frequency range. As we shall see, this fact necessarily follows as a consequence of the particular DtN-map kernel and of this approximation approach.

Consider the complex function

$$f(s) = \sqrt{s^2 + 1} - s \tag{4.28}$$

which has the same form of the regularized DtN-map kernel of the nondissipative constrained rod. Its asymptotic expansion for $s \to \infty$ is

$$\tilde{h}(s) \cong \frac{1}{s}\left(\frac{1}{2} - \frac{1}{8s^2} + \frac{1}{16s^4} + \dots\right). \tag{4.29}$$

We shall approximate the term in the bracket with rational functions. It contains only terms with even powers of $1/s$ whose coefficients are real. Assume that the term is approximated by a rational function $g_N(s)$ with denominator polynomial of degree $2N$. Its partial fraction form is

$$g_N(s) = \sum_{j=0}^{N} \frac{a_j}{s^2 - \lambda_j^2}. \tag{4.30}$$

Let us consider functions $g_N(s)$ which is analytic in C_+. That is, the roots of the denominators $s^2 - \lambda_j^2$ have negative real part: $Re(\pm\lambda_j) \leq 0$. Because the coefficients a_j have to be real the

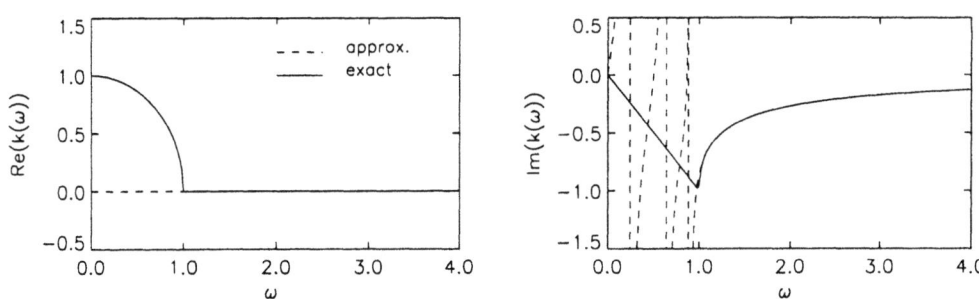

Figure 4.3: [5,6]-Padé-approximation of the function (4.28)

coefficients λ_j^2 must be real and negative. Hence, all the poles of a stable rational function with real coefficients fitting the asymptotic expansion (4.29) are located on the imaginary axis. Observe that this result is independent of the approximation method actually used.

Now assume that the approximations $g_N(s)$ converge absolutely for $|s| > 1$. Clearly, the modulus of the poles of the rational function must be smaller than unity: $|\lambda_j| \leq 1$. But since the function (4.28) is continuous on the imaginary axis, the approximation $g_N(s)$ will never converge uniformly for $|s| < 1$. Exactly this behaviour shows the [5,6]-Padé-approximation of (4.28) in Figure 4.3. It exhibits three poles on the imaginary axis beneath the cut-off frequency. Above the cut-off frequency the approximation is excellent. Apparently, the failure of the asymptotic approach seems to be related to the convergence radius of the asymptotic series (4.29) which is limited by the essential singularity located at the cut-off frequency.

Let us investigate further the effects of singularities on the approximation of DtN-maps. We consider the rational approximation of DtN-maps of the half-space with a semi-circular hole. In section 3.3.4 we have shown that the modal DtN-map kernels at $r = R$ are given by

$$Rh(\omega) = \mathrm{diag}\left(kR\frac{H_j^{(2)\prime}(kR)}{H_j^{(2)}(kR)}\right). \tag{4.31}$$

where $k = \omega/c$. The prime in $H_j^{(2)\prime}(kr)$ symbolizes the derivative with respect to the argument kr. Because it is rather cumbersome to develop the asymptotic expansion of the DtN-map for $\omega \to \infty$ using the exact solution (4.31) we apply a much easier approach which is based on the WKB-method [LS82]. With this method we try to construct an approximate solution of the differential equation (3.72.b) using the exponential form

$$u(r) = C\exp\left(\int_0^r \phi(z, k)dz\right). \tag{4.32}$$

Inserting the last equation in (3.72.b) we obtain a differential equation for $\phi(r, k)$

$$\phi'(r, k) + \phi^2(r, k) + \frac{1}{r}\phi(r, k) + k^2 - \frac{m^2}{r^2} = 0. \tag{4.33}$$

Now, assume that for $\omega \to \infty$ $\phi(r, k)$ has an asymptotic expansion of the form

$$\phi(r, k) = k\sum_{n=0}^{\infty} \phi_n(r)k^{-n}. \tag{4.34}$$

Combining the last equation with (4.33) and equating powers of k to zero we obtain an infinite set of equations defining the functions $\phi_n(r)$:

$$\phi_0^2(r) + 1 = 0$$

$$\phi_0'(r) + 2\phi_0(r)\phi_1(r) + \frac{\phi_0(r)}{r} = 0$$

$$\phi_1'(r) + \phi_1^2(r) + 2\phi_0(r)\phi_2(r) + \frac{\phi_1(r)}{r} - \frac{m^2}{r^2} = 0 \qquad (4.35)$$

$$\cdots$$

$$\phi_{n-1}'(r) + 2\phi_0(r)\phi_n(r) + \frac{\phi_{n-1}(r)}{r} + \sum_{j=1}^{n-1} \phi_j(r)\phi_{n-j}(r) = 0$$

Starting with the first equation we can compute recursively all the functions $\phi_n(r)$. In order to consider outgoing waves we choose the solution $\phi_0(r) = -i$. Next we obtain $\phi_1(r) = -1/2r$. The subsequent functions $\phi_n(r)$ depends on the parameter m. For $m = 0$ we obtain

$$\phi(r,k) = -ik - \frac{1}{2r} - \frac{i}{8kr^2} + \frac{1}{8k^2r^3} + O\left(\frac{1}{k^3}\right). \qquad (4.36)$$

Now, remember that the DtN-map is defined by $u_{,r}(r,\omega) + H(r,\omega)u(r,\omega) = 0$. Because of (4.32) we obtain the remarkable result that the differential equation (4.33) defines the DtN-map kernel. That is,

$$h(r,\omega) = -\phi(r,k) . \qquad (4.37)$$

The series (4.34) can be used to construct Padé approximations of the DtN-map kernel. These have the form

$$rh(r,\omega) = ikr + \frac{1}{2} + \frac{p(kr)}{q(kr)}, \qquad (4.38)$$

where $p(kr)$ is a polynomial of degree $n-1$ and $q(kr)$ is one of degree n. In the following we denote such a rational function as one of degree n. with Figure 4.4 shows the plots of several Padé approximations of the DtN-map kernel of the propagation mode $m = 0$. Note that the approximations have no singularities on the imaginary axis. In fact, the poles of the Padé approximations are real and negative. The approximations become more accurate increasing the degree of the polynomials. The largest errors are in the neighborhood of the singularity $kr = 0$ (the first derivative with respect to kr of the DtN-map kernel is infinite at $kr = 0$). Nevertheless, the result suggests that increasing the degree of the Padé approximations we may obtain a uniformly convergent approximation of the DtN-map kernels.

In this case, the direct asymptotic approach is successful although the DtN-map kernel has an essential singularity. However, this singularity of the DtN-map kernel is located at the origin. This suggests that the breakdown of the direct asymptotic approach experienced with the kernel function (4.28) is due to the fact that it has an essential singularity on point of the imaginary axis having a modulus larger than zero. In fact, it is well-known that Padé approximations of functions having an essential singularity exhibit a cluster of poles in the neighborhood of the singularity. Moreover, if the function to be approximated has a branch cut the poles and zeroes of Padé approximations are on a line radiating out from the origin and passing through the branch point. However, these effects are also found when other rational approximation methods are used provided that these are based on asymptotic expansions.

On the other hand, rational functions can be exactly recovered from their asymptotic series

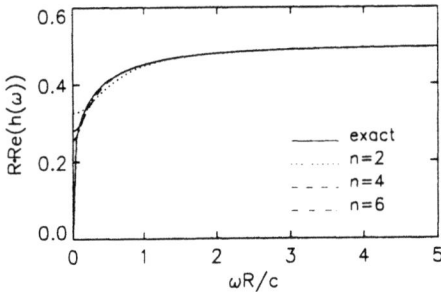 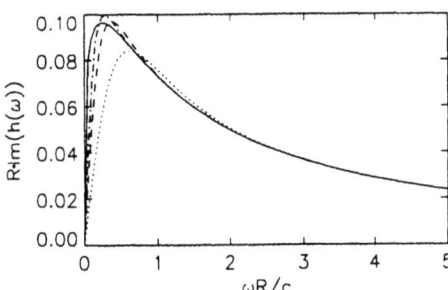

Figure 4.4: Padé approximations of the regularized DtN-map kernel (propagation mode $m = 0$).

although these series don't converge in the complete complex plane. The essential point is that the DtN-map kernel (4.28) is nonrational and has an essential singularitiy. Systems with nonrational kernels define so-called infinite-dimensional linear systems. If kernels of infinite-dimensional linear systems have no singularities the direct asymptotic approach can be very effective as is shown in Figure 4.5. The Padé approximation of degree 2 of the DtN-map kernel associated with the propagation mode $m = 2$ is extremely accurate and that of degree 4 is virtually identical with the kernel.

Hence, the direct asymptotic approach works well only under very special conditions. Clearly, the DtN-map kernels developed in chapter 3 don't share these conditions so that we need a more general approximation method. However, given a nonrational function analytic in C_+, does there exists a uniformly convergent approximation with stable rational functions? Fortunately, the answer is positive. The theorem of Coifman and Rochberg [CR80] assures its existence. Unfortunately, it doesn't suggest how to construct such an approximation.

Remarks:

- The WKB approach sketched in this section can be used to construct absorbing boundary conditions of nonhomogeneous wave-guides. For the nonhomogeneous elastic rod it is similar to the ray theory method described in [BBM88].

- The differential equation (4.33) defining the DtN-map kernel is similar to the differential equation which results applying the so-called consistent infinitesimal finite-element cell method [WS96]. This affinity suggests to use the asymptotic expansion at $\omega \to \infty$ combined with the Padé approximation technique to approximate the solution of the differential equation. This approach is numerically much more effective than any numerical integration method. In addition, the effectiveness of the direct asymptotic expansion approach in approximating DtN-maps of the half-space with a semi-circular hole shows that these class of problems are much easier to approximate than DtN-maps of wave-guides.

4.2.2 The approximation with Laguerre functions

To overcome the problems associated with the asymptotic series approach we have to consider a more general approximation method. This method should be able to uniformly approximate a DtN-map kernel defined on iR. That is, given a DtN-map kernel $h(\omega)$ and an error ε this approximation method should provide an approximation $\tilde{h}(\omega)$ so that $\left\| h(\omega) - \tilde{h}(\omega) \right\|_{L_\infty(iR)} < \varepsilon$. Moreover, the approximation method should be numerically robust and efficient but also general enough to handle a broad class of DtN-map kernels.

Approximation problems become particularly simple when they are formulated in separable Hilbert spaces because we can define an orthonormal basis. The orthogonal projection of the function being approximated to this basis defines the coefficients of the series expansion. Because DtN-map kernels are causal they are nonzero only on R_+. This suggests to use an ortho-

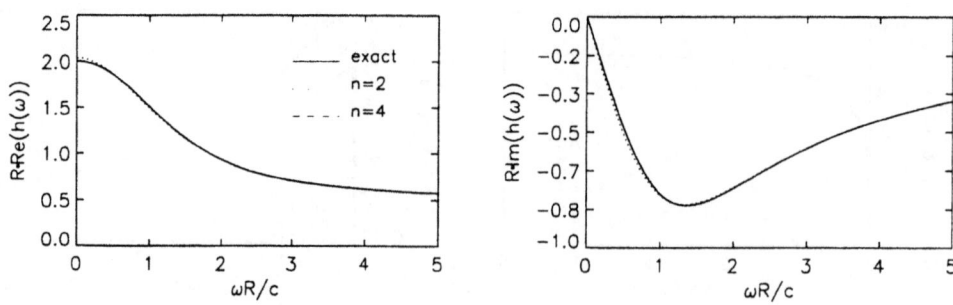

Figure 4.5: Padé approximations of the regularized DtN-map kernel (propagation mode $m = 2$).

normal basis of the Lebesgue space $L_2(R_+)$. A complete orthonormal basis of this space are the generalized Laguerre functions [Leb72, Sze75]. These are defined by

$$\phi_k(t) = \sqrt{2a}e^{-at}L_k(2at) \quad \text{for } k \geq 0 \tag{4.39}$$

where the functions $L_k(t)$ are the Laguerre polynomials

$$L_k(t) = \sum_{n=0}^{k} (-1)^n \binom{k}{n} k(k-1)...(n+1)t^n. \tag{4.40}$$

The Laguerre polynomials are generated by the forward recurrence relation

$$L_0(t) = 1 \qquad L_1(t) = 1 - 2t$$

$$L_k(t) = \frac{1}{k}\{(2k-1-2t)L_{k-1}(t) - (k-1)L_{k-2}(t)\} \tag{4.41}$$

Remarkably, the Laplace transforms of the Laguerre functions are an orthonormal basis of the Hilbert space $H_2(C_+)$. The transformed Laguerre functions are given by

$$\phi_k(s) = \sqrt{2a}\frac{(s-a)^k}{(s+a)^{k+1}} \quad \text{for} \quad k \geq 0. \tag{4.42}$$

The functions $\phi_k(s)$ are analytic and rational in $C_+ \cup iR$. Furthermore, with $s = i\omega$ the set (4.42) forms an orthonormal basis of $L_2(iR)$. Hence, we may try to represent DtN-map kernels by an infinite Fourier series convergent in $H_2(C_+)$

$$h(s) = \sum_{k=0}^{\infty} h_k\phi_k(s) = \sqrt{2a} \sum_{k=0}^{\infty} h_k\frac{(s-a)^k}{(s+a)^{k+1}}. \tag{4.43}$$

The Fourier coefficients h_k are found evaluating the inner product

$$h_k = \langle h, \phi_k \rangle = \frac{1}{2\pi i} \int_{-i\infty}^{i\infty} h(s)\phi_k(-s)ds \quad \text{with} \quad s = i\omega. \tag{4.44}$$

The Fourier series (4.43) with Fourier coefficients defined by (4.44) converges in $H_2(C_+)$ if and only if the sum of the square modulus of the Fourier coefficients converges:

$$\sum_{k=0}^{\infty} h_k\bar{h}_k = \sum_{k=0}^{\infty} |h_k|^2 < \infty. \tag{4.45}$$

In section 4.1.3 we pointed out that reasonable DtN-map kernels belongs to $L_1(R_+)$. However, using Laguerre functions we can approximate DtN-map kernels belonging to $L_2(R_+)$. That is, we capture only those kernels which are elements of the set $L_1(R_+) \cap L_2(R_+)$. Because the set $L_1(R_+) - (L_1(R_+) \cap L_2(R_+))$ is not empty, there are kernels belonging to $L_1(R_+)$ which can't be approximated with a series of Laguerre functions. Hence, we must assume that the regularized DtN-map kernels $h(t)$ belongs to $L_1(R_+) \cap L_2(R_+)$. Then the Laplace transform of $h(t)$ belongs to $H_2(C_+)$.

Let us consider the following approximation strategy. Given a function $h(\omega)$ in $L_2(iR)$ find an approximation $h_N(\omega)$ in the form of a finite series of Laguerre functions. Define an analytic extension $\hat{h}_N(s)$ of $h_N(\omega)$ on C_+. Then, because $h(t)$ belongs to $L_1(R_+) \cap L_2(R_+)$, $\hat{h}_N(s)$ belongs to $H_2(C_+)$ and is a causal and BIBO-stable function.

Remarks:

- Hagstrom [Hag95] proposed Laguerre functions and exponential polynomials (via Legendre polynomials) expansions to approximate the kernel of DtN-maps. Generally, Laguerre functions expansions are more accurate than exponential polynomials expansions.

- Mäkila [Mäk90] and Gu et al. [GKL89] uses Laguerre functions expansions to approximate a special class of linear infinite-dimensional systems (so-called delay systems).

4.2.3 A Möbius transformation

In this section, we discuss a conformal mapping ([MH87, Hen74]) which allows to associate normed spaces defined in C_+ with normed spaces defined in D. This mapping will not only elucidate several theoretical aspects of the approximation but will even greatly simplify the numerical computation of the Laguerre approximation.

Consider the Möbius transformation (also called fractional linear or bilinear transformation) $z = M(s)$ given by

$$z = \frac{s - (a + r)}{s + (a - r)} = \frac{s - p}{s + q}, \tag{4.46}$$

where s is an element of the complex plane. The positive real parameters a and r are defined according to Figure 4.6. Obviously, z is also an element of the complex plane. Therefore, the Möbius transformation (4.46) maps an element of the complex plane to another element of the complex plane. The Möbius transformation is invertible provided that $p + q \neq 0$. Therefore, it is bijective and onto. From (4.46) we deduce that the point at infinity $s \to \infty$ is mapped to $z = 1$. Furthermore, the element of the real axis $s_\infty = r - a$ is mapped to infinity. All other points of the complex plane are mapped to points of the complex plane with finite modulus. Note that the inverse transformation of (4.46) is also a Möbius transformation. It is given by

$$s = \frac{p + qz}{1 - z}. \tag{4.47}$$

Next, we investigate the geometrical properties of this Möbius transformation. Consider the mapping of a point on the vertical straight line L given by $s = r + iy$, $y \in R$ (Figure 4.6). Computing the modulus of $z = M(s)$ yields $|z| = 1$. Therefore, the vertical straight line is mapped to the unit circle T. Furthermore, the right half plane with respect to the vertical straight line L denoted by C_{L+} is mapped onto the unit disk D. In fact, given a point $s = x + iy$ in C_{L+} with $x > r$, we have

$$|z| = \frac{y^2 + (a - (x - r))^2}{y^2 + (a + (x - r))^2} < 1. \tag{4.48}$$

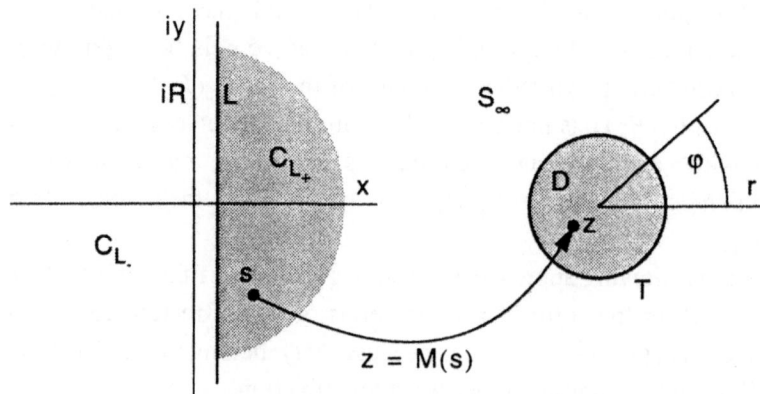

Figure 4.6: The Möbius transformation described by equation (4.46).

Since the Möbius transformation is bijective, the left half-plane with respect to the line L denoted by C_{L-} is mapped onto the open domain S_∞ given by $|z| > 1$. Observe that setting $r = 0$ in (4.46) the Möbius transformation maps iR to T, C_+ to D and C_ to S_∞. This transformation is given by

$$z = \frac{s-a}{s+a} \tag{4.49}$$

$$s = a\frac{1+z}{1-z} \tag{4.50}$$

The transformation (4.46) can be interpreted as a composition of elementary Möbius transformations. In fact, we have $z = M_3(M_2(M_1(s)))$, where M_1 is a translation given by

$$s' = M_1(s) = s - r, \tag{4.51}$$

M_2 is a stretching given by

$$s'' = M_2(s') = \frac{s'}{a} \tag{4.52}$$

and M_3 is the Möbius transformation

$$z = \frac{s-1}{s+1} \tag{4.53}$$

Now assume that $f(s)$ is a function defined in the complex plane C and analytic in C_+. The Möbius transformation relates $f(s)$ to a function $f_s(z)$ analytic in D. Define the function

$$f_s(z) = f(s) \quad\text{, where}\quad s = \frac{p+qz}{1-z}. \tag{4.54}$$

That is, the value of the function $f(s)$ is simply transplanted without modification from the point s to the point z of the complex plane C. The point z is defined by the Möbius transformation (4.46). An example: any function $f(\omega)$ defined on the imaginary axis is mapped by (4.49) to a function $f_s(e^{i\varphi})$ on the unit circle T.

4.2.4 Trigonometric Fourier series and Laguerre series

The Möbius transformation discussed before allows to compute the Fourier coefficients (4.44) very efficiently. Consider the transformation

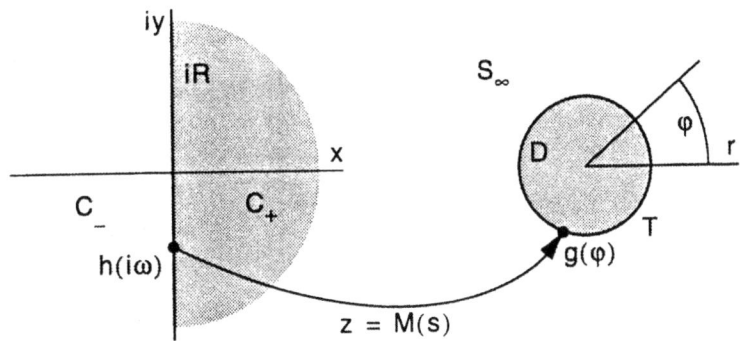

Figure 4.7: Transplantation of functions by way of the Möbius transformation (4.49)

$$s = M(z) = a\frac{1+z}{1-z}.$$ (4.55)

With (4.55) the integral (4.44) evaluated at $c = 0$ becomes

$$h_k = \frac{1}{2\pi i}\int_{|z|=1}\frac{\sqrt{2a}h(M(z))}{1-z}z^{-(k+1)}dz.$$ (4.56)

Hence, the coefficients h_k of the Laguerre series are the Fourier coefficients of the function

$$f(z) = \frac{\sqrt{2a}h(M(z))}{1-z}$$ (4.57)

with respect to the orthonormal basis of $L_2(T)$: the trigonometric Fourier basis. In fact, with $z = e^{i\theta}$ we obtain

$$h_k = \frac{1}{2\pi}\int_{-\pi}^{\pi}f(e^{i\theta})e^{-k\theta}d\theta = f_k \quad \text{for } k \geq 0.$$ (4.58)

Therefore, $h(s)$ given by (4.43) is in $L_2(iR)$ if and only if $f(z)$ given by (4.57) is in $L_2(T)$, because only in this case (4.45) holds. This result is a special case of a general theorem [Hof62]: Let $h(s)$ be an analytic function in C_+ and let $f(z)$ be the analytic function in D defined by $f(z) = h(a(1+z)/(1-z))$. Then $h(s)$ is in $H_p(C_+)$ if and only if $f(z) = (1-z)^{2/p}F(z)$, where $F(z)$ is in $H_p(D)$. In addition, we have $\|F(z)\|_{L_p(T)} = \|h(s)\|_{L_p(iR)}$. Equivalently, we can state that $f(z)$ is in $H_p(D)$ if and only if $h(s) = (a+s)^{2/p}H(s)$, where $H(s)$ is in $H_p(C_+)$. Moreover, $\|f(z)\|_{L_p(T)} = \|H(s)\|_{L_p(iR)}$.

The second part of this theorem suggests another method to compute the Fourier coefficients h_k. Let us assume that $h(t)$ belongs to $L_1(R_+) \cap L_2(R_+)$. Then $h(i\omega) \in L_2(iR)$ so that the Laguerre functions expansion (4.7) exists. Consider the function $f(s) = (a+s)h(s)$. This function is analytic in C_+ and bounded and continuous on iR. It is at most a constant at infinity because $h(s)$ is of the order $O(1/s)$ for $s \to \infty$. $f(s)$ may be written as

$$f(s) = \sqrt{2a}\sum_{k=0}^{\infty}h_k\frac{(s-a)^k}{(s+a)^k}.$$ (4.59)

Combined with the Möbius transformation (4.55) we obtain

$$f(z) = \sqrt{2a} \sum_{k=0}^{\infty} h_k z^k , \qquad (4.60)$$

$f(z)$ is bounded and continuous on T so that its Fourier series exists. The Fourier coefficients are defined by

$$h_k = \frac{1}{2\pi i} \int_{|z|=1} \frac{f(M(z))}{\sqrt{2a}} z^{-(k+1)} dz . \qquad (4.61)$$

The approximation of the DtN-map kernel of the undamped constrained rod with generalized Laguerre functions is shown in Figure 4.8. Contrary to the asymptotic approximation, the approximation with Laguerre functions is bounded on iR. The accuracy of the approximation $h_N(s)$ increases with the number of terms. However, the approximation converges slowly. The real and imaginary parts of the approximated DtN-map kernel in the neighborhood of the cut-off frequency is fairly crude even when 20 terms are considered (see Figure 4.8). This is because the function $h(t)$ is a slowly decaying oscillatory function in the time domain. On the other hand, Laguerre functions are polynomials so that many terms are needed to approximate an oscillatory function. In the time domain, however, the accuracy is much better (see Figure 4.8). The approximation with four terms ($n = 4$) is very rough. However, it already covers the essential features of this kernel. The approximation with ten terms ($n = 10$) is sufficiently accurate up to 5 s and that with twenty terms ($n = 20$) is very accurate up to 20 s. However, as we shall see, Laguerre series expansions are not optimal when approximating oscillatory functions.

Remarks:

- Generalized Laguerre functions have been applied to numerically invert Laplace transforms (Weeks method [Wee66]). The Möbius transformation (4.47) and various methods for trigonometric Fourier series expansion are used to determine the coefficients of the Laguerre series expansion.

- Because of the relation $h(-\omega) = \bar{h}(\omega)$ we obtain $h_z(-\theta) = \bar{h}_z(\theta)$ on the unit circle where $0 \le \theta \le \pi/2$. As a consequence, the Fourier coefficients h_k are real:

$$h_k = \frac{1}{2\pi} \int_{-\pi}^{\pi} h_z(e^{i\theta}) e^{-ik\theta} d\theta = \frac{1}{2\pi} \int_{-\pi}^{0} h_z(e^{i\theta}) e^{-ik\theta} d\theta + \frac{1}{2\pi} \int_{0}^{\pi} h_z(e^{i\theta}) e^{-ik\theta} d\theta$$

$$\qquad (4.62)$$

$$= \frac{1}{2\pi} \int_{0}^{\pi} \bar{h}_z(e^{i\theta}) \overline{e^{-ik\theta}} d\theta + \frac{1}{2\pi} \int_{0}^{\pi} h_z(e^{i\theta}) e^{-ik\theta} d\theta = a_k + \bar{a}_k = \mathrm{Re}(a_k)$$

- The sum of the square of the coefficients of functions belonging to $L_2(T)$ is finite. That is,

$$\sum_{k=0}^{\infty} |h_k|^2 < \infty \qquad (4.63)$$

A sequence of real or complex numbers $\{h_k\}$ satisfying (4.63) is said to belong to the normed space ℓ_2. ℓ_2 is a Hilbert space.

- Generally, the Fourier coefficients have to be computed using numerical Fourier transform techniques (preferably using FFT). These coefficients are approximations of the exact Fourier coefficients. However, these can be made arbitrarily accurate [Hen86].

- In practical applications, given an error ε we shall truncate the infinite series (4.43) after N_ε terms

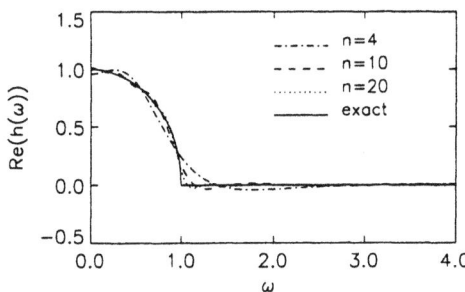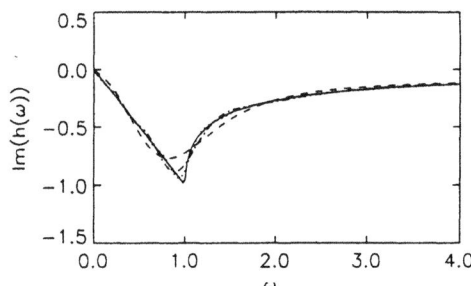

Figure 4.8: Approximations of the function (4.28) with Laguerre functions

$$h_N(s) = \sum_{k=0}^{N_\varepsilon} h_k \phi_k(s),$$ (4.64)

provided that $\left\| h_{N_\varepsilon} - h \right\|_{L_2(C_+)} < \varepsilon$. Observe that because of (4.42) $h_N(s)$ is a rational function with denominator polynomial of degree $N_\varepsilon + 1$ and analytic in $C_+ \cup iR$.

• Observe that the mapped function $f_s(z)$ given by (4.57) is not singular at $z = 1$ because the DtN-map kernels are regularized such that $f(s) \to 0$ for $s \to \infty$.

4.2.5 Uniform convergence of Laguerre-series

Laguerre series converge in spaces induced by quadratic norms $(L_2(iR))$. However, of much more concern is the question if Laguerre series expansions converge uniformly. That is, if the approximation error can be made arbitrarily small in $(L_\infty(iR))$. The answer to this question characterizes the DtN-map kernels we will be able to approximate with arbitrary accuracy.

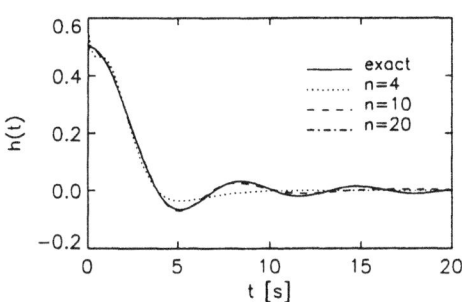

Figure 4.9: Time-domain representation of approximations of the kernel function (4.28) with series of Laguerre functions

The close relationship between Laguerre functions and trigonometric Fourier series on the unit circle suggests to analyse this question using the theory of trigonometric Fourier series on the unit circle T. Assume that $h(s) \in L_2(iR)$ so that the Laguerre series expansion exists. Define the function $f(s) = (s + a)h(s)$. This function is continuous if $h(s)$ is continuous and has the series expansion

$$f(s) = \sqrt{2a} \sum_{k=0}^{\infty} h_k \left(\frac{s-a}{s+a} \right)^k,$$ (4.65)

where the sequence $\{h_k\}$ are the Fourier coefficients of the Laguerre series expansion of $h(s)$. The Möbius transformation (4.55) transforms (4.65) to the Fourier series of the mapped function $f_s(z)$ on T. Because $f(s)$ is bounded and continuous on iR the mapped function $f_s(z)$ is bounded and continuous on T so that $f_s(z)$ belongs to $L_\infty(T)$. Furthermore, because of (4.65) the Fourier series expansion of $f_s(z)$ is

$$f_s(z) = \sqrt{2a} \sum_{k=0}^{\infty} h_k z^k.$$ (4.66)

Hence, if $f_s(z)$ converges in $L_\infty(T)$ then $f(s)$ converges in $L_\infty(iR)$ because of the relation $\left\| f_s(z) \right\|_{L_\infty(T)} = \left\| f(s) \right\|_{L_\infty(iR)}$. However, continuity is not enough to assure uniform convergence of a Fourier series in $L_\infty(T)$. In fact, let $S_n(f(z))$ be the nth partial sum of the Fourier series of $f(z)$ defined by

$$S_n(f_s(z)) = \sum_{k=-n}^{k=n} f_k e^{ik\theta}. \tag{4.67}$$

Then we have this remarkable theorem [Kör88]: Let $f(z)$ be continuous everywhere on T and df/dz be continuous and bounded except on a finite number of points. Then the Fourier series converges uniformly in $L_\infty(T)$. That is,

$$\left\| S_n(f_s(z)) - f_s(z) \right\|_{L_\infty(T)} \to 0 \quad \text{for} \quad n \to \infty. \tag{4.68}$$

Let us apply this theorem to the mapped function $f_s(z)$. It follows that $f(s)$ must be a continuous function on iR. That is, $h(s)$ must be continuous and $sh(s) \to K < \infty$ for $s \to \infty$. In other words, $h(s)$ is of the order $O(1/s)$ for $s \to \infty$. This property requires that the Fourier transform of the first derivative of $h(t)$ exists and that it is bounded for $t \to 0$. Next, let us consider the first derivative df_z/dz. Because of Leibniz's rule and of the Möbius transformation (4.55) we obtain

$$df_z/dz = (df/ds)(ds/dz) = -(h(s) + (s+a)(dh/ds)(s))\frac{(a+s)^2}{2a} \tag{4.69}$$

Hence, dh/ds have to be continuous on iR except on a finite number of points. Observe that the Fourier series converges uniformly even if $(dh/ds)s^2 \to \infty$ for $s \to \infty$. If dh/ds is continuous on iR then the only point of discontinuity is $s \to \infty$.

Example: Let us consider the DtN-map kernel of the dissipative constrained rod given in (4.28). Obviously the kernel is bounded and continuous on iR and $sh(s) = (1 - \varepsilon^2)/2$ for $s \to \infty$. Its first derivative is

$$\frac{dh}{ds} = \frac{s + \varepsilon - \sqrt{s^2 + 2\varepsilon s + 1}}{\sqrt{s^2 + 2\varepsilon s + 1}} \tag{4.70}$$

and is bounded and continuous on iR because the two poles are in the left half plane $s_{1,2} = -\varepsilon \pm i\sqrt{1 - \varepsilon^2}$. Hence the Laguerre series converge uniformly. The DtN-map kernel of the nondissipative rod has $\varepsilon = 0$. The first derivative is singular at the poles $s = \pm i$ so that the kernel has two points of singularity. Therefore, the Laguerre series converges uniformly in $L_\infty(iR)$.

The requirements on the derivative df/ds can be dropped if instead of the Fourier series (4.66) we consider its Cesaro means $s_n(z)$ defined by

$$s_n(z) = \frac{1}{n+1} \sum_{k=0}^{n} f_k(z), \tag{4.71}$$

where $f_n(z) = \sum_{k=0}^{n} h_k z^k$ are the nth partial sums of the Fourier series (4.66). For Cesaro means we have the following theorem [Hof62]: If the function $f(z)$ is continuous on T then the Cesaro means $s_n(z)$ converge uniformly to the function $f(z)$.

Hence, if $h(s) = O(1/s)$ for $s \to \infty$ the sequence of Laguerre-Cesaro functions

$$h_n(s) = \sqrt{2a} \sum_{k=0}^{n} \left(1 - \frac{k}{n+1}\right) h_k \frac{(s-a)^k}{(s+a)^{k+1}}, \tag{4.72}$$

where the coefficients h_k are those of the Laguerre series expansion of $h(s)$, converge uniformly to $h(s)$ on iR when $n \to \infty$. The Laguerre-Cesaro function $h_n(s)$ with $n = 10$ of the kernel function (4.28) is shown in Figure 4.10. The Laguerre-Cesaro function is less accurate than the

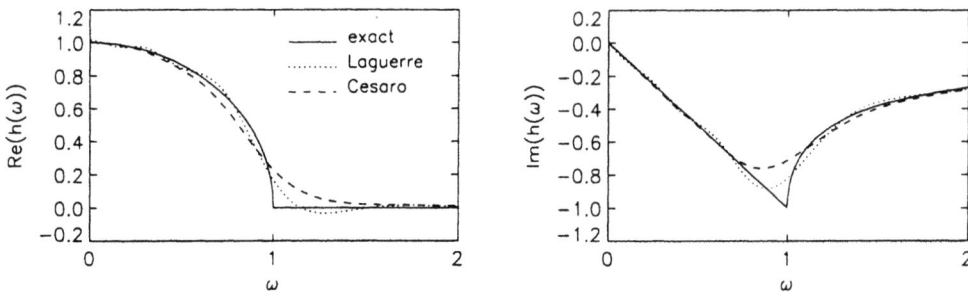

Figure 4.10: Laguerre and Laguerre-Cesaro approximations of the function (4.28)

Laguerre approximation. However, the last oscillates around the kernel function whereas the Laguerre-Cesaro function is much smoother. Nevertheless, the approach with Laguerre-Cesaro functions seems to be less effective than that with Laguerre functions. However, increasing the order of the approximation the difference between these two series becomes more and more insignificant provided that both converge uniformly.

The Fourier series expansion of $f_s(z)$ establishes the causality of the DtN-map kernel $h(s)$. Consider the extension of $f_s(z)$ to the complex plane by defining the function

$$f_s(z) = \sqrt{2a} \sum_{k = -\infty}^{k = \infty} h_k z^k, \tag{4.73}$$

where z is a complex number. This is the Laurent series expansion of $f_s(z)$ at $z = 0$. Let us split the last function in

$$f_+(z) = \sqrt{2a} \sum_{k = 0}^{k = \infty} h_k z^k. \tag{4.74}$$

and

$$f_-(z) = \sqrt{2a} \sum_{k = -\infty}^{k = -1} h_k z^k = \sqrt{2a} \sum_{k = 1}^{k = \infty} h_{-k} z^{-k}. \tag{4.75}$$

Obviously $f_s(z) = f_+(z) + f_-(z)$. Because of the Cauchy integral theorem, $f_+(z)$ is analytic in the unit disk and belongs to $H_\infty(D)$. On the other hand, $f_-(z)$ is singular at $z = 0$ so that it is not analytic in the unit disc. However, $f_-(z)$ is analytic in S_∞ and belongs to $H_\infty(S_\infty)$. These facts give a simple criterion to establish the analyticity of DtN-map kernels: If all coefficients of the Fourier series of $f_s(z)$ with negative indices vanish, then $h(s)$ is analytic in C_+, belongs to $H_2(C_+)$ and is therefore causal. This criterion is easy to apply in practice.

4.2.6 The approximation with trigonometric Fourier series on the unit circle

Finally, let us study an alternative approximation approach for DtN-map kernels $h(s)$. Let $h(s)$ be bounded and continuous on iR. Consider the mapped kernel $h_s(z)$ resulting when the Möbius transformation (4.55) is directly applied to $h(s)$. The function $h_s(z)$ is bounded and continuous on T. Therefore there exists a trigonometric Fourier series which converges to $h_s(z)$ in $L_2(T)$. Mapping this series back using (4.49) we obtain the series expansion

$$h(s) = \sum_{k = 0}^{\infty} \tilde{h}_k \frac{(s - a)^k}{(s + a)^k} \tag{4.76}$$

The functions

$$\psi_k(s) = \frac{(s-a)^k}{(s+a)^k} \qquad k \geq 0 \tag{4.77}$$

are bounded, continuous and analytic in $C_+ \cup iR$ provided that $a > 0$. Their modulus is exactly unity, that is, $|\psi_k(s)| = 1$. These functions are an orthonormal basis of $L_2(iR)$ with respect to the weight function

$$w(s) = \frac{2a}{a^2 - s^2}. \tag{4.78}$$

That is, the coefficients $\tilde{h}_k$ are defined by

$$\tilde{h}_k = \langle h, \psi_k \rangle = \frac{1}{2\pi i} \int_{-i\infty}^{i\infty} w(s)h(s)\psi_k(-s)ds \qquad s = i\omega. \tag{4.79}$$

Now, all the results about uniform convergence of the function $f(s) = (s+a)h(s)$ obtained before can be directly applied to the DtN-map kernel as well. That is,

If $h(s)$ is bounded and continuous and its first derivative $(dh/ds)(s)$ is continuous except at finitely many points then the series

$$\tilde{h}(s) = \sum_{k=-\infty}^{\infty} \tilde{h}_k \frac{(s-a)^k}{(s+a)^k} \tag{4.80}$$

converge uniformly to $h(s)$ on iR.

The coefficients $\tilde{h}_k$ are the Fourier coefficients of the Fourier series expansion of $h_s(z)$:

$$h_s(z) = \sum_{k=-\infty}^{k=\infty} \tilde{h}_k z^k. \tag{4.81}$$

Clearly, the result about the Cesaro means applies as well:

If $h(s)$ is bounded and continuous then the sequence of functions

$$s_n(s) = \sum_{k=-n}^{n} \left(1 - \frac{|k|}{n+1}\right) \tilde{h}_k \frac{(s-a)^k}{(s+a)^k}, \tag{4.82}$$

converges uniformly to $h(s)$ on iR. Furthermore, If all coefficients $\tilde{h}_k$ of the Fourier series of $h_s(z)$ with negative indices vanish, then $h(s)$ is analytic in C_+ and causal. The causality follows from the observation that the function $g(s) = h(s)/(s+a)$ belongs to $H_2(C_+)$. In fact, the sequence $\{\tilde{h}_k\}$ belongs to ℓ_2 because $g(s)$ has a Laguerre series expansion

$$g(s) = \frac{1}{2a} \sum_{k=0}^{\infty} h_k \frac{(s-a)^k}{(s+a)^{k+1}} \tag{4.83}$$

Hence, $g(s)$ is the Laplace transform of a causal function. Because of $h(s) = (s+a)g(s)$ we obtain in the time domain:

$$\begin{array}{ll} h(t) = ag(t) + \dot{g}(t) + g(0+) & \text{for} \quad t > 0 \\ h(t) = 0 & \phantom{\text{for}} \quad t < 0 \end{array} \tag{4.84}$$

so that $h(t)$ is causal.

If $h(s)$ belongs to $H_2(C_+)$ the coefficients h_k of the Laguerre series expansion (4.43) can be related to the coefficients of the series expansion (4.81). Applying the Möbius transformation (4.48) to both series we obtain

$$\frac{1}{\sqrt{2a}}(z-1)\sum_{k=0}^{\infty}h_k z^k = \frac{1}{\sqrt{2a}}\left(h_0 + \sum_{k=1}^{\infty}(h_k - h_{k-1})z^k\right) = \tilde{h}_0 + \sum_{k=1}^{\infty}\tilde{h}_k z^k. \tag{4.85}$$

Because Laurent series expansions are unique, the equality must hold for every power of z so that

$$\sqrt{2a}\tilde{h}_0 = h_0 \qquad \sqrt{2a}\tilde{h}_k = h_k - h_{k-1}, \quad k \geq 1. \tag{4.86}$$

This relations holds only if $h(s)$ is in $L_2(C_+)$. In fact, because of the Minkowsky inequality, if the sequence $\{h_k\}$ belongs to ℓ_2 the sequence $\{\tilde{h}_k\}$ belongs to ℓ_2 too. Hence, for kernels which are in $L_2(R_+)\cap L_1(R_+)$ the Laguerre and the Fourier series approach are essentially equivalent. That is, all DtN-map kernels which can be uniformly approximated by a Laguerre series expansion can be approximated uniformly by the Fourier series (4.80).

However, the reverse is not true. Consider the function $h(t) = e^{-t}/\sqrt{\pi t}$. It is unbounded at $t = 0$ and belongs to $L_1(R_+)$ but not to $L_2(R_+)$. Its Laplace transform is $h(s) = 1/\sqrt{s+1}$ which is bounded on the imaginary axis and in C_+. The mapped function $h(z) = \sqrt{(1-z)/2}$ is bounded and continuous on T and its first derivative $dh/dz = -1/\sqrt{4(1-z)}$ is continuous everywhere except at $z = 1$. Therefore, $h(z)$ has a Fourier series which converges uniformly on T. On the other hand, the Laguerre series doesn't converge uniformly because $h(s)$ is not in $L_2(C_+)$. In, fact the function $f(s) = (s+a)h(s)$ is unbounded when $s \to \infty$ so that $f_s(z)$ is unbounded at $z = 1$ and doesn't belong to $L_\infty(T)$. The effect on the Laguerre function approximation is shown in Figure 4.11. Therefore, the approach based on the series (4.80) allows to approximate a larger class of functions than that belonging to $L_2(R_+)\cap L_1(R_+)$.

Remarks:

- $h_+(s)$ vanish at $s \to \infty$ if $\sum_{k=0}^{\infty}\tilde{h}_k = 0$.
- In practical numerical computations, we must truncate the infinite series after a finite number of terms. Obviously, the truncated series $h_n(s)$ doesn't vanish when $s \to \infty$. In fact, $h_n(s)$ converge to a real constant $h_\infty \neq 0$ at infinity. However, the function $\tilde{h}_n(s) = h_n(s) - h_\infty$ vanish at infinity. Obviously, $h_n(s)$ is still a causal function because of $h_n(t) = \tilde{h}_n(t) + h_\infty\delta(t)$. In addition, $h_n(s)$ is analytic in $C_+ \cup iR$.
- The transfer function of a mechanical spring-mass system is given by the rational function $h(s) = 1/(s^2 + \Omega^2)$. This function doesn't belong to $L_\infty(iR)$ because it has two poles on the imaginary axis. It doesn't belong even to $L_2(iR)$ because the integral (4.9) diverges. As a consequence, we are unable to uniformly approximate the mapped function $h(z)$ with a Fourier series in T. However, the kernel function $h(s) = 1/(s^2 + \Omega^2)$ is still analytic in C_+. Hence, our approach fails to approximate a rational function analytic C_+. This inconsistency emerges from the fact that the kernels we try to approximate are nonrational functions analytic in C_+. On the other hand, $h(s) = 1/(s^2 + \Omega^2)$ is rational and therefore the kernel of a finite-dimensional linear system.

4.2.7   Laurent series and linear difference equations

So far, we have shown how to uniformly approximate DtN-map kernels defined on the imaginary axis iR. Now it's the moment to establish a link between these approximations and systems of linear differential equations. However, this link is not as easy to establish as that between Laurent

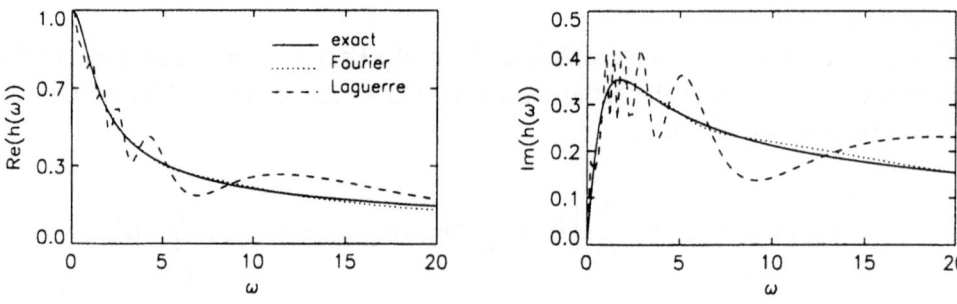

Figure 4.11: Approximations of the function $h(s) = 1/(\sqrt{s+1})$

series and systems of difference equations. Therefore, we will explore this relationship first. For this purpose, we use the theory of Z-transforms. Here, we give a very rough sketch of the Z-transform with two important results. Detailed presentations of this subject are found in [Oga87], [Ela96].

Consider a sequence of bounded real or complex numbers $\{f_k, -\infty < k < \infty\}$. To this sequence we associate a complex function $F(z)$, with complex z, defined by

$$F(z) = Z(f_k) = \sum_{k=-\infty}^{k=\infty} f_k z^{-k}. \tag{4.87}$$

This mapping is the so-called bilateral Z-transform of the sequence $\{f_k\}$. Equation (4.87) is the formal Laurent series of a complex function expanded at $z = 0$. On the unit circle T, (4.87) corresponds to the formal Fourier series

$$f(\theta) = \sum_{k=-\infty}^{k=\infty} f_k e^{-ik\theta}. \tag{4.88}$$

The inverse transform formula of Z-transforms uses (4.88) to recover the sequence $\{f_k\}$ of a function $F(z)$:

$$f_k = Z^{-1}(F(z)) = \frac{1}{2\pi i} \int_{|z|=1} f(z) z^{k-1} dz. \tag{4.89}$$

Without proof we list two important results of the theory of the Z-transform [Ela96]:

1) Consider the shifted sequence $\{f_{k+n}, -\infty < k < \infty\}$. Its Z-transform is given by

$$f_n(z) = Z(f_{k+n}) = z^n Z(f_k) = z^n f(z). \tag{4.90}$$

2) Consider the convolution of the sequences $\{f_k\}$ and $\{g_k\}$

$$h_n = \{f_k\} * \{g_k\} = \sum_{k=-\infty}^{k=\infty} f_k g_{n-k} = \sum_{k=-\infty}^{k=\infty} g_k f_{n-k}. \tag{4.91}$$

Its Z-transform is given by

$$h(z) = Z(h_n) = Z(\{f_k\} * \{g_k\}) = Z(f_k) Z(g_k) = f(z) g(z). \tag{4.92}$$

The Z-transform is a powerful analytical method to solve linear difference equation with constant coefficients. Consider the difference equation

$$x_{n+1} = ax_n + f_n, \tag{4.93}$$

where $\{f_n, n \geq 0\}$ is a bounded sequence. We try to find the solution of the difference equation with initial condition $x_0 = 0$. Applying the rule (4.90), the Z-transform is given by

$$zx(z) = ax(z) + f(z) \tag{4.94}$$

which yields the Z-transform of $x(z)$:

$$x(z) = \frac{f(z)}{z-a}. \tag{4.95}$$

The function $h(z) = 1/(z-a)$ is called the transfer function of (4.93). Because of (4.91) and (4.92) the solution of the difference equation is given by the convolution of $\{f_n\}$ with the inverse Z-transform of $h(z)$. This is obtained expanding the function $h(z)$ in a Laurent series at $z = 0$:

$$h(z) = \frac{1}{(z-a)} = \frac{1}{z} \sum_{n=0}^{\infty} (-1)^n \left(\frac{a}{z}\right)^n = \frac{1}{z} - \frac{a}{z^2} + \frac{a^2}{z^3} + \dots. \tag{4.96}$$

The sequence h_k is given by $\{h_0 = 0, \{h_n = (-1)^{n+1} a^n, n \geq 1\}\}$. With the convolution formula (4.91) the sequence $\{x_n\}$ is

$$x_n = \{f_k\} * \{h_k\} = \sum_{k=-\infty}^{k=\infty} h_k f_{n-k} = \sum_{k=1}^{k=\infty} h_k f_{n-k}. \tag{4.97}$$

Because $h_k = 0$ for $k \leq 0$ each element of the sequence $\{x_n\}$ is a linear combination of the subsequence of past inputs $\{f_k, k < n\}$. In analogy to the relation $h(t) = 0$ for $t < 0$ which defines causal convolution kernels of continuous-time systems, sequences $\{h_k\}$ with $h_k = 0$ for $k < 0$ define causal convolution kernels of discrete-time systems.

From the convolution (4.97) and the boundedness of the sequence $\{f_n\}$ we can deduce that the sequence $\{x_n\}$ defined in (4.97) is bounded if and only if the sequence $\{h_n\}$ is absolutely convergent. That is

$$\sum_{k=0}^{\infty} |h_n| < \infty. \tag{4.98}$$

With the last equation we deduce that all poles of the Z-transform $h(z)$ of the sequence $\{h_n\}$ must be in the unit disk D. In fact, $h(z)$ is bounded for $|z| \geq 1$:

$$|h(z)| = |Z(h_k)| = \left| \sum_{k=0}^{\infty} h_k z^{-k} \right| \leq \sum_{k=0}^{\infty} |h_k z^{-k}| < \sum_{k=0}^{\infty} |h_k| < \infty. \tag{4.99}$$

Furthermore, because of the convergence theorem of Laurent series, $h(z)$ is analytic inside its domain of convergence. In this case, this domain is $1 < |z| < \infty$, that is, S_∞. For the sequence given in (4.96) this means that $|a| < 1$.

Now, let us investigate how we can relate Laurent series defined by the approximation of DtN-map kernels to difference equations. In order to be related to a causal function these Laurent series have to be analytic in D. In contrast, the Laurent series associated with difference equations are analytic in S_∞. Hence, we have to transform Laurent series analytic in D to Laurent series analytic in S_∞. This provides the Möbius transformation

$$z = M_D(z') = -\frac{1}{z'} \tag{4.100}$$

which uniquely maps an element of D to an element of S_∞. In addition, (4.100) maps functions analytic in D to functions analytic in S_∞. Hence, the composed map $z = M_D(M(s))$, where $M(s)$ is given by (4.55), maps functions analytic in C_+ to functions analytic in S_∞. This map is

$$z = \frac{a+s}{a-s} \tag{4.101}$$

and its inverse is

$$s = a\frac{z-1}{z+1} \tag{4.102}$$

However, because of the additional mapping (4.100) we are obliged to redefine the causal subseries of Laurent series. Consider the extension to the complex plane of a trigonometric Fourier series on T by way of the Laurent series

$$f_s(z) = \sum_{k=-\infty}^{k=\infty} h_k z^k = f_+(z) + f_-(z) = \sum_{k=1}^{k=\infty} h_k z^k + \sum_{k=0}^{k=\infty} h_{-k} z^{-k}. \tag{4.103}$$

The subseries $f_+(z)$ represents now the anticausal and $f_-(z)$ the causal part.

Suppose that the causal part of (4.103) is the Laurent series of a rational transfer function $h(z)$. Then, the Z-transform relates $h(z)$ to a system of linear difference equations. A general theory of systems defined by linear difference equations is provided by the theory of linear discrete-time systems. This shall be the subject of the next chapter.

Remarks:

- A sequence of real or complex numbers $\{f_k\}$ with the property $\|h_k\|_{1_1} = \sum_{k=-\infty}^{\infty} |f_k| < \infty$ is said to be absolutely convergent or, equivalently, to belong to the sequence space ℓ_1.

- Because $h(z)$ is bounded on the unit circle T the Fourier series expansion (4.88) exists and we have

$$\|h_k\|_{\ell_2} = \sum_{k=0}^{\infty} |h_k|^2 = \|h(\theta)\|_{L_2(T)}. \tag{4.104}$$

Therefore the sequence h_k belongs to the Hilbert space ℓ_2 of infinite sequences of complex numbers such that

$$\sum_{k=0}^{\infty} |h_k|^2 < \infty. \tag{4.105}$$

Because of (4.98) the sequence h_k is an element of $\ell_1 \cap \ell_2$.

- $h(z)$ is bounded in the infinite open domain S_∞ because of

$$\lim_{z \to \infty} h(z) = h_0. \tag{4.106}$$

Chapter 5

Linear time-invariant systems

In this section, we give a brief overview of the theory of linear time-invariant discrete-time and continuous-time systems. This presentation deals only with those aspects of the theory which are relevant for this work. We refer to the multitude of excellent textbooks on this subject like [Bro70, Kai80] and [Che84] for readers feeling interested in more detailed presentations. The theory of discrete-time systems is thoroughly described in [Oga87].

Linear systems are represented by essentially two descriptions: The external or input-output description and the internal or state variable description. An example of the external or input-output description is the convolution integral discussed in the previous chapter. By this description, the properties of the system are defined by the kernel function. In linear systems theory, this function is called transfer function. On the other hand, the internal or state variable description defines a system by a set of difference (discrete-time systems) or differential (continuous-time systems) equations. The dynamics of these systems is described by so-called internal or state variables. In this chapter, we mainly investigate this description of linear systems.

5.1   Linear time-invariant discrete-time systems

5.1.1   The state-space description

Consider a linear time-invariant discrete-time system described by

$$
\begin{aligned}
x(k+1) &= Ax(k) + Bu(k) \\
v(k) &= Cx(k) + Du(k)
\end{aligned}
\tag{5.1}
$$

The first equation in (5.1) defines a difference equation relating the input $u(k)$ to the state $x(k)$. The second equation defines the output $v(k)$ by means of the input $u(k)$ and the state $x(k)$. The last is a $n \times 1$ column vector containing the scalar state variables $x_j(k)$. Similarly, the input $u(k)$ is a $m_u \times 1$ and the output $v(k)$ is a $m_v \times 1$ column vector. The state $x(k)$, the input $u(k)$ and the output $v(k)$ are elements of sequence spaces $(-\infty < k < \infty)$. The constant matrices A, B, C and D are called the system matrices. A is $n \times n$, B is $n \times m_u$, C is $m_v \times n$ and D is $m_v \times m_u$. The system defined by (5.1) is linear, time-invariant and causal.

If the dimension of the input $u(k)$ is equal to that of the output $v(k)$, that is $m = m_u = m_v$, the system (5.1) is called to be input-output symmetric. Transfer functions of input-output symmetric systems are described by square matrices. Because DtN-map kernels are usually input-output symmetric in the following we consider these type of systems. If $m > 1$ the system (5.1) is called to be a multiple-input/multiple-output system (MIMO system) and if $m = 1$ the system is called to be a single-input/single-output system (SISO system). DtN-map kernels are usually symmetric infinite-dimensional input-output MIMO systems.

Now let us introduce some basic concepts and results of linear systems theory.

1. When a system moves from the state $x(p)$ to the state $x(q)$, where $p < q$, we obtain from (5.1) the equation

$$x(q) = A^{q-p}x(p) + \sum_{j=p+1}^{q} A^{q-j}Bu(j-1).$$

(5.2)

This relation is called the state-transition equation of a linear discrete-time system. In particular, starting from the initial state $x(0) = 0$, the state $x(k)$ is given by

$$x(k) = \sum_{j=1}^{k} A^{k-j}Bu(j-1),$$

(5.3)

which is essentially a convolution equation. In this case, the state $x(k)$ is uniquely defined by the input sequence $u(k)$.

2. The input-output relation of a system is directly obtained from the state-space description. Combining (5.3) with the second equation of (5.1) we obtain

$$v(k) = CA^q x(0) + \sum_{j=1}^{k} CA^{k-j}Bu(j-1) + Du(k) = CA^k x(0) + \sum_{j=0}^{k} H_j u(k-j),$$

(5.4)

where the sequence H_j is given by

$$H_0 = D \qquad (H_j = CA^{j-1}B), j = 1...k.$$

(5.5)

The matrices H_j are called the Markov parameters of the linear system. A system, whose output is uniquely defined by the input is said to be relaxed. From (5.4) we obtain that a linear system is relaxed if and only if $x(0) = 0$. Because the output of DtN-maps are completely defined by its input, we will consider only relaxed systems. That is, systems with initial condition $x(0) = 0$.

3. A linear discrete-time system with space-description (5.1) is said to be controllable (or completely controllable) if the state vector can be brought from a state $y(p)$ to a state $y(q)$ in a finite number of steps by applying an appropriate input sequence $u(k)$. In other words, the system is controllable if there exists a sequence $\{u(k), k = p...q\}$ so that

$$y(q) = \sum_{j=p+1}^{q} A^{q-j}Bu(j-1),$$

(5.6)

A standard theorem of discrete-time linear systems states that a discrete-time system described by (5.1) is controllable if and only if the $n \times mn$ matrix

$$M_c = [B, AB, ..., A^{n-1}B]$$

(5.7)

has full rank. The matrix M_c is called the partial controllability matrix.

4. A discrete-time linear system with space-description (5.1) is said to be observable (or completely observable) if for each $k = p$ there exists a $q > p$, such that the state $x(p)$ is uniquely determined by the input-output information $\{(u(k), v(k)), (k = p...q)\}$. Similar as for controllability we obtain: A discrete-time system described by (5.1) is observable if and only if the $nm \times n$ matrix

$$M_o^T = [C, CA, ..., CA^{n-1}]^T \tag{5.8}$$

has full rank. The matrix M_o is called the partial observability matrix. By definition, if the system is observable we can reconstruct any particular state. As a consequence, every subsequent state is well-defined by the state-transition formula (5.2).

5. Applying the Z-transform we obtain the transfer function of (5.1) which is given by

$$\begin{aligned} zx(z) &= Ax(z) + Bu(z) \\ v(z) &= Cx(z) + Du(z) \end{aligned}, \tag{5.9}$$

where

$$x(z) = \sum_{k=0}^{\infty} x_k z^{-k} \qquad v(z) = \sum_{k=0}^{\infty} v_k z^{-k} \qquad u(z) = \sum_{k=0}^{\infty} u_k z^{-k}. \tag{5.10}$$

Eliminating the state $x(z)$ in (5.9) we obtain the input-output relation

$$v(z) = H(z)u(z), \tag{5.11}$$

The $n \times n$ matrix $H(z)$ is called the transfer function of the linear system and is given by

$$H(z) = C(zI - A)^{-1}B + D. \tag{5.12}$$

Another form of the transfer function is found expanding $(zI - A)^{-1}$ in a formal Neumann series

$$H(z) = \frac{C}{z}\left(I + \frac{A^1}{z} + \frac{A^2}{z^2} + ...\right)B + D = \sum_{k=1}^{\infty} CA^{k-1}Bz^{-k} + D = \sum_{k=0}^{\infty} H_k z^{-k}, \tag{5.13}$$

where the coefficients of the Neumann series H_k are given by

$$H_0 = D \qquad (H_k = CA^{k-1}B), k \geq 1. \tag{5.14}$$

These are exactly the Markov parameters H_k defined by the input-output relation (5.4). In particular, the columns $h_{k,j}$ of H_k are the response of the system to the unit impulse sequence $\{u_{0,j} = e_{(j)}, \{u_{k,j} = 0, k \geq 1\}\}$ where $e_{(j)}$ has a 1 as its ith component and 0's elsewhere. In fact, by the input-output relation (5.4) we obtain $v_k = h_{k,j}$. In addition, (5.11) shows that the output of the system to an arbitrary input sequence is defined by the convolution of the unit impulse response with the input sequence. The formal Neumann series converges to (5.12) by choosing z sufficiently large.

Using the adjoint formulation of the inverse of a matrix A we obtain a third representation of the transfer function

$$A^{-1} = \frac{\text{adj}(A)}{\det(A)}. \tag{5.15}$$

Applied to the transfer function representation (5.12) yields

$$H(z) = C\frac{\mathrm{adj}(zI - A)}{\det(zI - A)}B + D. \tag{5.16}$$

$\det(zI - A)$ is a polynomial in z of degree n with leading coefficient 1. The components of $\mathrm{adj}(zI - A)$ are also polynomials in z but of maximum degree $n - 1$. Therefore, every component of $H(z)$ is a rational function with $\det(zI - A)$ as denominator and with a numerator which is of a degree not larger than that of the denominator.

6. With the sequence of Markov parameters H_k we can form an infinite matrix Γ_H whose components are defined by

$$\Gamma_{H[i,j]} = H_{j+j-1} \quad , i, j \geq 1, \tag{5.17}$$

or more explicitly

$$\Gamma_H = \begin{bmatrix} H_1 & H_2 & H_3 & \dots \\ H_2 & H_3 & \dots & \dots \\ H_3 & \dots & \dots & \dots \\ \dots & \dots & \dots & \dots \end{bmatrix} \tag{5.18}$$

The matrix Γ_H is called the infinite block Hankel matrix. This matrix contains a lot of information about the discrete-time system. In particular, because the Markov parameters H_k are defined by (5.14), we obtain the following relationship between the infinite-dimensional observability and controllability matrices and the block Hankel matrix:

$$\Gamma_H = M_o M_c = \begin{bmatrix} C \\ CA \\ CA^2 \\ \dots \end{bmatrix} [B, AB, A^2 B, \dots]. \tag{5.19}$$

7. So far, we assumed that the system is defined by a state-space description. Using this description, we derived the transfer function of the system. Now let us have a look to the inverse problem. Suppose, we know the transfer function and want to find the associated state-space description. It turns out that the solution is not unique. In fact, to every realizable transfer function there exist an infinite number of state-space realizations. Particularly important are the realizations with smallest dimension of the state space (minimal number of degrees of freedom). Such a state-space description is called to be a minimal realization and its dimension is called the dimension of minimal realization. An important test for minimal realization is given by the following theorem: A state-space description of a linear system is a minimal realization if and only if both the partial controllability and partial observability matrices given in (5.7) and (5.8) have full rank.

This result suggests that the block Hankel matrix $\Gamma_H = M_o M_c$ contains information about the dimension of minimal realization. In fact, we obtain the following theorem: The dimension of minimal realization of a linear system with a transfer function $H(z)$ given by (5.12) is equal to the rank of the associated infinite block Hankel matrix Γ_H defined by (5.18).

Remarks:

- Because of the Cayley-Hamilton theorem the infinite-dimensional controllability and observability matrices given in (5.19) have the same ranks as the finite-dimensional partial controllability and observability matrices defined in (5.7) and (5.8).
- If the system is of minimal realization all components of $H(z) - D = C(zI - A)^{-1}B$ are

strictly proper rational functions. That is, the degree of the denominators of $H(z) - D$ is larger than the degree of the numerator.

- Observe that controllability, observability and minimal realization depends only on the system matrices $\{A, B, C\}$ and not on D.

Example: We illustrate the concepts discussed in this section investigating a small model system. Consider a discrete-time system given by the system matrices

$$A = \begin{bmatrix} a & b \\ c & -a \end{bmatrix} \qquad B = \begin{bmatrix} 1 \\ 0 \end{bmatrix} \qquad C = [1,0]. \qquad (5.20)$$

Computing the controllability and observability matrices we obtain

$$M_c = \begin{bmatrix} 1 & a \\ 0 & c \end{bmatrix} \qquad M_o = \begin{bmatrix} 1 & a \\ 0 & b \end{bmatrix}. \qquad (5.21)$$

Using the full rank criterions (5.7) and (5.8), the system is controllable if $c \neq 0$ and observable if $b \neq 0$. For $b = 0$ and $c = 0$ the system is neither controllable nor observable.

Now, let us consider the problem of controllability in detail. Let us assume that $c = 0$. Starting from the initial state $x(0) = 0$, we wish to reach the state $x(n)^T = [0,1]$ after a finite number of steps. According to the state transition formula the state $x(k)$ is given by (5.3). If $c = 0$ the sequence of vectors $A^{k-1}B$ is given by

$$x(k) = A^{k-1}B = \begin{bmatrix} a^{k-1} \\ 0 \end{bmatrix}. \qquad (5.22)$$

Obviously, for any input $u(k)$ we will never reach the state $x(n)^T = [0,1]$ because the state variable $x_2(k)$ will remain zero forever. Therefore, if $c = 0$ the state variable $x_2(k)$ cannot be influenced by any arbitrary input and is therefore uncontrollable.

Now, let us assume that $b = 0$. Starting from the initial state $x(0) = 0$ we compute the output $v(k)$ due to a unit impulse sequence $u(k)$. The output is given by the sequence

$$\{v(k) = a^{k-1}, k \geq 1\} \text{ or } \{1, a, a^2, a^3, ...\} \qquad (5.23)$$

and the state vectors are given by

$$\{x_k\} = \left\{ \begin{bmatrix} 1 \\ 0 \end{bmatrix}, \begin{bmatrix} a \\ c \end{bmatrix}, \begin{bmatrix} a^2 \\ 0 \end{bmatrix}, \begin{bmatrix} a^3 \\ a^2c \end{bmatrix}, \begin{bmatrix} a^4 \\ 0 \end{bmatrix}, \begin{bmatrix} a^5 \\ a^4c \end{bmatrix}, ... \right\}. \qquad (5.24)$$

Because of the relation $v(k) = Cx(k)$, where C is given by (5.20), the output sequence corresponds to $v(k) = x_1(k)$. Now let us assume that we guessed the state vector $\hat{x}(2)$ to be $\hat{x}(2)^T = [a,0]$. Starting from $\hat{x}(2)$ the sequence of states computed by the state-transition formula becomes

$$\{\hat{x}_k\} = \left\{ \begin{bmatrix} a \\ 0 \end{bmatrix}, \begin{bmatrix} a^2 \\ ca \end{bmatrix}, \begin{bmatrix} a^3 \\ 0 \end{bmatrix}, \begin{bmatrix} a^4 \\ a^3c \end{bmatrix}, \begin{bmatrix} a^5 \\ 0 \end{bmatrix}, ... \right\}. \qquad (5.25)$$

Obviously the latter sequence is different from the sequence (5.24). However, the input-output relation $\{v(k), k \geq 2\}$ is identical to (5.23). Therefore, if the system is not observable, the relation between the state and the output is not unique. A different sequence of states produce exactly the same output sequence.

Next we compute the transfer function $h(z)$ of the system (5.20). It is given by the rational function

$$h(z) = \frac{z+a}{z^2 - a^2 + bc}.$$ (5.26)

The denominator is a polynomial in z of degree 2 with leading coefficient 1 and the numerator is a polynomial in z of degree 1. Choosing $b = 0$ or $c = 0$ the transfer function becomes

$$h(z) = \frac{z+a}{z^2 - a^2} = \frac{z+a}{(z-a)(z+a)}.$$ (5.27)

The term $(z+a)$ appears in the denominator polynomial as well as in the numerator polynomial. Therefore, the rational function is not strictly proper and $(z+a)$ can be cancelled. This fact is called pole cancellation. As a consequence, the degree of both polynomials can be reduced by one and the transfer function becomes

$$h(z) = \frac{1}{z-a}$$ (5.28)

This suggests that the system (5.20) can be described by only one state variable instead of two. In fact, with $b = 0$ or $c = 0$ the system is either not controllable or not observable. Because of the minimal realization theorem, the realization (5.20) is not of minimal dimension. In fact, the Hankel matrix of (5.28) is

$$\Gamma_H = \begin{bmatrix} 1 & a & a^2 & \dots \\ a & a^2 & \dots & \dots \\ a^2 & \dots & \dots & \dots \\ \dots & \dots & \dots & \dots \end{bmatrix}$$ (5.29)

which has rank 1. With $b = 0$ or $c = 0$ the system is described by

$$A = [a] \qquad B = [1] \qquad C = [1].$$ (5.30)

Hence, if a system is controllable and observable it is uniquely defined.

5.1.2 Stability of discrete-time systems

One of the topics of major concern in system theory is stability. The usual definition of stability in system theory is that given in section 4.1.3: a system is said to be bounded input/bounded output (BIBO) stable if and only if given a bounded input sequence $u(k)$ the output sequence $v(k)$ is bounded too. Because the system (5.1) is vector valued we have to define specific norms for vectors and matrices in order to characterize the stability of these systems. A vector norm is a functional which associates to any vector $x \neq 0$ a real positive number $\|x\| \geq 0$, the vector norm of x. A collection of vectors together with a norm induces a normed vector space with distance function $d(x, y) = \|x - y\|$.

The usual vector norms on n-dimensional vector spaces are the ℓ_p-norms defined by

$$\|x\|_p = \left(\sum_{k=1}^{n} |x(k)|^p \right)^{1/p},$$ (5.31)

where $p \geq 1$. The most familiar vector norms are

• The Euclidian or ℓ_2 norm:

$$\|x\|_2 = (|x(1)|^2 + \ldots + |x(n)|^2)^{1/2}. \tag{5.32}$$

The Euclidian norm is induced by the inner product $(x, y) = \sum_{k=1}^{n} x(k) \bar{y}(k)$ and the associated normed space is therefore a Hilbert space.

• The sum or ℓ_1 norm:

$$\|x\|_1 = (|x(1)| + \ldots + |x(n)|) \tag{5.33}$$

• The max or ℓ_∞ norm:

$$\|x\|_\infty = \max(|x(1)|, \ldots, |x(n)|). \tag{5.34}$$

With vector norms we can define matrix norms. These matrix norms are said to be induced by vector norms and are called operator norms. They are defined by

$$\|A\| = \sup_{\|x\|_p = 1} \|Ax\|_p. \tag{5.35}$$

By its very definition, we obtain $\|Ax\| \leq \|A\| \|x\|$ and $\|I\| = 1$. The most famous operator norm is the spectral norm defined by

$$\|A\|_2 = \max \sqrt{\lambda} = \sup_{\|x\|_2 = 1} \|Ax\|_2, \tag{5.36}$$

where λ is the largest eigenvalue of the positive semi-definite Hermitian matrix A^*A.

The form of the stability criterions depends on how a system is described. For systems described by the input-output relation (5.4), a system is BIBO stable if and only if there exists a positive number M so that

$$\sum_{k=0}^{\infty} \|CA^k B\| = \sum_{k=0}^{\infty} \|H_{k+1}\| \leq M < \infty. \tag{5.37}$$

This criterion is similar to the ℓ_1-criterion for BIBO stable scalar systems mentioned in section 4.2.7. Because vector norms of finite-dimensional vector spaces are equivalent, (5.37) is valid with respect to any operator norm. From (5.37) we may deduce the sufficient condition $\|A\| < 1$ for BIBO stability. In fact, we obtain

$$\sum_{k=0}^{\infty} \|CA^k B\| < \|C\| \|B\| \sum_{k=0}^{\infty} \|A\|^k = \frac{\|C\| \|B\|}{1 - \|A\|} < \infty. \tag{5.38}$$

As a consequence, if the system is BIBO stable the convergence radius is unity. Therefore, to be BIBO stable, every component of the transfer function $H(z)$ is analytic in the open domain $z \geq 1$. Hence, all the zeroes of the denominator polynomial $\det(zI - A)$ must be in the open unit disk D. Because the zeroes of $\det(zI - A)$ are identical to the eigenvalues of the system matrix A, we obtain: the system (5.1) is BIBO stable if the eigenvalues of the system matrix A are of magnitude smaller than unity. That is,

$$|\lambda(A)| < 1, \; \lambda(A) \text{ is an eigenvalue of } A. \tag{5.39}$$

The latter condition is even necessary provided that the system (5.1) is controllable and observable and therefore of minimal realization. The condition for stability $|\lambda(A)| < 1$ is not necessary in general because of the eventuality of pole cancellation. To avoid pole cancellation, we must assume that the linear system (5.1) is of minimal dimension. Only under this assumption the condition $|\lambda(A)| < 1$ becomes a necessary condition for stability.

Example: Let us reconsider the model system described in the last section. The Markov parameters are given by

$$\{h_k\} = \{1, a, a^2 + bc, (a^2 + bc)a, (a^2 + bc)^2, (a^2 + bc)^2 a, ...\} . \tag{5.40}$$

According to (5.37) we obtain the stability condition

$$|a^2 + bc| < 1 . \tag{5.41}$$

In fact, the eigenvalues of the system matrix A given in (5.20) are

$$\lambda_1 = \sqrt{a^2 + bc} \qquad \lambda_2 = -\sqrt{a^2 + bc} . \tag{5.42}$$

Because the square roots of a complex number is $\sqrt{c} = \sqrt{|c|}e^{i\phi_c}$ the eigenvalue criterion (5.39) gives (5.41) too.

5.2   Linear time-invariant continuous-time systems

Similar as for linear time-invariant discrete-time systems, we may define linear time-invariant continuous-time systems. The most relevant difference between discrete-time and continuous-time systems is that the former is a map which relates an input sequence to an output sequence whereas the latter is a map which relates an input function to an output function. As before, we concentrate our presentation on the state-space description of a system. The input-output description of a time-invariant system is given by convolution integrals. This description has been analyzed in section 4.1. As will be evident in the following, discrete-time and continuous-time systems have many common features. Therefore, in this section, we will only emphasize those features of continuous-time systems which are really different from those of discrete-time systems.

5.2.1   The state-space description

The state-space description of linear time-invariant continuous-time systems is given by the pair of equations

$$\begin{aligned} \dot{x}(t) &= Ax(t) + Bu(t) \\ v(t) &= Cx(t) + Du(t) \end{aligned}, \tag{5.43}$$

The state $x(t)$ is a $n \times 1$ column vector containing the scalar state variables x_k. The input $u(t)$ and the output $v(t)$ are $m \times 1$ column vectors. The state $x(t)$, the input $u(t)$ and the output $v(t)$ are elements of function spaces. The constant matrices A, B, C and D are called the system matrices. A is $n \times n$, B is $n \times m$, C is $m \times n$ and D is $m \times m$. The first equation in (5.1) defines a differential equation relating the input $u(t)$ to the state $x(t)$. The second equation defines the output $v(t)$ by means of the state $x(t)$ and the input $u(t)$. The system defined by (5.43) is linear, time-invariant and causal.

Let us state some properties of linear time-invariant input-output systems

1. The set of all solutions of the differential equation $\dot{x}(t) = Ax(t)$ forms an n-dimensional vector space. Now we may use these solutions to construct a set of linearly independent vectors

$\psi_k(t)$ which define the so-called fundamental matrix $\Psi(t) = [\psi_1(t), \psi_2(t), ..., \psi_n(t)]$. This satisfies the matrix equation $\dot{\Psi}(t) = A\Psi(t)$. With the fundamental matrix we define the state transition matrix

$$\Phi(t, \hat{t}) = \Psi(t)\Psi^{-1}(t_0) \tag{5.44}$$

Choosing $\Psi(t) = e^{At}$ as fundamental matrix of $\dot{x}(t) = Ax(t)$ we obtain the state transition matrix

$$\Phi(t, t_0) = e^{A(t-t_0)}. \tag{5.45}$$

With the state transition matrix we can formulate the state $x(t)$ of a system starting from a state x_0 at t_0 by the equation

$$x(t) = e^{A(t-t_0)}x_0 + \int_{t_0}^{t} e^{A(t-s)}Bu(s)ds. \tag{5.46}$$

This relation is called the state-transition equation of a linear time-invariant continuous-time system. In particular, starting from the initial state $x_0 = 0$ the state $x(t)$ is given by

$$x(t) = \int_{t_0}^{t} e^{A(t-s)}Bu(s)ds. \tag{5.47}$$

Observe, that the latter equation has the structure of a convolution integral.

2. The input-output relation is obtained from the state-transition equation. In the following we will assume that the continuous-time system is relaxed. That is, $x_0 = 0$. Combining (5.47) with the second equation of (5.43) we obtain the input-output relation

$$v(t) = \int_{t_0}^{t} Ce^{A(t-s)}Bu(s)ds + Du(t). \tag{5.48}$$

3. The definitions of controllability and observability are equal to those of discrete-time systems except that they are formulated for the times t_1 and t_2 instead of the sequence indexes p and q. Because the state-space description of linear time-invariant discrete-time systems is formally identical to that of linear time-invariant continuous-time systems, the theorems on controllability and observability are exactly the same.

4. Performing the Laplace transform of the state-space equations (5.43) we obtain

$$\begin{aligned} sx(s) &= Ax(s) + Bu(s) \\ v(s) &= Cx(s) + Du(s) \end{aligned}, \tag{5.49}$$

With (5.49) the input-output relation is described by

$$v(s) = H(s)u(s), \tag{5.50}$$

where the $m \times m$ matrix $H(s)$ is called to be the transfer function of the linear system and is given by

$$H(s) = C(sI - A)^{-1}B + D. \tag{5.51}$$

Equation (5.50) shows that the output of the system to an arbitrary input function $u(t)$ is defined by the convolution of the transfer function with the input function. The transfer function may be represented using the adjoint formulation of the inverse of the matrix A given by

$$A^{-1} = \frac{\text{adj}(A)}{\det(A)}.$$ (5.52)

Applying (5.52) to the transfer function matrix (5.51) yields

$$H(s) = C\frac{\text{adj}(sI - A)}{\det(sI - A)}B + D.$$ (5.53)

Obviously $\det(sI - A)$ is a polynomial in s of degree n with leading coefficient 1. The components of $\text{adj}(sI - A)$ are also polynomials in s but of maximum degree $n - 1$. Therefore, every component of $H(s)$ is a rational function with $\det(sI - A)$ as a denominator and with a numerator which is of a degree not larger than that of the denominator.

Another form of the transfer function is found expanding $(sI - A)^{-1}$ in the formal Neumann series

$$H(s) = \frac{C}{s}\left(I + \frac{A^1}{s} + \frac{A^2}{s^2} + \dots\right)B + D = \sum_{k=1}^{\infty} CA^{k-1}Bs^{-k} + D = \sum_{k=0}^{\infty} H_k s^{-k}.$$ (5.54)

The Markov parameters H_k are given by

$$H_0 = D \qquad \{H_k = CA^{k-1}B, k \geq 1\}.$$ (5.55)

The formal von Neumann series converges provided that $s > \|A\|$. By virtue of the Laplace transform relation

$$\frac{1}{s^k} \leftrightarrow \frac{t^{k-1}}{(k-1)!}$$ (5.56)

we obtain

$$H(t) = \sum_{k=0}^{\infty} H_{k+1}\frac{t^k}{k!} + H_0\delta(t).$$ (5.57)

Therefore, the Markov parameters H_k for $k \geq 1$ are the coefficients of the Taylor expansion of the transfer function at $t = 0$. They are defined by the equation

$$H_k = \frac{d^k}{dt^k}(H(t) - H_0\delta(t))\bigg|_{t=0}.$$ (5.58)

In fact, using the input-output relation (5.48), the transfer function in the time domain is given by

$$H(t) = Ce^{At}B + D\delta(t)$$ (5.59)

Expanding the fundamental matrix e^{At} in a Taylor series at $t = 0$ we obtain

$$v(t) = \int_0^t Ce^{A(t-s)}B\delta(s)ds + D\delta(t) = \sum_{k=0}^{\infty} CA^k B\frac{t^k}{k!} + D\delta(t) = \sum_{k=0}^{\infty} H_{k+1}\frac{t^k}{k!} + H_0\delta(t)$$ (5.60)

The latter series converges for any finite t.

5. The results concerning the Hankel matrix and minimal realization which have been

obtained for discrete-time systems applies to continuous-time systems as well.

6. In chapter 5 we have investigated in detail the question of the stability of scalar input-output systems defined by convolution integrals. We found that the input-output relation is BIBO stable if and only if the kernel belongs to $L_1(R_+)$. As a consequence, we found that the Laplace transform of the kernel is analytic in the right half-plane C_+. Because the input-output relation of a linear continuous-time MIMO systems are defined by a convolution integral, each component of the output vector $v(t)$ is defined by a finite linear combination of the components of the transfer function with the components of the input vector $u(t)$. As a consequence, we obtain: A multivariable linear time-invariant system is BIBO stable if and only if each component $\hat{h}_{ij}(t)$ of the transfer function $\hat{H}(t) = H(t) - H_0\delta(t)$ belongs to $L_1(R_+)$.

Because any component $\hat{h}_{ij}(s)$ of the Laplace transform of the regularized transfer function $\hat{H}(t)$ is a proper rational function, provided that the system is of minimal realization, and the Laplace transform of a $L_1(R_+)$ is bounded on iR it follows: A multivariable linear time-invariant system of minimal realization is BIBO stable if and only if all the poles of every component $\hat{h}_{ij}(s)$ of the Laplace transform of the regularized transfer function matrix $\hat{H}(s) = H(s) - H_0$ are on the left half-plane C_-. Furthermore, since by (5.53) the denominator of every component $\hat{h}_{ij}(s)$ is given by the polynomial $\det(sI - A)$ the poles of every component $\hat{h}_{ij}(s)$ are a subset of the eigenvalues of the system matrix A. Therefore: A multivariable linear time-invariant system of minimal realization is BIBO stable if and only if all the eigenvalues of the system matrix A are on the left half-plane C_-.

Remarks:

- Because the poles of finite-dimensional systems are one the left half plane C_-, $\hat{H}(s)$ is analytic even on iR. On the other hand, we have seen that the regularized DtN-map kernel of the nondissipative constrained rod is not analytic on iR. Nevertheless, it is still BIBO stable. The essential fact is that the linear system (5.43) is of finite dimension and has therefore a rational transfer function. The DtN-map kernel of the nondissipative constrained rod is the kernel of an infinite-dimensional linear time-invariant system and therefore nonrational.

- If the system matrix A have a purely imaginary eigenvalue the linear system (5.43) is obviously not BIBO stable.

5.3 Further properties of linear time-invariant systems

5.3.1 State-space realizations of linear systems

So far, we assumed that linear systems are defined by a state-space descriptions. With this description, we deduced many elementary properties of linear systems. In this section, we assume that a system is described by a transfer function $H(s)$ and try to construct a state-space description with matrices A, B, C and D so that the transfer function $H(s)$ given by (5.12) or (5.51) is identical to the transfer function $H(s)$. Then, we have a so-called state-space realization of the system. Every rational transfer function $H(s)$ has a finite-dimensional state-space realization. In this section, we will work mainly with continuous-time systems. However, because of the formal equivalence between discrete-time and continuous-time linear systems, all results apply for discrete-time systems as well.

Next we introduce some important state-space realizations. Consider the transfer function

$$H(s) = \frac{N(s)}{p(s)} + D, \tag{5.61}$$

where $N(s)$ is a polynomial matrix of maximum degree n-1, that is

$$N(s) = \sum_{k=0}^{n-1} N_k s^k, \tag{5.62}$$

and $p(s)$ is a polynomial of maximum degree n with leading coefficient equal to one:

$$p(s) = \sum_{k=0}^{n} p_k s^k. \tag{5.63}$$

Then, it is easy to verify that the following state-space description is a realization of $H(s)$

$$A = \begin{bmatrix} -p_{n-1}I_n & -p_{n-2}I_n & \cdots & -p_0I_n \\ I_n & 0 & \cdots & 0 \\ 0 & I_n & \cdots & 0 \\ \cdots & \cdots & \cdots & \cdots \\ 0 & 0 & I_n & 0 \end{bmatrix} \quad B = \begin{bmatrix} I_n \\ 0 \\ 0 \\ \cdots \\ 0 \end{bmatrix} \quad C^T = \begin{bmatrix} N_{n-1} \\ N_{n-2} \\ N_{n-3} \\ \cdots \\ N_0 \end{bmatrix} \tag{5.64}$$

This realization is called the block controller form. Evaluating the partial controllability matrix (5.7), it can be easily shown that this form is controllable. However, nothing can be said about its observability without more informations about the matrices N_k.

Similarly, we can define the so-called block observer form by

$$A = \begin{bmatrix} -p_{n-1}I_n & I_n & 0 & \cdots & 0 \\ -p_{n-2}I_n & 0 & I_n & \cdots & 0 \\ \cdots & \cdots & \cdots & \cdots & I_n \\ -p_0I_n & 0 & 0 & \cdots & 0 \end{bmatrix} \quad B = \begin{bmatrix} N_{n-1} \\ N_{n-2} \\ N_{n-3} \\ \cdots \\ N_0 \end{bmatrix} \quad C^T = \begin{bmatrix} I_n \\ 0 \\ 0 \\ \cdots \\ 0 \end{bmatrix}, \tag{5.65}$$

which is observable but not necessarily controllable.

A third and, as we will see, for our purpose very important realization is obtained if the transfer function can be expanded in a partial fractions expansion:

$$H(s) = \sum_{k=0}^{q} \frac{R_k}{s - \lambda_k}, \tag{5.66}$$

where the set of complex numbers $\lambda_k, k = 1 \ldots q$ are the distinct roots of the denominator polynomial (5.63) and the matrices R_k are the residue matrices defined by the limit

$$R_k = \lim_{s \to \lambda_k} (s - \lambda_k) H(s). \tag{5.67}$$

The roots of the denominator polynomial are exactly the eigenvalues of the system matrix A. These systems are called to be diagonalizable. A $n \times n$ matrix is diagonalizable if and only if A has n linearly independent eigenvectors. Hence, defining a matrix $T = [x_1, x_2, \ldots, x_n]$ whose columns are the eigenvectors of A, the similarity transformation with T yields the state-space description

$$\bar{A} = \Lambda = T^{-1}AT \qquad \bar{B} = T^{-1}B \qquad \bar{C} = CT \qquad \bar{D} = D, \tag{5.68}$$

where the diagonal matrix Λ contains the eigenvalues of A. Observe, that the matrix A need not to have n distinct eigenvalues. In fact, any eigenvalue of A may have a geometric multiplicity which is larger than one. Therefore, the number of summands of the partial fraction expansion (5.66) may be smaller than n, say $q \leq n$. Obviously, if we denote by r_k the geometric multiplicity of the eigenvalue λ_k we must have $r_1 + r_2 + \ldots + r_q = n$. The state-space description defined in (5.68) can be partitioned into the block structure

$$\Lambda = \begin{bmatrix} \lambda_1 I_{r_1} & 0 & \ldots & 0 \\ 0 & \lambda_2 I_{r_2} & \ldots & 0 \\ \ldots & \ldots & \ldots & \ldots \\ 0 & 0 & \ldots & \lambda_p I_{r_q} \end{bmatrix} \qquad \bar{B} = \begin{bmatrix} B_1 \\ B_2 \\ \ldots \\ B_p \end{bmatrix} \qquad \bar{C}^T = \begin{bmatrix} C_1^T \\ C_2^T \\ \ldots \\ C_p^T \end{bmatrix}. \qquad (5.69)$$

where the matrices B_k are $r_k \times m$ and the C_k are $m \times r_k$. As a consequence the $m \times m$ residue matrices R_k are given by

$$R_k = C_k B_k, \qquad (5.70)$$

with $\text{rank}(R_k) \leq r_k$.

This description enables to formulate a simple criterion to verify the controllability and observability of linear systems: A diagonalizable system is controllable if and only if $\text{rank}(B_k) = r_k$ and observable if and only if $\text{rank}(C_k) = r_k$. In other words, a diagonalizable system is controllable and observable, and therefore of minimal realization, if and only if every residue matrices have $\text{rank}(R_k) = r_k$.

Remarks:
- If the matrix A have n distinct eigenvalues then it is diagonalizable and the partial fraction expansion contains n terms with residue matrices R_k of rank at most one. If all the residue matrices has rank one then the diagonalizable realization is a minimal realization.
- If the linear system is real the denominator polynomial (5.63) has real coefficients. Therefore, the spectrum of A is composed of real and of pairs of conjugate complex numbers. Furthermore, the residue matrices R_k associated with a real eigenvalues are real and those associated with pairs of conjugate complex eigenvalues are conjugate complex.
- If the transfer function is symmetric, say $H^T(s) = H(s)$, the residue matrices R_k must be symmetric.
- In the time domain, the transfer function of a diagonalizable system is given by

$$H(t) = \sum_{k=0}^{p} R_k e^{\lambda_k t}. \qquad (5.71)$$

- We know from matrix analysis that those $n \times n$ matrices which have less than n independent eigenvectors are not diagonalizable by a similarity transformation. These matrices can be transformed to the so-called Jordan canonical form. Because of the approximation process, the finite-dimensional models of DtN-maps will hardly be nondiagonalizable. Therefore, we don't investigate the Jordan canonical form further.

Next, let us consider the nonuniqueness of state-space realizations. Consider a system described by

$$\begin{aligned} sx &= Ax + Bu \\ v &= Cx + Du \end{aligned} \qquad (5.72)$$

Let us transform the state x to the state y by way of the similarity transformation

$$x = Ty,$$ (5.73)

where T is an arbitrary nonsingular matrix. Combining (5.73) with (5.72) we obtain

$$sy = T^{-1}ATy + T^{-1}Bu$$
$$v = CTy + Du$$ (5.74)

This is still a linear system. The system matrices are defined by

$$\bar{A} = T^{-1}AT \qquad \bar{B} = T^{-1}B \qquad \bar{C} = CT \qquad \bar{D} = D.$$ (5.75)

Because there are infinitely many nonsingular matrices T there are infinitely many different but equivalent state-space realizations. Note the transfer functions are invariant with respect to a similarity transformation:

$$\bar{H}(s) = \bar{C}(sI - \bar{A})^{-1}\bar{B} + \bar{D} = CT(sT^{-1}T - T^{-1}AT)^{-1}T^{-1}B + D$$
$$= CTT^{-1}(sI - A)^{-1}TT^{-1}B + D = C(sI - A)^{-1}B + D = H(s)$$ (5.76)

Hence, the denominator polynomial $\det(sI - A)$, the numerator matrix $N(s) = C\mathrm{adj}(sI - A)B$ and the Markov parameters H_k are also invariant with respect to a similarity transformation. Moreover, because the eigenvalues of a matrix are invariant with respect to a similarity transformation, that is, $\lambda(T^{-1}AT) = \lambda(A)$, the stability of a linear system is also invariant with respect to similarity transformations. Similarly, controllability, observability, minimal realization and the Hankel matrix are also invariant with respect to similarity transformations.

Example: Let us reconsider the simple model system of section 5.1.1. From the transfer function (5.26) we conclude that the input normal form is

$$A = \begin{bmatrix} 0 & -(a^2 + bc) \\ 1 & 0 \end{bmatrix} \qquad B = \begin{bmatrix} 1 \\ 0 \end{bmatrix} \qquad C^T = \begin{bmatrix} 1 \\ a \end{bmatrix}$$ (5.77)

and the output normal form is

$$A = \begin{bmatrix} 0 & 1 \\ -(a^2 + bc) & 0 \end{bmatrix} \qquad B = \begin{bmatrix} 1 \\ a \end{bmatrix} \qquad C^T = \begin{bmatrix} 1 \\ 0 \end{bmatrix}.$$ (5.78)

In diagonal form, the model system is given by

$$A = \begin{bmatrix} \sqrt{a^2 + bc} & 0 \\ 0 & -\sqrt{a^2 + bc} \end{bmatrix} \qquad B = \begin{bmatrix} \dfrac{c}{2\sqrt{a^2 + bc}} \\ -\dfrac{c}{2\sqrt{a^2 + bc}} \end{bmatrix} \qquad C^T = \begin{bmatrix} \dfrac{\sqrt{a^2 + bc} + a}{c} \\ \dfrac{\sqrt{a^2 + bc} - a}{c} \end{bmatrix}$$ (5.79)

For $c = 0$ the matrix B vanish and the system is not controllable. For $b = 0$ the matrix C becomes $C = [2a/c, 0]$ so that the second mode is not observable.

5.3.2 Stability and Lyapunov equations

An alternative and very elegant approach to analyze the stability of linear systems is provided by the stability theory of Lyapunov. For an introduction to Lyapunov stability theory we refer to [LL61, Mer97] and for deeper results to [Hah63]. This theory states that an equilibrium state x_∞ of a linear or nonlinear dynamical system is stable in the sense of Lyapunov if there exists a continuously differentiable positive definite scalar function $V(x)$, the Lyapunov function, whose derivative with respect to the time along the trajectory of the state $x(t)$ is negative semi-definite

$$\dot{V}(x) = \frac{dV}{dt} = \frac{dV}{dx}\frac{dx}{dt} \leq 0. \tag{5.80}$$

For linear time-invariant systems, the condition (5.80) can be reduced to an algebraic equation. Consider the ODE

$$\dot{x}(t) = Ax(t). \tag{5.81}$$

with initial condition $x(t_0) = x_0$. Let us define the positive definite Lyapunov function $V(x)$ given by

$$V(x) = x^T Q x \tag{5.82}$$

where the $n \times n$ matrix Q is positive definite ($Q > 0$). Combining (5.82), (5.81) and (5.80) we obtain

$$\dot{V}(x) = x^T(A^T Q + QA)x. \tag{5.83}$$

Therefore, the stability condition (5.80) becomes

$$A^T Q + QA = -R, \tag{5.84}$$

where R is an arbitrary positive semi-definite $n \times n$ matrix. The latter equation is the famous Lyapunov matrix equation. The system (5.81) is said to be stable if for any positive semi-definite matrix R there exists a solution of (5.84) with positive definite matrix Q. If there exists a positive definite solution Q of (5.84) with R positive definite then the system is said to be asymptotically stable. That is, all eigenvalues of the matrix A have negative real part.

Therefore, the linear time-invariant system (5.43) is BIBO stable if and only if given a positive definite matrix R there exists a positive definite matrix Q that satisfies the Lyapunov matrix equation (5.84). Moreover, if the linear system is observable the matrix R can even be positive semi-definite. In particular, if $R = C^T C$ we obtain: The solution Q of the Lyapunov matrix equation (5.84) is positive definite if and only if the linear time-invariant system (5.43) is observable and BIBO stable. A dual theorem is obtained analyzing the controllable system $\dot{x}(t) = A^T x(t)$ with positive definite Lyapunov function $V(x) = x^T P x$ where the right-hand side is defined by $R = B^T B$.

Combining these two results we obtain: The observable and controllable linear time-invariant system (5.43) (and therefore of minimal realization) is BIBO stable if and only if there exist symmetric positive definite matrices P and Q such that the Lyapunov matrix equations

$$AP + PA^T = -BB^T$$
$$A^T Q + QA = -C^T C \tag{5.85}$$

are satisfied.

A remarkable fact is that the solutions P and Q of (5.85) can be given analytically. In fact, the matrices

$$W_c = \int_0^\infty e^{At} BB^T e^{A^T t} dt \qquad W_o = \int_0^\infty e^{A^T t} C^T C e^{At} dt \qquad (5.86)$$

known as controllability and observability grammians are the solutions of the Lyapunov matrix equations (5.85). This is proved by inspection. Setting $P = W_c$ and inserting the left equation of (5.86) into the first equation of (5.85) yields

$$AP + PA^T = \int_0^\infty A e^{At} BB^T e^{A^T t} dt + \int_0^\infty e^{At} BB^T e^{A^T t} A^T dt = \int_0^\infty \frac{d}{dt}(e^{At} BB^T e^{A^T t}) dt$$

$$= e^{At} BB^T e^{A^T t} \Big|_{t=0}^{t=\infty} = -BB^T \qquad (5.87)$$

Similarly we obtain $Q = W_o$ as solution of the second equation of (5.85).

The controllability and observability grammians have a nice physical meaning. Consider a system with initial conditions $x(0) = x_0$ and with input function $u(t) = 0$ for $t \geq 0$. Then the output of the system is given by

$$v(t) = C e^{At} x_0 \text{ for } t \geq 0. \qquad (5.88)$$

The norm of $v(t)$ in $L_2(R_+)$ is

$$\|v(t)\|_{L_2(R_.)} = \int_0^\infty v^T(t) v(t) dt = \int_0^\infty x^T_0 e^{A^T t} C^T C e^{At} x_0 dt = x^T_0 W_o x_0. \qquad (5.89)$$

Hence, the observability grammian defines the energy of the output. The interpretation of the controllability grammian is related to the optimization problem: What is the minimum input energy $\|u(t)\|_2$ which brings a linear system from the state $x(-\infty) = 0$ to the state $x(0) = x_0$? The answer is quite suggestive and says

$$\min\|u(t)\|_{L_2(R_.)} = x^T_0 W_c^{-1} x_0. \qquad (5.90)$$

Now, suppose we find a state-space realization which diagonalize the observability grammian. Because it is positive definite all components must be positive. The output energy is then given by

$$\|v(t)\|_2 = \sum_{k=0}^n \sigma_k y_k^T(0) y_k(0). \qquad (5.91)$$

This form is very useful to analyze how much each state contributes to the output energy. State variables with large σ_k contribute much to the output energy. In contrast, state variables with a small σ_k have a also small impact on the output. In particular, any nonobservable state variable must have $\sigma_k = 0$. These state variables can be deleted from the state-space description of a system without affecting its input-output behavior.

Equivalently, if we find a state-space realization which diagonalize the controllability grammian we conclude that those state variables with large σ_k need less input energy to be reached. On the other hand, those state variables with small σ_k need a lot of input energy to be reached. Clearly, noncontrollable state variables ($\sigma_k = 0$) need an infinite input energy.

Remarks:

- The stability of linear systems depends only on the matrix A. Hence, (5.85) is a means to investigate the location of the eigenvalues of the matrix A, and, because A and A^T have the same eigenvalues, even those of A^T.

- The matrices P and Q are uniquely defined by (5.85) provided that no two eigenvalues of A add up to zero, that is $\lambda_i(A) + \lambda_j(A) = 0$. Each Lyapunov matrix equation is a linear algebraic system of equations for the $n(n-1)/2$ unknown (symmetry). Its efficient numerical solution is provided by the algorithms of Bartels and Stewart [BS72] or Hammarling [Ham82].

- The controllability and observability grammians exist if the eigenvalues of the system matrix A belong to half space C_-. These matrices gave a means to compute the solution of the Lyapunov equation without solving (5.85) directly. A detailed presentation of various numerical methods to solve the Lyapunov matrix equation is found in [GQ95].

The stability theory of Lyapunov can be formulated for discrete-time systems as well. Consider the linear time-invariant discrete-time system

$$x(k+1) = Ax(k) \qquad x(k_0) = x_0 \tag{5.92}$$

Given a positive definite Q we define a positive definite Lyapunov function $V(x(k))$ by

$$V(x(k)) = x(k)^T Q x(k). \tag{5.93}$$

For discrete-time systems the condition (5.80) transforms to

$$\Delta V(x(k)) = V(x(k+1)) - V(x(k)) \leq 0. \tag{5.94}$$

Combining (5.94) with (5.92) and (5.93) yields the matrix equation

$$Q - A^T Q A = R. \tag{5.95}$$

where R is a positive semi-definite matrix. Similarly, we can deduce an analogous equation when analyzing discrete-time systems defined by $(x(k+1) = A^T x(k))$, $(x(k_0) = x_0)$ so that we obtain: The observable and controllable linear time-invariant discrete-time system (5.1) (and therefore of minimal realization) is BIBO stable if and only if there exist symmetric positive definite matrices P and Q such that the Lyapunov matrix equations

$$\begin{aligned} APA^T - P &= -BB^T \\ A^T Q A - Q &= -C^T C \end{aligned} \tag{5.96}$$

are satisfied.

Similarly as for continuous-time systems, the solutions P and Q of the Lyapunov matrix equations (5.96) are given by the controllability and observability grammians of discrete-time systems W_c and W_o defined by

$$P = W_c = M_c M_c^T = \sum_{k=0}^{\infty} A^k BB^T (A^T)^k \tag{5.97.a}$$

$$Q = W_o = M_o^T M_o = \sum_{k=0}^{\infty} (A^T)^k C^T C A^k \tag{5.97.b}$$

In fact, combining (5.97.a) with the first equation in (5.96) we obtain

$$\sum_{k=1}^{\infty} A^k BB^T (A^T)^k - \sum_{k=0}^{\infty} A^k BB^T (A^T)^k = -BB^T. \tag{5.98}$$

Similarly we may verify that (5.97.b) is the solution of the second equation in (5.96).

Remarks:

- The controllability and observability grammians W_c and W_o converge provided that the operator norm of A is smaller than unity.

- Because of $\text{rank}(A^T A) = \text{rank}(AA^T) = \text{rank}(A) = \text{rank}(A^T)$ for any square matrix A, the controllability and observability grammians W_c and W_o given by (5.97.a) and (5.97.b) must have the same rank as the controllability and observability matrices M_c and M_o.

- The Lyapunov equations (5.96) can be brought in the form (5.85) by a so-called fractional transformation. In this form, they may be solved numerically provided that no two eigenvalues of the system matrix A have product equal to one. Clearly, a solution of the Lyapunov equations can be found evaluating the controllability and observability grammians W_c and W_o.

- The physical interpretations of the controllability and observability grammians of discrete-time systems W_c and W_o are similar to those given for continuous-time systems.

Example: We want briefly discuss the relationship between controllability and observability and the solutions of the Lyapunov equations investigating the model system of section 5.1.1. Let us assume that the free parameters are $a = 2/3$, $b = 1/3$ and $c = 2/3$. We know that the system is controllable and observable. The spectrum of the matrix A is $\sigma(A) = \{\sqrt{2/3}, -\sqrt{2/3}\}$ and is entirely in the unit disk. Therefore the system is BIBO stable. In fact, the controllability and observability grammians are

$$W_c = \frac{1}{5}\begin{bmatrix} 13 & 4 \\ 4 & 4 \end{bmatrix} \qquad W_o = \frac{1}{5}\begin{bmatrix} 13 & 2 \\ 2 & 1 \end{bmatrix} \tag{5.99}$$

and therefore positive definite. Now let us choose the following set of parameters $a = 2/3$, $b = 1/3$, $c = 0$. From section 5.1.1 we know that with these parameters the model system is stable but not controllable. In fact, the controllability and observability grammians are

$$W_c = \frac{1}{5}\begin{bmatrix} 9 & 0 \\ 0 & 0 \end{bmatrix} \qquad W_o = \frac{1}{65}\begin{bmatrix} 117 & 18 \\ 18 & 117 \end{bmatrix}. \tag{5.100}$$

The observability grammian is still positive definite. However, the controllability grammian is obviously positive semi-definite. Choosing $b = 0$, the observability grammian becomes positive semi-definite whereas the controllability grammian is positive definite because the model system is not observable. Now, what happens if we choose $a = 2/3$, $b = 1/3$, $c = 5/3$? The system is still observable and controllable. However, the spectrum of the matrix A is $\sigma(A) = \{1, -1\}$ so that the system is stable but not BIBO stable. By inspection, we will find that the Lyapunov equations have no solutions.

5.3.3 Normal and balanced realizations

In section 5.3.1 we have seen that the state-space description of linear systems is unique up to a similarity transformation. We could use this freedom to bring a system to a realization which is particularly convenient for their analysis. In this section, we shall derive several such useful realizations.

In the following, we shall assume that the linear systems are of minimal realization and BIBO stable. Therefore, the controllability and observability grammians W_c and W_o are positive definite. Because positive definite matrices are diagonalizable, we may find a similarity transformation which diagonalize both grammians simultaneously. First, we investigate how the controllability and observability grammians W_c and W_o transform with respect to a similarity transformation. For discrete-time systems the analysis is straightforward. Combining (5.97.a) and (5.97.b) with (5.75) we obtain

$$\overline{W}_c = \sum_{k=0}^{\infty} \overline{A}^k \overline{B} \overline{B}^T (\overline{A}^T)^k = \sum_{k=0}^{\infty} T^{-1} A^k B B^T (A^T)^k T^{-T} = T^{-1} W_c (T^{-1})^T$$

$$\overline{W}_o = \sum_{k=0}^{\infty} (\overline{A}^T)^k \overline{C}^T \overline{C} \overline{A}^k = \sum_{k=0}^{\infty} T^T (A^T)^k C^T C A^k T = T^T W_o T$$

(5.101)

Note that the grammians are not invariant with respect to similarity transformations. The transformed grammians $\overline{W}$ are T-congruent to those formulated with respect to the original basis. With Sylvester's law of inertia we deduce that the transformed grammians $\overline{W}$ are positive definite too.

To analyze the transformation of the grammians of continuous-time systems we consider the identity $e^{\overline{A}t} = T^{-1} e^{At} T$ and obtain

$$\overline{W}_c = \int_0^{\infty} e^{\overline{A}t} \overline{B} \overline{B}^T e^{\overline{A}^T t} dt = \int_0^{\infty} T^{-1} e^{At} B B^T e^{A^T t} (T^{-1})^T dt = T^{-1} W_c (T^{-1})^T$$

$$\overline{W}_o = \int_0^{\infty} e^{\overline{A}^T t} \overline{C}^T \overline{C} e^{\overline{A}t} dt = \int_0^{\infty} T^T e^{A^T t} C^T C e^{At} T dt = T^T W_o T$$

(5.102)

Hence, the grammians of continuous-time systems transforms like those of discrete-time systems so that in the following we don't distinguish between discrete- and continuous-time systems.

Now consider the controllability grammian W_c. Because it is positive definite, there exists a unitary matrix U_c so that $W_c = U_c \Lambda_c U_c^T$. Λ_c is diagonal and contains the eigenvalues of W_c (spectral representation of W_c). Define the matrix $T_1 = U_c \Lambda_c^{1/2}$ which is obviously nonsingular. By a similarity transformation with T_1 we obtain

$$\overline{W}_c = T_1^{-1} W_c (T_1^{-1})^T = \Lambda_c^{-1/2} U_c^T U_c \Lambda_c U_c^T U_c \Lambda_c^{-1/2} = I$$

(5.103)

Hence, a similarity transformation with $T_1 = U_c \Lambda_c^{1/2}$ transforms W_c to the identity matrix. The observability matrix

$$\overline{W}_o = T_1^T W_o T_1 = \Lambda_c^{1/2} U_c^T W_o U_c \Lambda_c^{1/2},$$

(5.104)

is in general not diagonal. Because $\overline{W}_o$ is positive definite it has the spectral decomposition

$$\overline{W}_o = V_{\bar{o}} \Sigma_{\bar{o}}^2 V_{\bar{o}}^T$$

(5.105)

with unitary $V_{\bar{o}}$. Therefore, a similarity transformation with the matrix

$$T_c = U_c \Lambda_c^{1/2} V_{\bar{o}}$$

(5.106)

diagonalize both grammians. In fact, evaluating (5.101) we obtain the transformed grammians

$$\overline{W}_c = I \qquad \overline{W}_o = \Sigma_{\bar{o}}^2 \qquad\qquad (5.107)$$

A system, whose controllability and observability grammians W_c and W_o are given by (5.107) is called to be in the input-normal realization form.

Similarly, we can start from the observability matrix W_o whose spectral representation is given by $W_o = U_o \Lambda_o U_o^T$. Using the nonsingular transformation matrix $T_2 = U_o \Lambda_o^{-1/2}$ we obtain

$$\overline{W}_c = T_2^{-1} W_c (T_2^{-1})^T = \Lambda_o^{1/2} U_o^T W_c U_o \Lambda_o^{1/2} = V_{\bar{c}} \Sigma_{\bar{c}}^2 V_{\bar{c}}^T$$
$$\overline{W}_o = I \qquad\qquad (5.108)$$

Defining then the nonsingular transformation matrix

$$T_o = U_o \Lambda_o^{-1/2} V_{\bar{c}} \qquad\qquad (5.109)$$

yields

$$\overline{W}_c = \Sigma_{\bar{c}}^2 \qquad \overline{W}_o = I. \qquad\qquad (5.110)$$

A system, whose controllability and observability grammians W_c and W_o are given by (5.107) is called to be in the output-normal realization form.

On the other hand using (5.109) we may define the transformation matrix

$$\tilde{T}_o = U_o \Lambda_o^{-1/2} V_{\bar{c}} \Sigma_{\bar{o}}^{-2} \qquad\qquad (5.111)$$

to derive the output-normal state-space realization form (5.110). It is easy to show that $\tilde{T}_o = T_o$. Furthermore, because the matrix defined by the product of the controllability grammian with the observability grammian is subjected to a similarity transform

$$\overline{W}_c \overline{W}_o = T^{-1} W_c W_o T \qquad\qquad (5.112)$$

their eigenvalues are invariant with respect to any transformation. Therefore, we obtain

$$\Sigma = \Sigma_{\bar{o}} = \Sigma_{\bar{c}}. \qquad\qquad (5.113)$$

Another more symmetric form of the transformed grammians is obtained if instead of (5.106) we define the transformation matrix

$$T_b = U_c \Lambda_c^{1/2} V_{\bar{o}} \Sigma^{-1/2}. \qquad\qquad (5.114)$$

In fact, under this transformation the transformed grammians are equal

$$\overline{W}_c = \overline{W}_o = \Sigma. \qquad\qquad (5.115)$$

A system whose controllability and observability grammians W_c and W_o are equal is called to be in the balanced realization form.

Remarks:

• The controllability and the observability matrices transform like

$$\overline{M}_c = T^{-1} M_c \qquad \overline{M}_o = M_o T. \qquad\qquad (5.116)$$

• The input normal realization can be obtained even if the system is controllable but not observable. Similarly, the output normal realization can be obtained if the system is observable but not controllable. In contrast, the balanced realization can only be obtained if the system is as controllable as it is observable. In other words $\text{rank}(W_c) = \text{rank}(W_o)$.

- From matrix analysis we know that two matrices are simultaneously diagonalizable if and only if they commute. That is, $W_c W_o = W_o W_c$. As a consequence, the eigenvalues of $W_c W_o$ are equal to those of $W_o W_c$ and are given by Σ^2.

- Defining the transformation matrix $T_\alpha = U_c \Lambda_c^{1/2} V_{\bar{o}} \Sigma^{-\alpha}$ we obtain the pair of grammians

$$\overline{W}_{c,\alpha} = \Sigma^{2\alpha} \qquad \overline{W}_{o,\alpha} = \Sigma^{2(1-\alpha)}. \tag{5.117}$$

This realization is called an intermediate realization. The input normal, output normal and balanced realizations are given by $\alpha = 0$, $\alpha = 1/2$, $\alpha = 1$, respectively.

An alternative way to find the balanced realization is suggested by the fact that Σ^2 are the eigenvalues of $W_o W_c$. For any eigenvalue λ not equal to zero we obtain

$$W_o W_c x = M_o^T M_o M_c M_c^T x = M_o^T \Gamma_H M_c^T x = \lambda x. \tag{5.118}$$

Multiplying the latter equation from the left with M_c^T yields

$$M_c^T M_o^T \Gamma_H M_c^T x = \Gamma_H^T \Gamma_H (M_c^T x) = \lambda (M_c^T x). \tag{5.119}$$

Hence, Σ^2 are the eigenvalues of $\Gamma_H^T \Gamma_H$. Clearly Σ^2 are also the eigenvalues of $\Gamma_H \Gamma_H^T$ too. Σ are the so-called singular values of the Hankel matrix Γ_H. They are obtained by the singular value decomposition of Γ_H

$$\Gamma_H = U \Sigma V^T. \tag{5.120}$$

We refer to [HJ90, HJ91] for the reader not being familiar with the theory of singular value decomposition of matrices. For the computational aspects we refer to [GL86].

Because of $\Gamma_H = M_o M_c$ we can factor (5.120) in the form

$$M_0 = U \Sigma^{1-\alpha} \qquad M_c = \Sigma^\alpha V^T \tag{5.121}$$

which gives the intermediate realizations

$$W_c = M_c M_c^T = \Sigma^{2\alpha} \qquad W_o = M_o^T M_o = \Sigma^{2(1-\alpha)}. \tag{5.122}$$

Example: Let us discuss these concepts with a simple example. Consider a single degree of freedom (SDOF) system described by

$$\ddot{x} + 2\xi\Omega\dot{x} + \Omega^2 x = \frac{f}{m} = u \tag{5.123}$$

If the velocity is taken as the output variable we obtain the realization

$$A = \begin{bmatrix} 0 & 1 \\ -\Omega^2 & -2\xi\Omega \end{bmatrix} \qquad B = \begin{bmatrix} 0 \\ 1 \end{bmatrix} \qquad C = [0 \ 1] \qquad D = 0, \tag{5.124}$$

where x and $y = \dot{x}$ have been chosen as state variables. The controllability and observability matrices of this system are

$$W_c = \frac{1}{4\xi}\begin{bmatrix} \Omega^{-3} & 0 \\ 0 & \Omega^{-1} \end{bmatrix} \qquad W_o = \frac{1}{4\xi}\begin{bmatrix} \Omega & 0 \\ 0 & \Omega^{-1} \end{bmatrix} \qquad (5.125)$$

Remarkably, they are diagonal so that the system is in a intermediate realization form. The singular values are the eigenvalues of $W_c W_o$ and are given by

$$\Sigma = \frac{1}{4\xi}\begin{bmatrix} \Omega^{-1} & 0 \\ 0 & \Omega^{-1} \end{bmatrix} \qquad (5.126)$$

Observe, that both are equal and that the singular values of the undamped SDOF system are infinite. To obtain the balanced realization form we have to transform the state-space realization (5.124) with (5.75) using the similarity matrix

$$T = \begin{bmatrix} \Omega^{-1} & 0 \\ 0 & 1 \end{bmatrix} \qquad (5.127)$$

This gives the state-space realization

$$A = \begin{bmatrix} 0 & \Omega \\ -\Omega & -2\xi\Omega \end{bmatrix} \qquad B = \begin{bmatrix} 0 \\ 1 \end{bmatrix} \qquad C = [0\ 1] \qquad D = 0, \qquad (5.128)$$

where the new state variables are

$$y_1 = \Omega x \qquad y_2 = \dot{x}. \qquad (5.129)$$

Observe, that both state variables have the dimension of a velocity.

5.3.4   The Möbius transformation again

In section 4.2.7 we have shown that there are Möbius transformations which map rational functions analytic in S_∞ to rational functions analytic in the right half-plane C_+. Hence, with these Möbius transformations we can relate discrete-time systems to continuous-time systems and vice versa. Therefore, for any state-space description of a discrete-time system there exists a related state-space description of a continuous-time system and vice versa.

The Möbius transformations given in section 4.2.3 have to be subjected to a successive Möbius transformation which maps elements belonging to the unit disk D to elements belonging to the open domain S_∞

$$z = M_4(z') = -\frac{1}{z'}, \qquad (5.130)$$

Clearly, the elements on the unit circle T are mapped to the unit circle T. Therefore, the Möbius transformation mapping rational functions analytic in C_L to rational functions analytic in S_∞ is given by

$$z = \frac{(a-r)+s}{(a+r)-s} = \frac{q+s}{p-s}. \qquad (5.131)$$

It maps the right half-plane $\text{Re}(s) > r$ to the unbounded domain S_∞ and the left half-plane $\text{Re}(s) < r$ to the unit disk D. The vertical line L is mapped to the unit circle T. The inverse

Möbius transformation is given by

$$s = \frac{zp - q}{z + 1}. \tag{5.132}$$

The Möbius transformation which maps C_+ to S_∞, C_- to D and iR to T is

$$z = \frac{a + s}{a - s}, \tag{5.133}$$

and its inverse is

$$s = a\frac{z - 1}{z + 1}. \tag{5.134}$$

Now let us investigate how the system matrices of a state-space description transforms under a Möbius transformation. Consider the transfer function

$$H(z) = C(zI - A)^{-1}B + D. \tag{5.135}$$

Let us assume that the map is given by the general Möbius transformation

$$z = \frac{\alpha s + \beta}{\gamma s + \delta}. \tag{5.136}$$

Now we combine (5.136) with (5.135) and compute

$$\tilde{H}(s) = C\left(\frac{\alpha s + \beta}{\gamma s + \delta}I - A\right)^{-1}B + D = (\gamma s + \delta)C((\alpha I - \gamma A)s - (\delta A - \beta I))^{-1}B + D$$
$$= (\gamma s + \delta)C(s - (\alpha I - \gamma A)^{-1}(\delta A - \beta I))^{-1}(\alpha I - \gamma A)^{-1}B + D \tag{5.137}$$

where the last step is obtained by the factorization

$$((\alpha I - \gamma A)s - (\delta A - \beta I))^{-1} = ((\alpha I - \gamma A)(s - (\alpha I - \gamma A)^{-1}(\delta A - \beta I)))^{-1}. \tag{5.138}$$

Next, define the matrices

$$\begin{aligned}\tilde{A} &= (\alpha I - \gamma A)^{-1}(\delta A - \beta I)\\ \tilde{B} &= (\alpha I - \gamma A)^{-1}B\end{aligned} \tag{5.139}$$

These will be the system matrices A and B of the mapped system. Then (5.137) becomes

$$\tilde{H}(s) = (\gamma s + \delta)C(sI - \tilde{A})^{-1}\tilde{B} + D. \tag{5.140}$$

However, the transfer function (5.140) is not in the standard form. The remaining system matrices are obtained expanding (5.140) in a von Neumann-series

$$\tilde{H}(s) = (\gamma s + \delta)C(sI - \tilde{A})^{-1}\tilde{B} + D = \sum_{k=0}^{\infty} \frac{\gamma C\tilde{A}^k\tilde{B}}{s^k} + \sum_{k=1}^{\infty} \frac{\delta C\tilde{A}^{k-1}\tilde{B}}{s^k} + D. \tag{5.141}$$

Taking the limit $s \to \infty$ we obtain the transformed $\tilde{D}$ matrix

$$\tilde{D} = D + \gamma C(\alpha I - \gamma A)^{-1}B. \tag{5.142}$$

The matrix $\tilde{C}$ is obtained observing that (5.141) may be written as

$$\tilde{H}(s) = (\gamma s + \delta)C(s - \tilde{A})^{-1}\tilde{B} + D = \sum_{k=1}^{\infty} \frac{C(\gamma\tilde{A} + \delta I)\tilde{A}^{k-1}\tilde{B}}{s^k} + \tilde{D}. \qquad (5.143)$$

This yields

$$\tilde{C} = C(\gamma\tilde{A} + \delta I) = C(\gamma(\alpha I - \gamma A)^{-1}(\delta A - \beta I) + \delta(\alpha I - \gamma A)^{-1}(\alpha I - \gamma A))$$
$$= (\alpha\delta - \beta\gamma)C(\alpha I - \gamma A)^{-1} \qquad (5.144)$$

Finally, we obtain the standard form of the mapped transfer function

$$\tilde{H}(z) = \tilde{C}(zI - \tilde{A})^{-1}\tilde{B} + \tilde{D} \qquad (5.145)$$

The system matrices are given by

$$\tilde{A} = (\alpha I - \gamma A)^{-1}(\delta A - \beta I) = (\delta A - \beta I)(\alpha I - \gamma A)^{-1}$$
$$\tilde{B} = (\alpha I - \gamma A)^{-1}B$$
$$\tilde{C} = (\alpha\delta - \beta\gamma)C(\alpha I - \gamma A)^{-1} \qquad (5.146)$$
$$\tilde{D} = D + \gamma C(\alpha I - \gamma A)^{-1}B$$

Usually, in this work, we will have to transform a discrete-time system to the related continuous-time system using the Möbius transformation (5.131). Therefore, we will specialize the transformed matrices for this particular transformation. Setting $\alpha = 1$, $\beta = q$, $\gamma = -1$, $\delta = p$ we obtain

$$\tilde{A} = (I + A)^{-1}(pA - qI) = (pA - qI)(I + A)^{-1}$$
$$\tilde{B} = (I + A)^{-1}B$$
$$\tilde{C} = (p + q)C(I + A)^{-1} \qquad (5.147)$$
$$\tilde{D} = D - C(I + A)^{-1}B$$

The transformed matrices for the Möbius transformation (5.133) is

$$\tilde{A} = a(I + A)^{-1}(A - I) = a(A - I)(I + A)^{-1}$$
$$\tilde{B} = (I + A)^{-1}B$$
$$\tilde{C} = 2aC(I + A)^{-1} \qquad (5.148)$$
$$\tilde{D} = D - C(I + A)^{-1}B$$

Remarks:

- The mapped system exists only if the matrix $(\alpha I - \gamma A)$ is invertible or, in other words, if no eigenvalue of $(\alpha I - \gamma A)$ is equal to zero. This occurs if α/γ is an eigenvalue of A. The mapped systems (5.147) and (5.148) exists provided that -1 is not an eigenvalue of A.
- The second formula for $\tilde{A}$ in (5.146) is readily obtained using the factorization

$$((\alpha I - \gamma A)s - (\delta A - \beta I))^{-1} = ((s - (\delta A - \beta I)(\alpha I - \gamma A)^{-1})(\alpha I - \gamma A))^{-1} \qquad (5.149)$$

- As can be shown, the controllability and observability properties of the continuous-time system

$\{\tilde{A}, \tilde{B}, \tilde{C}, \tilde{D}\}$ are the same of its related system $\{A, B, C, D\}$. Therefore, if $\{A, B, C, D\}$ is of minimal realization then $\{\tilde{A}, \tilde{B}, \tilde{C}, \tilde{D}\}$ is of minimal realization too.

Let us assume that the discrete-time system has an intermediate realization $\{A_\alpha, B_\alpha, C_\alpha, D_\alpha\}$. What can be stated about the realization of the related continuous-time system? Consider the Lyapunov equation of the continuous-time system $\{\tilde{A}, \tilde{B}, \tilde{C}, \tilde{D}\}$ given by

$$\tilde{A}P + P\tilde{A}^T = -\tilde{B}\tilde{B}^T. \tag{5.150}$$

Combining the last equation with (5.146) yields

$$(\alpha I - \gamma A)^{-1}(\delta A - \beta I)P + P(\delta A - \beta I)^T(\alpha I - \gamma A)^{-T} + (\alpha I - \gamma A)^{-1}BB^T(\alpha I - \gamma A)^{-T}$$
$$= (\alpha I - \gamma A)^{-1}((\delta A - \beta I)P(\alpha I - \gamma A)^T + (\alpha I - \gamma A)P(\delta A - \beta I)^T + BB^T)(\alpha I - \gamma A)^{-T} \tag{5.151}$$
$$= (\alpha I - \gamma A)^{-1}((\alpha\delta + \beta\gamma)(AP + PA^T) - 2\delta\gamma APA^T - 2\alpha\beta P + BB^T)(\alpha I - \gamma A)^{-T} = 0$$

If $(\alpha I - \gamma A)^{-1}$ exists we finally obtain

$$(\alpha\delta + \beta\gamma)(AP + PA^T) - 2\delta\gamma APA^T - 2\alpha\beta P = -BB^T. \tag{5.152}$$

Similarly we will find for the Lyapunov equation $\tilde{A}^T P + P\tilde{A} = -\tilde{C}^T\tilde{C}$

$$(\alpha\delta + \beta\gamma)(A^T Q + QA) - 2\delta\gamma A^T QA - 2\alpha\beta Q = -(\alpha\delta - \beta\gamma)^2\tilde{C}^T\tilde{C}. \tag{5.153}$$

Therefore, if the following equations holds

$$\alpha\delta + \beta\gamma = 0 \qquad \alpha\beta + \delta\gamma = 0, \tag{5.154}$$

the solutions of the Lyapunov matrix equations, that is the controllability and observability grammians, are invariant (except for a multiplication with a positive constant) with respect to Möbius transformations.

Now consider the Möbius transformation (5.131) with $\alpha = 1, \beta = q, \gamma = -1, \delta = p$. Clearly, the first of the equations in (5.154) gives $p = q$. This yields the condition $r = 0$. The second condition is then automatically fulfilled. Therefore, the Möbius transformations which maps the imaginary axis iR to the unit circle T fulfils the criterions (5.154). The controllability and observability grammians are transformed according to

$$\tilde{P} = 2aP \qquad \tilde{Q} = \frac{1}{2a}Q. \tag{5.155}$$

As a consequence, the controllability and observability grammians preserve their positive-definiteness only if $a > 0$. Furthermore, because $\tilde{P}\tilde{Q} = PQ$ the singular values are invariant with respect to this class of Möbius transformations. Using the more "symmetric" transformation

$$\begin{aligned}
\tilde{A} &= a(I + A)^{-1}(A - I) = a(A - I)(I + A)^{-1} \\
\tilde{B} &= \sqrt{2a}(I + A)^{-1}B \\
\tilde{C} &= \sqrt{2a}C(I + A)^{-1} \\
\tilde{D} &= D - C(I + A)^{-1}B
\end{aligned} \tag{5.156}$$

the controllability and observability grammians are even invariant with respect to this transform. That is $\tilde{P} = P$ and $\tilde{Q} = Q$. As a consequence, a discrete-time system which is in the intermediate realization form is mapped by (5.156) to a continuous-time system which is also in the intermediate realization form. In addition, the parameter α describing the intermediate realizations, see the equation (5.117), is equal for both systems. Therefore, if the discrete-time system is in the input-normal, output-normal or balanced realization form then the related continuous-time system is in the input-normal, output-normal or balanced realization form, respectively.

Chapter 6

Approximation of scalar DtN-maps II

The Laguerre and the Fourier series expansions of DtN-map kernels developed in chapter 4 solve in principle the approximation problem. With these expansions we can define a truncated Laurent series which, by way of the Z-transform, can be associated with linear finite-dimensional discrete-time systems. However, these systems are not optimal with regard to numerical efficiency because they have many state variables with little influence on the output. These states could be deleted without relevant consequences on the accuracy of the approximation. In system theory, this deletion of state variables is called model reduction. So far, many methods have been developed to reduce the number of state variables of linear systems. An overview of these techniques may be found in [FNG92]. In this chapter, we study three techniques to reduce the degrees of freedom of discrete-time systems. A classical one, the Padé approximation method, and two more advanced methods: the truncated balanced and the Hankel-norm approximation methods.

6.1   State-space description of truncated Laurent series

Let us assume that a mapped scalar DtN-map kernel has been approximated by a Fourier series of degree n. With this series we define a truncated Laurent series of degree n analytic in S_∞:

$$h(z) = \sum_{k=0}^{n} h_k z^{-k} = h_0 + \sum_{k=1}^{n} \frac{h_k z^{n-k}}{z^n}. \tag{6.1}$$

Now let us interpret $h(z)$ as a transfer function of a linear finite-dimensional discrete-time system. Applying (5.64) we find that a controllable realization is given by the controller form

$$A = \begin{bmatrix} 0 & 0 & \dots & 0 \\ 1 & 0 & \dots & 0 \\ 0 & 1 & \dots & 0 \\ \dots & \dots & \dots & \dots \\ 0 & 0 & 1 & 0 \end{bmatrix} \quad B = \begin{bmatrix} 1 \\ 0 \\ 0 \\ \dots \\ 0 \end{bmatrix} \quad C^T = \begin{bmatrix} h_1 \\ h_2 \\ h_3 \\ \dots \\ h_n \end{bmatrix} \quad D = h_0 \tag{6.2}$$

Because the transfer matrix has finitely many terms, the systems (6.2) is necessarily BIBO-stable. In fact, the matrix A has one eigenvalue of algebraic multiplicity n. This eigenvalue is zero. As a consequence, the system matrix A is nilpotent of order n, that is $A^k = 0$ for $k \geq n$. The matrix A is called the forward shift matrix because of its effect on the elements of the standard basis of finite dimensional vector spaces. That is, $A^k e_j = e_{k+j}$.

Now let us investigate the controllability and observability matrices of the system (6.2) in order to establish under which conditions it is a minimal realization. As is readily seen, the $n \times n$ partial controllability matrix is the identity matrix

$$M_c = [B, AB, ..., A^{n-1}B] = \begin{bmatrix} 1 & 0 & ... & 0 \\ 0 & 1 & ... & 0 \\ ... & ... & ... & ... \\ 0 & ... & ... & 1 \end{bmatrix} \tag{6.3}$$

so that the system is controllable. The $n \times n$ partial observability matrix is the Hankel matrix

$$M_o = \Gamma_H = \begin{bmatrix} C \\ CA \\ ... \\ CA^{n-1} \end{bmatrix} = \begin{bmatrix} h_1 & h_2 & ... & h_n \\ h_2 & h_3 & ... & 0 \\ ... & ... & ... & ... \\ h_n & 0 & ... & 0 \end{bmatrix} \tag{6.4}$$

Hence, the observability depends on the rank of the Hankel matrix and in particular on the Markov parameters h_k. Now, it is easily seen that the Hankel matrix (6.4) has full rank if and only if $h_n \neq 0$. Therefore, given a DtN-map kernel, we can always define a truncated Laurent series with $h_n \neq 0$ such that the associated system is of minimal dimension.

Next we compute the controllability and observability grammians. This yields

$$W_c = M_c M_c^T = I \qquad \text{and} \qquad W_o = M_o^T M_o = \Gamma_H^T \Gamma_H \tag{6.5}$$

Therefore, the realization (6.2) is almost in the input-normal form. What remains to do is to transform the observability grammian W_o in the diagonal form. Because the Hankel matrix Γ_H has full rank, W_o is positive definite and can be represented in the spectral form as $W_o = V\Sigma^2 V^T$, where Σ are the singular values of the Hankel matrix Γ_H and V is a unitary matrix which corresponds to the right unitary matrix of the SVD of Γ_H. Defining the new set of state variables $x = Vy$ we obtain the input-normal realization

$$\bar{A} = V^T A V \qquad \bar{B} = V^T B \qquad \bar{C} = CV \qquad \bar{D} = D. \tag{6.6}$$

Finally, with the similarity transformation $x = V\Sigma^{-1/2}y$ we obtain the balanced realization

$$\bar{A} = \Sigma^{1/2} V^T A V \Sigma^{-1/2} \qquad \bar{B} = \Sigma^{1/2} V^T B \qquad \bar{C} = CV\Sigma^{-1/2} \qquad \bar{D} = D. \tag{6.7}$$

Remarks:

- Because the matrix A is nilpotent of order n we have $A^k B = 0$ and $CA^k = 0$ for $k \geq n$. Therefore, the infinite controllability and observability matrices are equal to the associated finite-dimensional partial controllability and observability matrices.

- Because the Hankel matrix is symmetric the columns of the right and left unitary matrices of the SVD $\Gamma_H = U\Sigma V^T$ are equal up to the sign. The change of sign occurs because Γ_H may have negative eigenvalues (Γ_H is generally not positive definite).

- Instead of the controller form (6.2) we could have used the observer form

$$A = \begin{bmatrix} 0 & 1 & 0 & \dots & 0 \\ 0 & 0 & 1 & \dots & 0 \\ \dots & \dots & \dots & \dots & 1 \\ 0 & 0 & 0 & \dots & 0 \end{bmatrix} \qquad B = \begin{bmatrix} h_1 \\ h_2 \\ h_3 \\ \dots \\ h_n \end{bmatrix} \qquad C^T = \begin{bmatrix} 1 \\ 0 \\ 0 \\ \dots \\ 0 \end{bmatrix} \qquad D = h_0 \qquad (6.8)$$

to describe the system. Both forms are completely equivalent if the system (6.2) is of minimal realization. With the observer form the observability matrix is the identity matrix and the controllability matrix is equal to the Hankel matrix. The matrix A is called the backward shift matrix.

- The state-space realization described above can also be used for truncated Laurent series generated by Laguerre series expansions. With a Laguerre series of degree n we obtain a state-space realization with $n + 1$ state variables. As is easily shown, the related discrete-time system is given by

$$A = \begin{bmatrix} 0 & 0 & \dots & 0 \\ 1 & 0 & \dots & 0 \\ 0 & 1 & \dots & 0 \\ \dots & \dots & \dots & \dots \\ 0 & 0 & 1 & 0 \end{bmatrix} \qquad B = \begin{bmatrix} 1 \\ 0 \\ 0 \\ \dots \\ 0 \end{bmatrix} \qquad C^T = \frac{1}{\sqrt{2a}} \begin{bmatrix} h_0 - h_1 \\ -h_1 + h_2 \\ h_2 - h_3 \\ \dots \\ (-1)^n h_n \end{bmatrix} \qquad D = h_0 \qquad (6.9)$$

6.2 The Padé approximation method

The Padé approximation method is undoubtedly the most popular method to construct rational approximations. It is used in many areas of applied mathematics. A throughout description of the method and its fields of application may be found in [BG96]. The goal of this section is to give a brief description of the basic ideas and to explain why the Padé approximation method is generally not adequate for the approximation of DtN-map kernels.

Consider the formal Laurent series expansion of a function $f(z)$ given by

$$f(z) = \sum_{k=0}^{\infty} f_k z^{-k} = f_0 + \sum_{k=1}^{\infty} f_k z^{-k}. \qquad (6.10)$$

A $[M,N]$-Padé-approximations is a rational function

$$r(z) = \left(\sum_{k=0}^{M} a_k z^k \right) \Big/ \left(\sum_{k=0}^{N} b_k z^k \right) \qquad (6.11)$$

whose numerator and denominator are polynomials in z of degree M and N. The coefficients a_k and b_k are defined by imposing that the first $M + N + 1$ coefficients of the formal Laurent series agree with those of the asymptotic expansion of the rational function (6.10). We set $b_N = 1$ in the denominator of (6.11). Usually, we will approximate the formal Laurent series

$$\tilde{f}(z) = f(z) - f_0 = \sum_{k=1}^{\infty} f_k z^{-k} \tag{6.12}$$

This series vanish for $z \to \infty$ and is at most of the order $O(1/z)$ when $z \to \infty$. The rational function (6.11) exhibits the same asymptotic behavior for $z \to \infty$ if the degree of the numerator is $M = N - 1$. Therefore, given the degree of the denominator N, the coefficients a_k and b_k are determined by the condition

$$\sum_{k=1}^{\infty} f_k z^{-k} \cdot \sum_{k=0}^{N} b_k z^k = \sum_{k=0}^{N-1} a_k z^k + O((1/z)^{N+1}). \tag{6.13}$$

Comparing the coefficients of the terms with equal degree in z on the left and on right-hand side of (6.13) we obtain a set of $2N$ equations

$$\sum_{j=1}^{N-k} b_{k+j} f_j = a_k \qquad \text{for} \qquad k = 1, ..., N \tag{6.14.a}$$

$$\sum_{j=0}^{N} b_j f_{k+j} = 0 \qquad \text{for} \qquad k = 1, ..., N \tag{6.14.b}$$

These equations define the coefficients b_k. Because of $b_N = 1$ (6.14.b) may be rewritten as

$$\sum_{j=1}^{N} b_j f_{k+j} = -f_{N+k} \quad \text{for} \quad k = 1, ..., N \tag{6.15}$$

Arranging the coefficients b_k and $-f_k$ in the vectors b and f (6.15) is written in the matrix form:

$$Hb = f \tag{6.16}$$

with the real $N \times N$ Hankel-matrix H defined by

$$H = \begin{bmatrix} f_1 & f_2 & \cdots & f_N \\ f_2 & f_3 & \cdots & f_{N+1} \\ \cdots & \cdots & \cdots & \cdots \\ f_N & f_{N+1} & \cdots & f_{2N} \end{bmatrix}. \tag{6.17}$$

If H is nonsingular, the linear system of equations (6.16) gives a unique solution for the coefficients of the denumerator b_k. The coefficients a_k, $k = 0, ..., N - 1$, of the numerator polynomial are computed straightforwardly using the set of equations (6.14.a).

The Padé-approximation method is conceptually simple and numerically straightforward. In fact, once we have chosen the degree N of the denominator, the Padé-approximation can be computed with standard techniques of numerical linear algebra provided that H is nonsingular. However, for large degree N the Hankel matrix H is nearly singular. As a consequence, the coefficients of the Padé-approximations may become sensitive to small perturbations because of the round-off errors which occurs when the solution of (6.16) is computed. A detailed error analysis shows that given a perturbed formal Laurent series $f_\varepsilon = f + \varepsilon g$, where g is a formal Laurent series and ε is the perturbation parameter, the relative error of the denominator coefficients is

$$\frac{\|b_\varepsilon - b\|_2}{\|b\|_2} \leq Cs_1 \frac{\varepsilon}{s_n}, \tag{6.18}$$

where b refers to the exact solution, b_ε to the perturbed solution, C is a finite positive constant and s_1, s_n are the largest and smallest singular values of the Hankel matrix (6.17), respectively. Therefore, if the magnitude of the perturbation parameter is of the same order of the smallest singular value, the fraction ε / s_n is of the order $O(1)$ and the relative error becomes very large. Because the magnitude of the smallest singular value s_n depends on the size of the Hankel matrix, the degree N of the Padé-approximations must be chosen with care in order to avoid large errors. For nearly singular matrices it is convenient to use the singular value decomposition (SVD) of H to compute the solution vector b. In fact, writing

$$Hx = USV^T b = f \tag{6.19}$$

the solution b is given by

$$b = VS^{-1}U^T f = \sum_{j=1}^{n} \frac{v_j u_j^T f}{s_j}. \tag{6.20}$$

A numerically less sensitive method to construct Padé approximations of a given degree may be obtained from Padé approximations of smaller degree using recursive methods (see [MG96] for further details).

However, little can be said about the stability, minimality and accuracy of the rational approximation. Clearly, by its very definition, taking the limit $N \to \infty$ the Padé approximations converges to the function $f(z)$ defined by the formal Laurent series because the first $2N$ coefficients of the asymptotic expansion of the Padé approximation (6.11) are by definition equal to those of the Laurent series expansion (6.10). As we have seen in chapter 4, Padé approximations may even converge in a domain which is larger than that of the underlying formal Laurent series. However, Padé approximations may not be stable. That is, the denominator may have poles located in the right half-plane. In fact, unstable states may arise especially when the denominator polynomial is of large degree [FNG92]. In these cases, special algorithms are used to construct stable Padé approximation [FNG92]. Moreover, approximations with large degree exhibit often nearly pole cancellation. The associated linear system has therefore states which are nearly non-controllable and/or nonobservable.

6.3   The truncated balanced realization method

The most common techniques to approximate a given system with one of lower dimension are the methods of model reduction by truncation. The goal of this approach is to remove all state variables which are assumed to be negligible for the solution of a given problem. In structural dynamics, this is explicitly done by neglecting all those modes whose eigenfrequencies are larger than a certain, usually problem dependent, frequency (modal truncation method). On the other hand, truncation of modes with high eigenfrequency may be performed implicitly by the use of time integration algorithms with strong numerical damping at large frequencies. Implicitly, because in this case the high frequency modes are not really removed, that is, the algebraic dimension of the system is not reduced. In order to be reliable, a truncation scheme must preserve the stability of the system. Moreover, the error induced by the truncation must be bounded with respect to an appropriate norm.

In this section, we describe a conceptually simple procedure, the truncated balanced realization method, which allows to construct a stable reduced-order model. The truncated balanced realization method is somehow similar to the classical mode truncation procedure. In fact, we delete those components of the state vector $x(t)$ which have little influence on the energy transfer from the input $u(t)$ to the output $v(t)$. The error induced by the truncated balanced realization method is bounded with respect to the spaces $L_\infty(T)$ and $L_\infty(iR)$.

6.3.1 The truncated balanced realization method

Let us consider the following problem to introduce the truncated balanced realization method. Consider a linear system driven by an arbitrary input function $u(t)$ belonging to $L_2(R)$ from the initial state $x_{-\infty} = 0$ at $t = -\infty$ to the state x_0 at $t = 0$. There, we stop the input and measure the energy of the output $\|v(t)\|_2$ for $t \geq 0$. Next, we divide the output energy by the input energy and take the supremum to form the norm of $h(\omega)$ with respect to $L_\infty(iR)$. Using the interpretation of the controllability and observability grammians given in section 5.3.3 we obtain

$$\|h(\omega)\|_{L_\infty(iR)} = \sup \frac{\|v(t)\|_{L_2(R_-)}}{\|u(t)\|_{L_2(R_-)}} = \max \frac{x^T_0 W_o x_0}{x^T_0 W_c^{-1} x_0}. \tag{6.21}$$

Now, suppose we have an input-normal representation. Then, (6.21) becomes

$$\|h(\omega)\|_{L_\infty(iR)} = \max \frac{y^T_0 \Sigma^2 y_0}{y^T_0 y_0}. \tag{6.22}$$

The last equation suggests to delete those state variables with small singular values, because they have little influence on the norm of $h(\omega)$ with respect to $L_\infty(iR)$. This idea for model reduction via balanced truncation realization was first proposed by Moore [Mo81]. A textbook presentation of the subject is found in [GL96].

Let us assume that the system $\{A, B, C\}$ is in the balanced realization form and let us partition the controllability and observability grammians as

$$\Sigma = \begin{bmatrix} \Sigma_1 & 0 \\ 0 & \Sigma_2 \end{bmatrix}, \tag{6.23}$$

with

$$\Sigma_1 = \begin{bmatrix} \sigma_1 I_{r_1} & \cdots & 0 \\ \cdots & \cdots & \cdots \\ 0 & \cdots & \sigma_k I_{r_k} \end{bmatrix} \qquad \Sigma_2 = \begin{bmatrix} \sigma_{k+1} I_{r_{k+1}} & \cdots & 0 \\ \cdots & \cdots & \cdots \\ 0 & \cdots & \sigma_p I_{r_p} \end{bmatrix} \tag{6.24}$$

where $\sigma_k > \sigma_{k+1}$ and r_k is the dimension of the subspace associated with the singular value σ_k. Let us partition the system matrices conformably with (6.23) as

$$A = \begin{bmatrix} A_{11} & A_{12} \\ A_{21} & A_{22} \end{bmatrix} \qquad B = \begin{bmatrix} B_1 \\ B_2 \end{bmatrix} \qquad C = \begin{bmatrix} C_1 & C_2 \end{bmatrix}, \tag{6.25}$$

We define the system $\{A_{11}, B_1, C_1, D\}$ as the balanced truncation of $\{A, B, C\}$.

Next, let us give some properties of this balanced truncated system. One can easily verify that the system $\{A_{11}, B_1, C_1, D\}$ obeys the Lyapunov equations

$$A_{11}\Sigma_1 A_{11}^T - \Sigma_1 = -B_1 B_1^T$$
$$A_{11}^T \Sigma_1 A_{11} - \Sigma_1 = -C_1^T C_1$$
(6.26)

That is, (6.25) is formally in the balanced realization form. Furthermore, the system is stable because Σ_1 is positive definite but not necessarily BIBO stable because the system (6.25) may be unobservable or uncontrollable or both. However, in Pernebo and Silverman [PS82] proved that any balanced truncation of a BIBO stable system is aso BIBO stable. Hence, the balanced truncated system $\{A_{11}, B_1, C_1, D\}$ is also BIBO stable. That is, the system $\{A_{11}, B_1, C_1, D\}$ is of minimal realization with McMillan degree $n = r_1 + \ldots + r_k$. Moreover, the approximation error with respect to $L_\infty(iR)$ is bounded by twice the sum of the singular values associated with the state variables that have been deleted [GL96]. That is,

$$\|h(\omega) - h(\omega)\|_\infty \leq 2(\sigma_{k+1} + \ldots + \sigma_p),$$
(6.27)

where p is the number of distinct singular values.

Remarks:
- In (6.22) we used instead of the balanced the input-normal realization because (6.22) is easier to interpret. Clearly, the same results apply for balanced realizations too. In addition, the state-space representation in the balanced form is more symmetric.
- Observe that because of (6.22) we may delete the nonobservable state variables without influencing the norm of $h(\omega)$ with respect to $L_\infty(iR)$.
- If only one state with associated singular value σ is deleted the approximation error (6.27) becomes exactly $\|h(\omega) - h(\omega)\|_\infty = 2\sigma$ and the maximum error occurs at $\omega = 0$ [GL96].
- Because the system matrix D is not affected by the truncation procedure, the reduced-order model has perfect matching at infinity, that is $h(\infty) = \hat{h}(\infty) = D$. Therefore, the reduced-order model exhibits good high-frequency fidelity. Little can be said about the variation of the error at low and intermediate frequencies. However, by the remark above, if all singular values are distinct we expect that the largest approximation error occurs at low frequencies.
- We have formulated the theory of the truncated balanced realization method in the context of discrete-time systems. However, it can be applied without any substantial modification to continuous-time systems.

Next, let us consider how we can apply the truncated balanced realization method for the approximation of DtN-map kernels. In section 6.1 we have shown that to every truncated Laurent series with $h_n \neq 0$ there exists a BIBO stable discrete-time system of minimal realization. This can be transformed to the balanced realization form. Now, if the discrete-time system and its related continuous-time system are associated via the Möbius transformation which maps the unit circle T to the imaginary axis iR, then, according to section 5.3.4, the related continuous-time system is in the balanced realization form. In addition, both systems have exactly the same singular values.

Hence, we can proceed along two ways to approximate DtN-map kernels: perform the model reduction on the discrete-time system and then map it to the related continuous-time system or first map the discrete-time system to the related continuous-time system and afterwards perform the model reduction of the continuous-time system. The error estimate (6.27) is valid for both approximations. However, as we shall see, the approximations will not be equal. In this section, we discuss the approach via truncating the discrete-time system. This approach is similar to that proposed by Gu et al. [GKL89] for delay systems. The approach via the mapped continuous-time system will be discussed in chapter 7.

Next we propose an algorithm to compute reduced-order models using the balanced truncation method:

1. Choose an upper bound of the approximation error, say ε.
2. Setup the observability matrix $W_o = \Gamma_H^T \Gamma_H$ and compute their eigenvalues Λ and eigenvectors V. Sort the eigenvalues and eigenvectors such that $\lambda_1 \geq \lambda_2 \geq \ldots \geq \lambda_n$.
3. Find the index k such that $2(\sigma_k + \ldots + \sigma_p) > \varepsilon \geq 2(\sigma_{k+1} + \ldots + \sigma_p)$. This is the index of the last state variable which will not be deleted.
4. Compute the truncated state-space realization of McMillan degree k by

$$\bar{A}_{11} = \Sigma_k^{1/2} V_k^T A V_k \Sigma_k^{-1/2} \qquad \bar{B}_1 = \Sigma_k^{1/2} V_k^T B \qquad \bar{C}_1 = C V_k \Sigma_k^{-1/2} \qquad \bar{D} = D, \quad (6.28)$$

where the $n \times k$ matrix V_k contains the k-first eigenvectors and Σ_k contains the k-first square roots of the eigenvalues of the observability matrix.

5. Map the discrete-time system to the related continuous-time system using (5.156).

Another algorithm to compute reduced-order models is obtained when we consider that the Hankel matrix of the shifted Markov parameters is given by

$$\hat{\Gamma}_H = \begin{bmatrix} h_2 & h_3 & \ldots & 0 \\ h_3 & h_4 & \ldots & 0 \\ \ldots & \ldots & \ldots & \ldots \\ 0 & 0 & \ldots & 0 \end{bmatrix} = M_o A M_c \qquad (6.29)$$

Note that the Hankel matrix is $\Gamma_H = M_o M_c$. Now we can proceed as follows:

1. Choose an upper bound of the approximation error, say ε.
2. Setup the Hankel matrix $\Gamma_H = M_o$ and compute its singular value decomposition $\Gamma_H = U \Sigma V^T$.
3. Find the index k such that $2(\sigma_k + \ldots + \sigma_p) > \varepsilon \geq 2(\sigma_{k+1} + \ldots + \sigma_p)$. This is the last state variable not being deleted. Define $M_o = U_k \Sigma_k^{1/2}$ and $M_c = \Sigma_k^{1/2} V_k^T$. The $n \times k$ matrix U_k and V_k contains the k-first left and right singular vectors and Σ_k contains the k-first singular values of the Hankel matrix.
4. Compute the truncated realization of McMillan degree k by

$$A_{11} = \Sigma_k^{-1/2} U_k^T \hat{\Gamma}_H V_k \Sigma_k^{-1/2} \qquad (6.30.a)$$

$$B = \text{first column of } M_c = \Sigma^{1/2} V^T \qquad (6.30.b)$$

$$C = \text{first row of } M_o = U \Sigma^{1/2}. \qquad (6.30.c)$$

5. Map the discrete-time system to the related continuous-time system via (5.156).

Both algorithms give exactly the same reduced-order model.

Finally, let us prove that the balanced truncation method can approximate a scalar DtN-map to any imposed accuracy with respect to the infinity norm. Let $h(z)$ be the scalar DtN-map, $\hat{h}_m(z)$ an approximation of degree m and ε an error bound. The error with respect to the infinity norm is $\|h(z) - \hat{h}_m(z)\|_{L_\infty(T)}$. Let $h_n(z)$ be the truncated Laurent series of degree n. Then we have

$$\|h(z) - \hat{h}_m(z)\|_{L_\infty(T)} \leq \|h(z) - h_n(z)\|_{L_\infty(T)} + \|h_n(z) - \hat{h}_m(z)\|_{L_\infty(T)}. \qquad (6.31)$$

Let us assume that $h_n(z)$ converges to $h(z)$ in $L_\infty(T)$ then there exists an index n_ε so that $\|h(z) - h_{n_\varepsilon}(z)\|_{L_\infty(T)} \leq \varepsilon/2$. Furthermore, because of the error bound (6.27) we can find a a system of McMillan degree $m_\varepsilon(n_\varepsilon) < n_\varepsilon$ so that $\left\| h_n(z) - \hat{h}_{m_\varepsilon}(z) \right\|_{L_\infty(T)} \leq \varepsilon/2$. Hence, we obtain $\left\| h(z) - \hat{h}_m(z) \right\|_{L_\infty(T)} \leq \varepsilon$. Because the error bound ε is arbitrary the prove is complete.

Remarks:

- When other Möbius transformations than (5.156) are applied we must first map the discrete-time system to a continuous-time system and then apply the balanced truncation procedure.

- Because the Hankel matrix (6.4) is symmetric we could use an eigenvalue analysis procedure to construct its the singular value decomposition. The singular values are the absolute value of the eigenvalues. Defining the eigenvectors as the left singular vectors the right singular vectors are given by

$$v_i = \text{sign}(\lambda_i) u_i. \tag{6.32}$$

- The algorithm via singular value decomposition of the Hankel matrix has been used by Weber [Web94] to obtain approximations of DtN-map kernels. The approximation was constructed using the following Hankel matrix

$$\Gamma_H = \begin{bmatrix} h_1 & h_2 & \dots & h_n \\ h_2 & h_3 & \dots & h_{n+1} \\ \dots & \dots & \dots & \dots \\ h_n & h_{n+1} & \dots & h_{2n} \end{bmatrix} \tag{6.33}$$

However, if the system to be approximated is infinite-dimensional, this Hankel matrix can't be associated with any BIBO stable finite-dimensional system. Therefore, the reduced-order model constructed via the singular value decomposition of the Hankel matrix (6.33) may be unstable. Weber's proposal was to delete the unstable state variables. However, this may induce an additional approximation error. Moreover, Weber's method needs generally a singular value decomposition to construct reduced-order models. This is numerically less effective than an eigenvalue analysis of a positive definite matrix. In the context of the theory developed in this work, Weber's procedure could be interpreted as an approximation. In fact, if we consider the first $2n + 1$ terms of the truncated Laurent series we obtain

$$\Gamma_{H_{2n+1}} = \begin{bmatrix} \Gamma_{H_n} & \Gamma_{H_{(n+1,2n+1)}} \\ \Gamma_{H_{(n+1,2n+1)}} & 0 \end{bmatrix}. \tag{6.34}$$

Therefore, Weber's approach picks out the left upper block to construct the reduced-order model whereas in this work we use the complete Hankel matrix. Shortly, Weber approximates a truncated Hankel matrix of an infinite-dimensional system. In this, work we approximate infinite-dimensional systems with stable finite-dimensional systems and then apply model reduction techniques to obtain smaller systems.

6.4 The Hankel-norm approximation method

The truncated balanced realization method is a conceptually simple, numerically efficient and very accurate method to construct a BIBO stable reduced-order model. However, the reduced-order model is not optimal with respect to any norm. In this section we describe the Hankel-norm approximation method. Contrary to the truncated balanced realization method, the Hankel-norm approximation method provides approximations which are optimal with respect to a specific norm: the Hankel norm. In this section, we develop some results of the theory of the Hankel-norm approximation of discrete-time systems. This, because the theory is much simpler than that of continuous-time systems. The Hankel-norm approximation of continuous-time systems will be described in the next chapter. A throughout description of this method is found in [CC92]. An introduction to Hankel operators is found in [Par90].

6.4.1   Hankel operator, Hankel norm and optimal Hankel-norm approximation

Consider the Fourier series on the unit circle T given by

$$f(z) = \sum_{n=-\infty}^{\infty} f_n z^{-n},$$ (6.35)

where $z = e^{i\theta}$. To this Fourier series we associate an infinite-dimensional Hankel matrix defined by $\Gamma_f[j, k] = [f_{j+k-1}]$ for $j, k \geq 1$, that is

$$\Gamma_f = \begin{bmatrix} f_1 & f_2 & f_3 & \cdots \\ f_2 & f_3 & \cdots & \cdots \\ f_3 & \cdots & \cdots & \cdots \\ \cdots & \cdots & \cdots & \cdots \end{bmatrix}.$$ (6.36)

Now let us consider Γ_f as a linear operator on the sequence space ℓ_2. Then, the Hankel norm of $f(z)$ is defined to be the spectral norm of the Hankel operator Γ_f. That is,

$$\|f\|_\Gamma = \|\Gamma_f\|_s = \sup_{\|x\|_{\ell_2} = 1} \|\Gamma_f x\|_{\ell_2}.$$ (6.37)

Therefore, $\|f\|_\Gamma < \infty$ if and only if the Hankel operator Γ_f is a bounded operator on ℓ_2. That is, $\|f\|_\Gamma$ is bounded if for any sequence x in ℓ_2 the sequence $y = \Gamma_f x$ is in ℓ_2. A definite answer about the boundedness of Hankel operators Γ_f was proved by Nehary [Neh57]: Let $f(z)$ be any function in $L_\infty(T)$ and $g(z)$ be any function in $H_\infty(D)$ then

$$\|f\|_\Gamma = \|\Gamma_f\|_s = \inf\|f - g\|_{L_\infty(T)}$$ (6.38)

An immediate consequence of this theorem is that every function $f(z)$ belonging to $L_\infty(T)$ has finite Hankel norm. Furthermore, a function $f(z)$ not belonging to $L_\infty(T)$ has finite Hankel norm if and only if there exists a function $g(z)$ belonging to $H_\infty(D)$ such that $f(z) - g(z)$ belongs to $L_\infty(T)$. In this work, the DtN-map kernels of concern are bounded functions. Therefore, they are in $L_\infty(T)$ and have a bounded Hankel operator. Moreover, because the DtN-map kernels are continuous the Hankel operator Γ_f is even compact [Neh57]. Moreover, if the Hankel norm is compared with the L_2 and L_∞ norms on the unit circle T we obtain

$$\|f\|_{L_2} \leq \|f\|_\Gamma \leq \|f\|_{L_\infty}.$$ (6.39)

Now suppose we have an optimal approximation $\hat{f}(z)$ with respect to the Hankel norm. The last equation states that the approximation error with respect to $L_2(T)$ will be smaller or at most equal to the error with respect to the Hankel norm. On the other hand, the approximation error with respect to $L_\infty(T)$ will be larger or at most equal to the error with respect to the Hankel norm.

Now let us formulate the approximation problem with respect to the Hankel norm. Given a function $f(z)$ find a rational function $\hat{r}_m(z)$ out of the class of strictly proper rational functions $R_m(S_\infty)$ of degree m and analytic in S_∞ (all poles are in the unit disk D) such that the error with respect to the Hankel norm is minimized. That is,

$$\|f - \hat{r}_m\|_\Gamma = \inf_{r_m \in R_m(S_\infty)}\|f - r_m\|_\Gamma. \tag{6.40}$$

Now, Kronecker's theorem [CC92] states that to any rational function $r_m(z)$ of degree m we can associate an infinite Hankel matrix Γ_{r_m} of finite rank m. Hence, because of the very definition of the Hankel norm, the approximation problem (6.40) can be reformulated as

$$\|\Gamma_f - \Gamma_{\hat{r}_m}\|_s = \inf_{\Gamma_{r_m} \in \Gamma_m}\|\Gamma_f - \Gamma_{r_m}\|_s. \tag{6.41}$$

That is, given a Hankel matrix Γ_f of rank $n > m$ find the Hankel matrix $\Gamma_{\hat{r}_m}$ of rank at most m such that the spectral norm is minimized.

The singular value decomposition provides the solution of a similar problem: Find an optimal approximation of a given Hankel matrix S of finite or infinite rank with a matrix S_m of finite rank m. The solution is given by

$$S_m = U_m \Sigma_m V_m^T, \tag{6.42}$$

where $U_m = \begin{bmatrix} u_1 & \ldots & u_m \end{bmatrix}$ and $V_m = \begin{bmatrix} v_1 & \ldots & v_m \end{bmatrix}$ are the unitary matrices containing the m-first left and right singular vectors and Σ_m is a diagonal matrix containing the m-first singular values of the Hankel matrix Γ_f. The approximation error with respect to the spectral norm is given by the singular value s_{m+1}

$$\|S - S_m\|_s = s_{m+1}. \tag{6.43}$$

However, if the matrix S is a Hankel matrix S_m is generally not a Hankel matrix so that $\|\Gamma_f - \Gamma_{\hat{r}_m}\|_s \geq s_{m+1}(\Gamma_f)$. Remarkably, Adamjan, Arov and Krein [AAK71] proved that there exists a rational function $r_m(z)$ such that $\|\Gamma_f - \Gamma_{\hat{r}_m}\|_s = s_{m+1}(\Gamma_f)$. This is the famous AAK-theorem: Let $f(z)$ be a function in $L_\infty(T)$ and $r_m(z)$ a rational function in $R_m(S_\infty)$. Then

$$\|f - \hat{r}_m\|_\Gamma = \inf_{r_m \in R_m}\|f - r_m\|_\Gamma = \inf_{h_m \in H_\infty^m}\|f - \hat{h}_m\|_{L_\infty(T)} = s_{m+1}(\Gamma_f). \tag{6.44}$$

H_∞^m is the class of function defined by $h_m(z) = r_m(z) + h(z)$, where $h(z)$ belongs to $H_\infty(D)$.

Even more remarkable as the existence of the solution of the extremum problem (6.44) is the fact that $\hat{r}_m(z)$ is the so-called singular part of $\hat{h}_m(z)$. This function belongs to H_∞^m and is the best approximation with respect to $L_\infty(T)$. Remarkably, $\hat{h}_m(z)$ can be even formulated explicitly. Let $[\hat{h}_m(z)]_s$ be the singular part of the rational function $\hat{h}_m(z)$ defined by

$$\hat{r}_m(z) = [h_m(z)]_s, \tag{6.45}$$

$\hat{h}_m(z)$ is given by

$$\hat{h}_m(z) = f(z) - s_{m+1}(\Gamma_f)\frac{v(z)}{u(z)} \tag{6.46}$$

where

$$u(z) = \sum_{k=1}^{\infty} u_k z^{k-1} \qquad v(z) = \sum_{k=1}^{\infty} v_k z^{-k}. \tag{6.47}$$

$u_{m+1} = [u_1, u_2, \ldots]^T$ and $v_{m+1} = [v_1, v_2, \ldots]$ are the left and right singular vectors associated with the singular value $s_{m+1}(\Gamma_f)$.

Remarks:

- Because the rational functions $r_m \in R_m(S_\infty)$ are strictly proper, R_0 contains only the function $r_0 = 0$. In this case, (6.44) becomes (6.38) such that the AAK-theorem is a generalization of Nehari's theorem for $f(z)$ belonging to $L_\infty(T)$.

- Observe that the AAK-theorem assures that $\hat{r}_m(z)$ is a stable rational function.

- Nehary [Neh57] proved that if $f(z)$ is continuous, the associated Hankel operator Γ_f is compact. Therefore, the accumulation point of the singular values is zero. That is, $s_m(\Gamma_f) \to 0$ for $m \to \infty$. As a consequence, $\|f - \hat{r}_m\|_\Gamma \to 0$ for $m \to \infty$. That is, every continuous function $f(z)$ can be approximated with arbitrary accuracy in the Hankel norm with a rational function $r_m \in R_m(S_\infty)$.

- Generally, the direct application of the AAK-theory to infinite-dimensional matrices with infinite rank is virtually impossible because it is very difficult to find an explicit singular value decomposition. Therefore, in order to be computable, the infinite Hankel matrix must be approximated by a Hankel matrix Γ_f^m of finite rank m.

6.4.2 The optimal Hankel-norm approximation method

In this section, we apply the optimal Hankel-norm approximation theory to the system defined by the truncated Laurent series. The associated Hankel matrix is given by (6.4). The method is contained as a special case of the procedures presented by Kung [Kun80] and Silvermann and Bettayeb [SB80] which uses the AAK-theory in the context of model reduction for finite-dimensional systems. Independently, a similar technique was proposed by Trefethen [Tre81] and later generalized by Gutknecht [Gut84] for the near best $L_\infty(T)$ approximation of polynomials. It has been called Carathéodory-Féjer method (CF-method). In [GST83] this method was applied to digital filter design.

In the following, we use the AAK-theorem to construct the optimal Hankel approximation $\hat{r}_m^n(z)$ of the truncated Laurent series (6.1). First we show that the rational function $\hat{h}_m^n(z)$ belonging to H_∞^m and given by (6.46) fulfils the equality

$$\|f - \hat{h}_m^n\|_{L_\infty(T)} = s_{m+1}(\Gamma_H^n). \tag{6.48}$$

Because the Hankel matrix Γ_H^n is symmetric the left and right singular eigenvectors differ only in the sign, that is, $v_{m+1} = \pm u_{m+1}$. Hence we have

$$
\begin{bmatrix} h_1 & h_2 & \dots & h_n \\ h_2 & h_3 & \dots & 0 \\ \dots & \dots & \dots & \dots \\ h_n & 0 & \dots & 0 \end{bmatrix}
\begin{bmatrix} u_1 \\ u_2 \\ \dots \\ u_n \end{bmatrix}
= \pm s_{m+1}
\begin{bmatrix} u_1 \\ u_2 \\ \dots \\ u_n \end{bmatrix}. \tag{6.49}
$$

Combining this fact with (6.46) we obtain

$$f(z) - h_m^n(z) = \pm s_{m+1}(\Gamma_H^n)\frac{v(z)}{u(z)}. \tag{6.50}$$

The rational function $v(z)/u(z)$ is given by

$$v(z)/u(z) = \frac{u_1 z^{-1} + u_2 z^{-2} + \dots + u_n z^{-n}}{u_1 + u_2 z + \dots + u_n z^{(n-1)}} = z\frac{u_1 + u_2 z^{-1} + \dots + u_n z^{-(n-1)}}{u_1 + u_2 z + \dots + u_n z^{(n-1)}}. \tag{6.51}$$

On the unit circle T we have $z = e^{i\theta}$ so that

$$v(z)/u(z) = e^{i\theta}\frac{u_1 + u_2 e^{-i\theta} + \dots + u_n e^{-i(n-1)\theta}}{u_1 + u_2 e^{-i\theta} + \dots + u_n e^{i(n-1)\theta}} = e^{i\theta}\frac{\bar{u}(z)}{u(z)}. \tag{6.52}$$

Hence

$$|v(z)/u(z)| = |e^{i\theta}|\left|\frac{\bar{u}(z)}{u(z)}\right| = 1 \text{ for } z = e^{i\theta} \tag{6.53}$$

and, as a consequence,

$$\|f - h_m^n\|_\infty = s_{m+1}(\Gamma_H^n)|v(z)/u(z)| = s_{m+1}(\Gamma_H^n). \tag{6.54}$$

Next we construct the explicit form of $\hat{h}_m^n(z)$. From (6.46) it is obvious that

$$\hat{h}_m^n(z) = \frac{p(z)}{u(z)} \tag{6.55}$$

where $u(z) = u_1 + u_2 z + \dots + u_n z^{(n-1)}$ and with $p(z) = h_m^n(z)u(z) - s_{m+1}(\Gamma_H^n)v(z)$. Now we will show that $p(z)$ is a polynomial of degree $n - 2$. In fact, because of (6.49) we have

$$h^n(z)u(z) - s_{m+1}v(z) = \left(\sum_{k=1}^{n} h_k z^{-k}\right)\left(\sum_{l=1}^{n} u_l z^{l-1}\right) \mp s_{m+1}\sum_{k=1}^{n} u_k z^{-k}$$

$$= \left(\sum_{k=1}^{n} h_k z^{-k}\right)\left(\sum_{l=1}^{n} u_l z^{l-1}\right) - \sum_{k=1}^{n}\sum_{l=1}^{n} h_{k+l-1}u_l z^{-k} \tag{6.56}$$

$$= \sum_{k=1}^{n}\sum_{l=k+1}^{n} h_{l-k}u_{l+1}z^k + \sum_{k=1}^{n}\sum_{l=1}^{n} h_{k+l-1}u_l z^{-k} - \sum_{k=1}^{n}\sum_{l=1}^{n} h_{k+l-1}u_l z^{-k}$$

The last two summands cancel out and therefore we obtain

$$p(z) = h_m^n(z)u(z) - s_{m+1}v(z) = \sum_{k=1}^{n-2}\sum_{l=k+1}^{n-1} h_{l-k}u_{l+1}z^k \tag{6.57}$$

which is a polynomial of degree at most $n - 2$. The polynomial $p(z)$ may be expressed in the matrix form:

$$p(z) = z^T T u_{m+1}, \tag{6.58}$$

where $z^T = \begin{bmatrix} 1 & z & \dots & z^{n-1} \end{bmatrix}$ and T is the $n \times n$ Toeplitz matrix

$$T = \begin{bmatrix} 0 & h_1 & h_2 & \dots & h_{n-1} \\ 0 & 0 & h_1 & \dots & h_{n-2} \\ \dots & \dots & \dots & \dots & \dots \\ 0 & 0 & 0 & \dots & h_1 \\ 0 & 0 & 0 & \dots & 0 \end{bmatrix}. \tag{6.59}$$

Finally, we may express $\hat{h}_m^n(z)$ as a strictly proper polynomial of degree $n - 1$

$$\hat{h}_m^n(z) = \frac{p(z)}{u(z)} = \frac{z^T T u_{m+1}}{z^T u_{m+1}}. \tag{6.60}$$

According to the AAK-theory, the best Hankel-norm approximation $\hat{r}_m^n(z)$ is obtained by the stable part of (6.60). The usual way to compute the stable part is to express (6.60) in a partial fraction expansion. Equivalently, we may diagonalize the controller realization of $\hat{h}_m^n(z)$

$$A = \begin{bmatrix} -\tilde{u}_{n-1} & -\tilde{u}_{n-2} & \cdots & -\tilde{u}_1 \\ 1 & 0 & \cdots & 0 \\ 0 & 1 & \cdots & 0 \\ \cdots & \cdots & \cdots & \cdots \\ 0 & 0 & 1 & 0 \end{bmatrix} \quad B = \begin{bmatrix} 1 \\ 0 \\ 0 \\ \cdots \\ 0 \end{bmatrix} \quad C^T = \begin{bmatrix} \tilde{p}_{n-1} \\ \tilde{p}_{n-2} \\ \tilde{p}_{n-3} \\ \cdots \\ \tilde{p}_0 \end{bmatrix} \tag{6.61}$$

where $\tilde{u}_k = u_k/u_n$ and $\tilde{p}_k = p_k/u_n$ for $k = 1 \ldots n-1$ and partition it according to

$$\Lambda = \begin{bmatrix} \Lambda_s & 0 \\ 0 & \Lambda_i \end{bmatrix} \quad \tilde{B} = \begin{bmatrix} B_s \\ B_i \end{bmatrix} \quad \tilde{C}^T = \begin{bmatrix} C_s^T \\ C_i^T \end{bmatrix} \tag{6.62}$$

where the diagonal $m \times m$ matrix Λ_s contains the eigenvalues of A being into the unit disk D and the diagonal $(n-m-1) \times (n-m-1)$ matrix Λ_i contain those being in the domain S_∞. The system related to the best Hankel-norm approximation $\hat{r}_m^n(z)$ is that given by $\{\Lambda_s, B_s, C_s\}$.

By the AAK-theorem, the approximation error of the optimal approximation with respect of the Hankel norm is given by the $m+1$ singular value $s_{m+1}(\Gamma_H^n)$. Of much more interest is to know the approximation error with respect to $L_\infty(T)$. The inequalities (6.39) suggest that this approximation error is in general larger than $s_{m+1}(\Gamma_H^n)$. Glover [Glo84] proves that the approximation error of the best Hankel-norm approximation with respect to $L_\infty(T)$ is

$$s_{m+1} + s_{m+2} + \ldots + s_n \leq \|h^n - \hat{r}_m^n\|_{L_\infty(T)} \leq 2(s_{m+1} + s_{m+2} + \ldots + s_n). \tag{6.63}$$

The lower bound is achieved defining an "optimal" system matrix D. The upper bound is equal to the approximation error of the truncated balanced realization with respect to $L_\infty(T)$. Therefore, the best Hankel-norm approximation with "optimal" system matrix D is generally the better approximation with respect to $L_\infty(T)$. However, even if the system matrix D is not optimal, the approximation error of the best Hankel-norm approximation is likely to be smaller than that of the truncated balanced realization.

Let us consider an algorithm to compute the optimal Hankel-norm approximation of systems defined by a truncated Laurent series:

1. Choose an upper bound of the approximation error, say ε.
2. Setup the observability matrix $W_o = \Gamma_H^T \Gamma_H$ and compute their eigenvalues Λ and eigenvectors U. Sort the eigenvalues and eigenvectors such that $\lambda_1 \geq \lambda_2 \geq \ldots \geq \lambda_n$.
3. Find the index m such that $2(\sigma_m + \ldots + \sigma_p) > \varepsilon \geq 2(\sigma_{m+1} + \ldots + \sigma_p)$.
4. Compute the coefficients of the polynomial $p(z)$ of $\hat{h}_m^n(z)$ according to (6.58).
5. Setup and diagonalize the system (6.61). Partition it according to (6.62) and extract the stable system $\{\Lambda_s, B_s, C_s\}$.
6. Map the discrete-time system to the related continuous-time system using (5.156).

Remarks:

• Let $\hat{r}_m(z)$ be the strictly proper best Hankel-norm approximation defined by (6.44). Chui, Li and Ward [CLW91] proved that if the function $h(z)$ belongs to $L_\infty(T)$ and is continuously differentiable then

$$\lim_{n \to \infty} \left\| \hat{r}_m^n(z) - \hat{r}_m(z) \right\|_\infty \to 0 \tag{6.64}$$

provided that $s_{m+1}(\Gamma_H)$ is simple. Therefore DtN-maps kernels are approximated with arbitrary accuracy by the optimal Hankel-norm approximation method.

- Because the CF-method was devised to construct a near-best solution in the $L_\infty(T)$ norm, in [Gut84] several proposals has been outlined to find a near-best approximation $r_m^{cf}(z)$. The original proposal of Trefethen [Tre81] is equivalent to the optimal Hankel-norm solution. A second approach is to perform a Padé approximation of $h(z)$ with fixed denominator $u(z)$. A third method, which is worth to mention, is to perform a least-squares approximation of $h(z)$ over the unit circle using as approximations the rational function $r(z) = p(z)/u(z)$ but letting $u(z)$ fixed. In practice all three methods give similar results.
- The step 2 of the foregoing algorithm may be replaced by the computation of the singular value decomposition of the Hankel matrix Γ_H.

6.5   Applications

6.5.1   A small model system

In this section, we shall study the behavior of the three approximation methods discussed so far considering a simple model problem. The model problem has been appositely chosen to emphasize certain important aspects of these approximation methods. This will give us useful insights how these approximation methods works.

Consider the discrete-time system given by the transfer function

$$h(z) = 1 + z^{-1} - z^{-2} + z^{-3} \tag{6.65}$$

It has the structure of a truncated Laurent series and is therefore stable and of McMillan degree three. The associated Hankel matrix is given by

$$\Gamma_H = \begin{bmatrix} 1 & -1 & 1 \\ -1 & 1 & 0 \\ 1 & 0 & 0 \end{bmatrix} \tag{6.66}$$

with singular values $s(\Gamma_H) = \{2.247, 0.802, 0.555\}$. We approximate the model system with one of McMillan degree at most two using the Padé-approximation, the balanced truncation and the optimal Hankel-norm approximation methods. The result is shown in Figure 6.1

The most remarkable fact is that the Padé-approximation gives a system of degree one. In fact, the linear system of equations defining the parameters of the denominator is

$$Hb = \begin{bmatrix} 1 & -1 \\ -1 & 1 \end{bmatrix} \begin{bmatrix} b_0 \\ b_1 \end{bmatrix} = \begin{bmatrix} 1 \\ 0 \end{bmatrix} \tag{6.67}$$

Because the Hankel matrix H is singular (6.65) has no solution. This is a consequence of the very definition of the Padé approximation method. It requires that its Laurent series expansion have to agree up to the term of order $O(z^{-3})$ with the transfer function (6.65). The terms of order $O(z^{-n})$

Method	Transfer function	Poles
Padé	$g(z) = \dfrac{1}{1+z} + 1$	$z_1 = -1$
truncated balanced	$g(z) = \dfrac{0.940z + 0.591}{z^2 + 0.677 + 0.194} + 1$	$z_1 = -0.339 + i0.281$ $z_2 = -0.339 - i0.281$
optimal Hankel-norm	$g(z) = \dfrac{z + 0.246}{z^2 + 1.247 + 0.555} + 1$	$z_1 = -0.623 + i0.408$ $z_2 = -0.623 - i0.408$

Figure 6.1: Approximations of the model system (6.65)

with $n > 3$ are irrelevant for the computation of the Padé approximation. In fact, the transfer function $\tilde{h}(z) = 1 + z^{-1} - z^{-2} + z^{-3} + az^{-4}$, with a arbitrary, has the same Padé approximation as (6.65). This behavior of the Padé approximation method may be a severe drawback when it is applied to approximate DtN-map kernels. Note that the approximation error with respect to $L_\infty(T)$ is infinite because of the singularity of the Padé approximation at $z = -1$.

Contrary to the Padé-approximation method, the truncated balanced and the optimal Hankel-norm methods produce strictly proper rational approximations of degree 2. However, the coefficients of the associated rational functions are completely different (Figure 6.1). The truncated balanced approximation method is exact at $\theta/\pi = 1$ and very accurate in the interval $\theta/\pi \in [0.7, 1.0]$ (Figure 6.2). Because the error curve of the optimal Hankel-norm approximation is a circle (Figure 6.3) the approximation error is uniform over the complete interval. Therefore, the Hankel-norm approximation is optimal with respect to $L_\infty(T)$ too. In fact, the error is equal to the third singular value of the Hankel matrix (6.66), that is $\|e\|_{L_\infty(T)} = s_3 = 0.555$. This result can be generalized because the optimal Hankel-norm approximation of McMillan degree $m = n - 1$ is exactly $\hat{h}_{n-1}^n(z)$ which is optimal with respect to the $L_\infty(T)$ norm. The error is given by $\|e\|_{L_\infty(T)} = s_n$. In our example, this error is smaller than that of the approximant obtained via the truncated balanced method which is approximately $\|e\|_{L_\infty(T)} = 0.813$. This error smaller than the error bound (6.27) which is $2s_3 = 1.11$. Observe, that truncating the transfer function (6.65) after the term of order $O(z^{-2})$ we obtain another rational approximation of degree 2

$$g(z) = 1 + \frac{z - 1}{z^2}. \qquad (6.68)$$

This is the optimal approximation with respect to $L_2(T)$ with denominator polynomial z^2. However, its error with respect to $L_\infty(T)$ is larger than that of the truncated balanced and the optimal Hankel-norm approximation. In fact, $|e(z)| = |z^{-3}|$ so that $\|e\|_{L_\infty(T)} = 1$.

6.5.2 The DtN-map of the elastically restrained rod

First, we investigate the approximation of the regularized DtN-map kernel of the nondissipative restrained rod given by

$$\tilde{k}(s) = \sqrt{s^2 + 1} - s. \qquad (6.69)$$

This gives us the opportunity to study the behavior of the approximation in the neighborhood of the cut-off frequency. The mapping from the imaginary axis to the unit circle T and back is provided by the pair of Möbius transformations

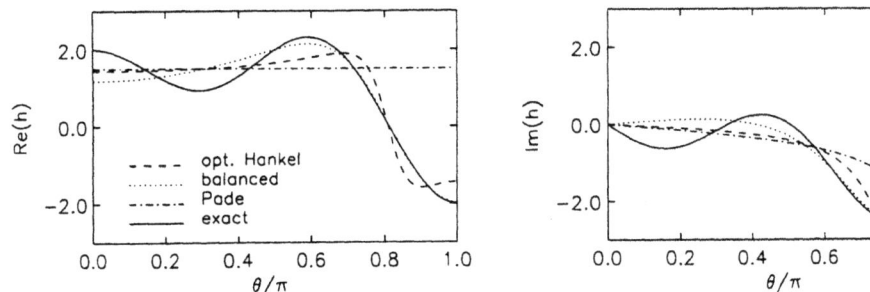

Figure 6.2: Plots of the transfer function of the model system (6.65) and its approximations

$$z = -\frac{s+1}{s-1} \quad \text{and} \quad s = \frac{z-1}{z+1}. \tag{6.70}$$

Combining (6.69) with the first equation of (6.70) yields the mapped DtN-map kernel

$$\tilde{k}(z) = \frac{\sqrt{2}\sqrt{1+z^2}+1-z}{1+z^2} \tag{6.71}$$

with Laurent series expansion

$$\tilde{k}(z) = (\sqrt{2}-1) + \frac{2-\sqrt{2}}{z} + \frac{(3\sqrt{2})/2-2}{z^2} + \frac{-(3\sqrt{2})/2+2}{z^3} + O\!\left(\frac{1}{z^4}\right) \tag{6.72}$$

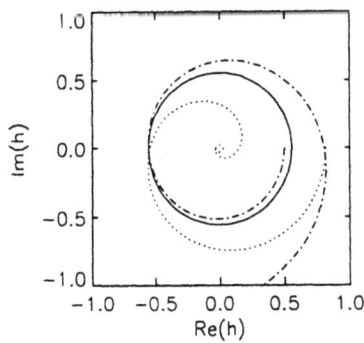

Figure 6.3: Error curves of the approximations plotted in Figure 6.2

First, we truncate the Laurent series after the term of order $O(z^{-12})$. This approximation is shown in Figure 6.4 and the associated error curve is plotted in Figure 6.5. It is apparent from both figures that the error $\|e_{12}\|_{L_\infty(T)} = 0.226$ is located at the cut-off frequency. Moreover, the error rapidly decreases with the distance to the cut-off frequency. This fact doesn't change if the number of terms of the Laurent series is increased to 24 or 36 (Figure 6.5). However, the approximation error is reduced to $\|e_{24}\|_{L_\infty(T)} = 0.161$ and $\|e_{36}\|_{L_\infty(T)} = 0.132$, respectively. Moreover, the domain exhibiting a significant error shrinks with increasing number of terms.

Now we apply the Padé, the truncated balanced and the optimal Hankel-norm approximation methods to construct rational approximations of degree four. The result is plotted in Figure 6.6 and shows no significant differences between the three methods. The error curves plotted in Figure 6.7, left, manifest a slightly better accuracy of the Padé approximation in the neighborhood of the cut-off frequency. Remarkably, the error with respect to $L_\infty(T)$ is approximately equal to that of the truncated Laurent series. In fact, the error of the approximant obtained with the truncated balanced method is $\|e_{12}\|_{L_\infty(T)} = 0.217$. This suggests that in the neighborhood of the cut-off frequency the error of the rational approximation is mainly determined by the degree of the truncated Laurent series. In fact, as is shown in Figure 6.7 right, using the optimal Hankel-norm and of the truncated balanced approximation methods no substantial improvement in accuracy is obtained when the degree of the approximations is increased to six. On the other hand, increasing the degree of the truncated Laurent series the error with respect to $L_\infty(T)$ of the truncated balanced approximations of degree four decreases from $\|e_{12}\|_{L_\infty(T)} = 0.217$ to $\|e_{24}\|_{L_\infty(T)} = 0.159$ and finally to $\|e_{36}\|_{L_\infty(T)} = 0.141$. Hence, to

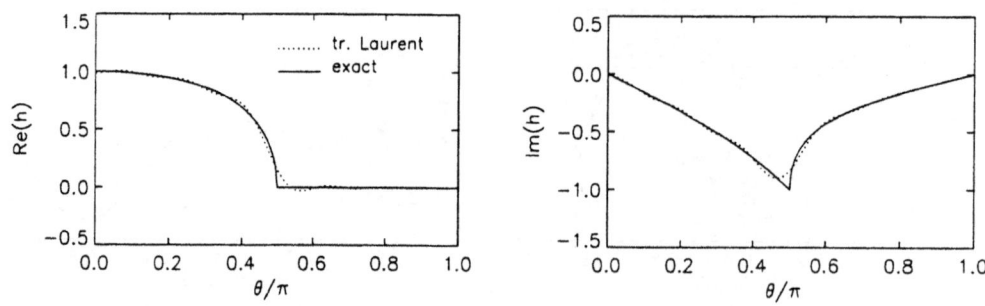

Figure 6.4: Approximation of (6.69) with a truncated Laurent series of degree 12

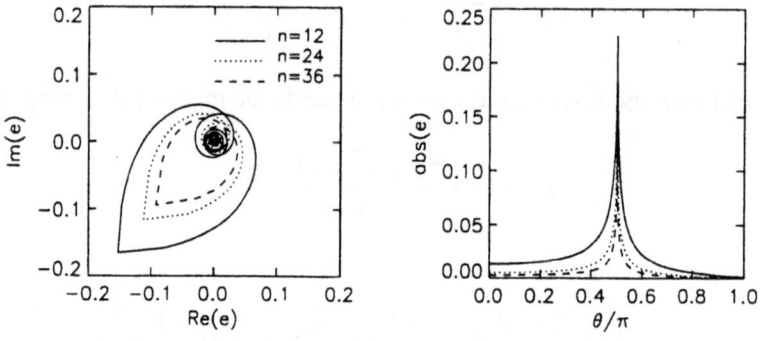

Figure 6.5: Error curves of the truncated Laurent series of degree 12, 24 and 36

achieve a small error in the neighborhood of the cut-off frequency the size of the system associated with the truncated Laurent series has to be large enough. In contrast, the error of the Padé approximation of degree six is reduced to $\|e\|_{L_-(T)} = 0.171$. This is because the Padé approximation of degree six considers four additional terms of the Laurent series expansion. That is, increasing the degree of a Padé approximation automatically increases the degree of the underlying truncated Laurent series.

These observations are confirmed if we analyze the distribution of the singular values. The magnitude of the first 12 singular values of the Hankel matrices associated with the three truncated Laurent series are shown in Figure 6.8. Note that these singular values are an estimate of the exact ones which are defined by the infinite-dimensional Hankel matrix. The magnitude of the singular values of the Laurent series with 12 terms decrease quickly up to the fourth singular value. The decrease of the following singular values is much more modest. This means that increasing the degree from 4 to 6 has only little impact on the accuracy of the approximation. This changes when the number of terms of the Laurent series are increased. An approximation with a rational function of degree 6 is more promising with a Laurent series which considers 36 terms. However, this approximation will modify only the second digit of the error because the fifth and sixth singular values are of order $1/100$. A further significant reduction of the error can only be obtained with Laurent series and approximations with large degrees.

Figure 6.9 shows the errors of two approximants of degree 4. Both approximants have been obtained using the balanced truncation method. One is based on the Fourier series and the other on the Laguerre series expansion of degree 12 of the DtN-map. The truncation of that based on the Laguerre series expansion was performed with the continuous-time system obtained via the mapping (5.156). The errors with respect to $L_\infty(T)$ of both approximants are virtually equal. However, that based on the Laguerre series has perfect matching at infinity. That is, the error vanish at $\theta/\pi = 1$. It is therefore more accurate at high frequencies. In contrast, it is less accurate at

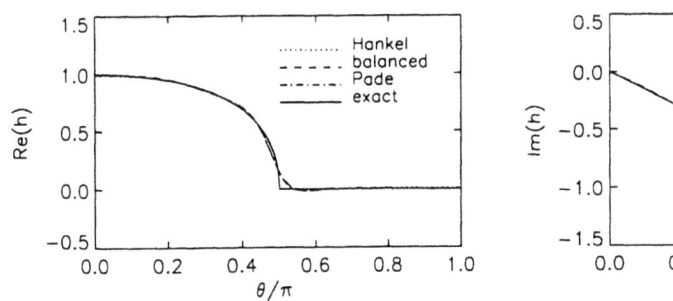
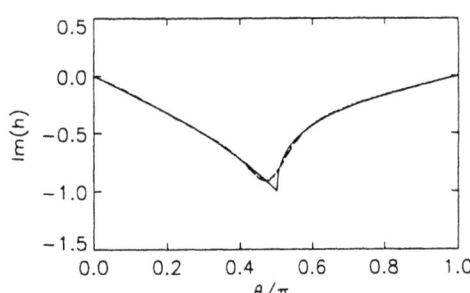

Figure 6.6: Approximations of (6.69) with different methods

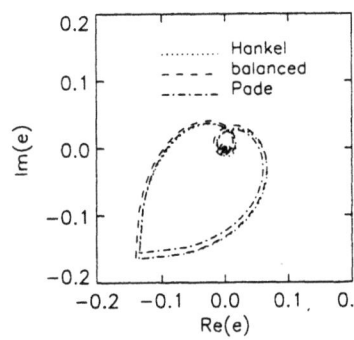
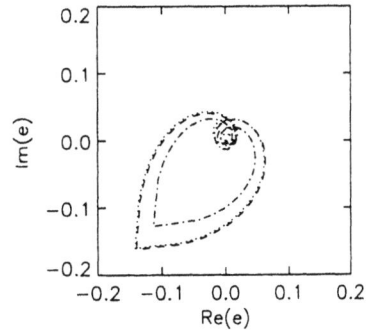

Figure 6.7: Error curves of approximations of (6.69) with different methods.
Left: approximations of degree four. Right: approximations of degree six.

small frequencies than that obtained with the Fourier series expansion. This effect is quite general. High accuracy in a subdomain is paid by lower accuracy in the complementary subdomain. The error with respect to $L_\infty(T)$, however, remains generally unchanged. Observe that to enforce perfect matching at infinity, we have to use Laguerre series expansion and the truncation must be applied to the mapped continuous-time system.

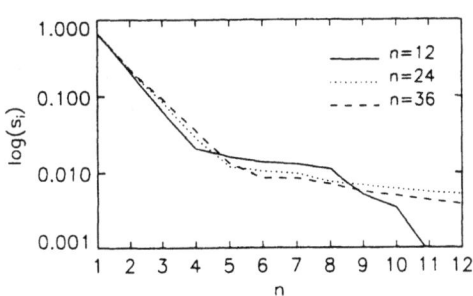

Figure 6.8: The first 12 singular values of the truncated Laurent series of degree 12, 24 and 36

The magnitude of the error with respect to $L_\infty(T)$ is dramatically reduced when we consider a dissipative restrained rod. The regularized DtN-map kernel of the nondissipative restrained rod is given by

$$\tilde{k}_\varepsilon(s) = \sqrt{s^2 + 2\varepsilon s + 1} - \varepsilon - s, \qquad (6.73)$$

where ε is the parameter which determines the magnitude of the dissipation and is given by $\varepsilon = 0.02$. This corresponds to a small dissipation. In fact, as is shown in Figure 6.10, the shape of the regularized DtN-map kernel $\tilde{k}_\varepsilon(s)$ differs only slightly from that of the nondissipative rod. In particular the sharp edge at the cut-off frequency is smoothed out. We consider as approximants rational functions of degree four. Similarly to the nondissipative rod, the Padé, the truncated balanced and the optimal Hankel-norm approximations are similar, see Figure 6.10. Observe, that we have used the Laurent series of degree 24 to construct the truncated balanced and the optimal Hankel-norm approximations.

The error curves shown in Figure 6.12 provide us with more details. The most significant fact is that the Padé approximation is clearly less accurate than the other two approximations. Its error is

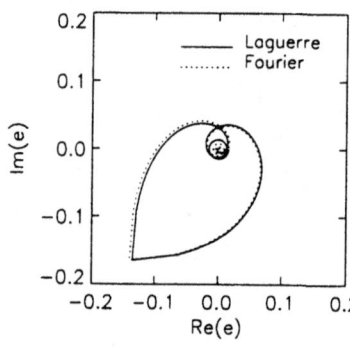 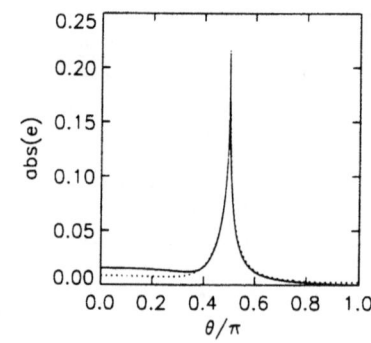

Figure 6.9: Approximations via balanced truncation based on Fourier and Laguerre series (n=12)

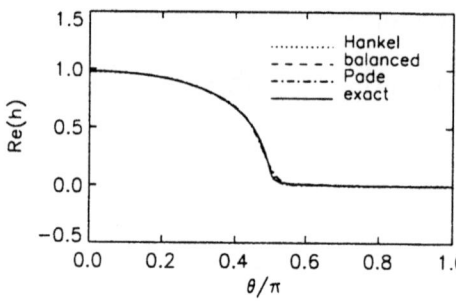 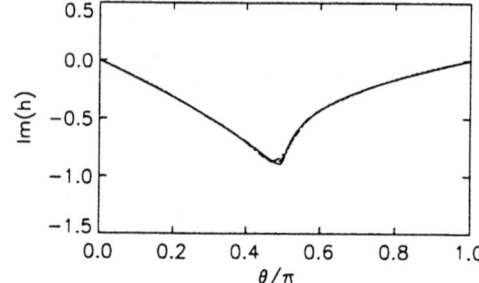

Figure 6.10: Approximations of the regularized DtN-map of the dissipative restrained rod

$\|e_{24}\|_{L_\infty(T)} = 0.0663$ whereas that of the approximant obtained with the truncated balanced method is $\|e_{24}\|_{L_\infty(T)} = 0.0352$ and that obtained via optimal Hankel-norm approximation is $\|e_{24}\|_{L_\infty(T)} = 0.0364$. The error curves are shown Figure 6.12. Similarly as for the nondissipative rod, the $L_\infty(T)$ errors are determined by the behavior of the approximations in the neighborhood of the cut-off frequency. In the remaining interval the approximation is at least 5 times smaller.

Finally, we compare the approximation obtained with the truncated balanced method with that obtained using Weber's approximation method [Web94]. Figure 6.11 shows that both approximations are very similar. That obtained by Weber's method is more accurate in the neighborhood of the cut-off frequency ($\|e_{24}\|_{L_\infty(T)} = 0.0297$). On the other hand, that obtained with the method described in this work is slightly more accurate in the remaining part of the domain.

To conclude we may reassume the most important facts about the applicability of the methods discussed in this chapter for the approximation DtN-map kernels:

- The Padé approximation method is certainly the numerically more efficient method provided that the degree of the rational approximation is given a priori. However, its approximation error is generally appreciably larger than that obtained with the truncated balanced or optimal Hankel-norm approximation method. Furthermore, if an error bound is given, then the Padé approximation method requires an iterative procedure to construct the approximant because we have no formula to estimate the approximation error a priori.

- The truncated balanced and the optimal Hankel-norm approximation methods are robust methods to achieve rational approximations of DtN-map kernels. The magnitude of the approximation errors is similar for both methods. For a given degree of the rational approximation the magnitude of the error depends essentially on the degree of the truncated Laurent series.

6.5.3   A remark to the approximation error in the time domain

The approximation methods described in the last two sections produce rational approximations

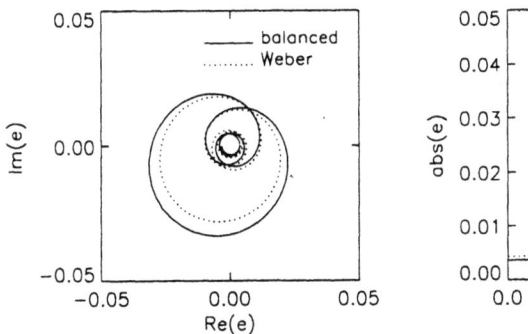

Figure 6.11: Error curves of the approximations by balanced truncation and Weber's method

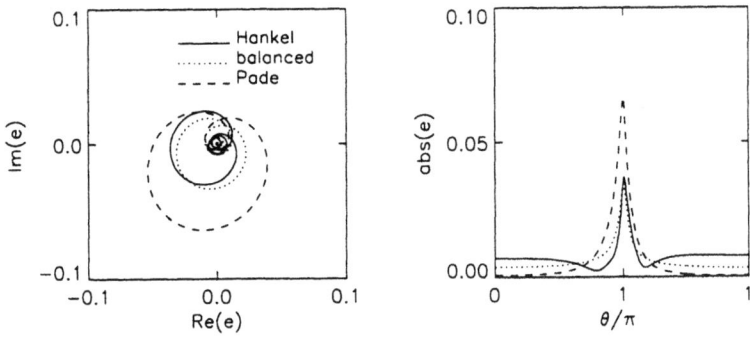

Figure 6.12: Error curves of different approximations

analytic in C_+ whose error can be made arbitrarily small with respect to $\|\cdot\|_{L_-(iR)}$. What implications follow from this result with regard to the error in the time domain?

Let us assume that $\hat{h}(s)$ is an approximation of the DtN-map kernel $h(\omega)$. Observe, that $\hat{h}(s)$ is continuous in $C_+ \cup iR$ and that $\hat{h}(t)$ belongs to $L_1(R_+)$. The error of the output function can be measured with respect to different norms. The most natural are the $\|\cdot\|_{L_-(R.)}$ and $\|\cdot\|_{L_2(R.)}$ norms. First, let us consider the error with respect to the infinity norm $\|\cdot\|_{L_-(R.)}$. The absolute output error at the time t is

$$|\varepsilon_y(t)| = |y(t) - \hat{y}(t)| = \left|\int_0^t (h(\tau) - \hat{h}(\tau))u(t-\tau)d\tau\right|, \qquad (6.74)$$

where $u(t)$ is a bounded input function. A simple estimate of the error is

$$\varepsilon_y(t) \le \sup_{\tau \in [0,t]}|u(\tau)|\int_0^t|h(\tau) - \hat{h}(\tau)|d\tau < \sup_{\tau \in [0,t]}|u(\tau)|\int_0^\infty|h(\tau) - \hat{h}(\tau)|d\tau \qquad (6.75)$$

Because this relation is valid for any t we have

$$\|\varepsilon_y(t)\|_{L_-(R.)} \le \|\varepsilon_h(t)\|_{L_1(R.)}\|u(t)\|_{L_-(R.)}, \qquad (6.76)$$

where $\varepsilon_h(t) = h(t) - \hat{h}(t)$. Apparently, the maximum error depends essentially on the accurateness of the approximation with respect to $L_1(R_+)$. Now, because $h(t)$ as well as $\hat{h}(t)$ belongs to $L_1(R_+)$ the integral in the last equation is finite. Hence, the error $\|\varepsilon_y(t)\|_{L_-(R.)}$ is finite for every bounded input $u(t)$. However, the computation of the error with respect to $L_1(R_+)$ requires that

$h(t)$ is given in the time domain. But this is generally not the case when working in the frequency domain so that an extra effort is needed to compute $h(t)$.

A more appropriate estimate of the approximation error when working in the frequency domain is obtained analyzing the output error with respect to $L_2(R_+)$. That is

$$\left\|\varepsilon_y(t)\right\|_{L_2(R_.)} = \left(\int_0^\infty |\varepsilon_y(t)|^2\right)^{1/2}. \tag{6.77}$$

This integral exists provided that the input function belongs to $L_2(R_+)$. Using the isometry induced by the Fourier transform we find

$$\left\|\varepsilon_y(t)\right\|_{L_2(R.)} = \left\|\varepsilon_y(\omega)\right\|_{L_2(iR)} = \left\|\varepsilon_h(\omega)u(\omega)\right\|_{L_2(iR)} \leq \left\|\varepsilon_h(\omega)\right\|_{L_\infty(iR)}\left\|u(\omega)\right\|_{L_2(iR)}. \tag{6.78}$$

That is,

$$\left\|\varepsilon_y(t)\right\|_{L_2(R.)} \leq \left\|\varepsilon_h(\omega)\right\|_{L_\infty(iR)}\left\|u(t)\right\|_{L_2(iR)}. \tag{6.79}$$

Hence, with the error $\left\|\varepsilon_h(\omega)\right\|_{L_\infty(iR)}$ we obtain an estimate of the error of the output energy. This error is defined in the frequency domain, is easy to estimate and fits very well in the approximation scheme we have developed so far.

Chapter 7

Approximation of matrix-valued DtN-maps

In chapter three we have shown that matrix-valued DtN-maps are the rule than the exception. Inhomogeneous strata, solid strata or fluid strata with frequency dependent boundary conditions produce matrix-valued DtN-maps. In these cases, the propagation modes of a system are coupled. The strength of this coupling is generally frequency dependent: Homogeneous solid strata exhibit a strong coupling at low frequencies and uncouple at high frequencies. In applications where the energy content of the loading is dominated by high frequency components, it might be possible to obtain accurate results neglecting the coupling of the propagation modes. In these cases, we can approximate the diagonal components of the DtN-map kernel using the methods developed in the previous chapter. However, if the energy content of the loading has dominant low frequency components, it is generally necessary to approximate the fully coupled DtN-map kernels. In this chapter, we present two methods to successfully approximate matrix-valued DtN-map kernels with matrix-valued rational functions: the truncated balanced and the Hankel-norm approximation method. These are generalizations of the methods presented in the chapter before.

7.1 Matrix-valued DtN-maps

Matrix-valued DtN-maps transform vector-valued input functions $u(t)$ to vector-valued output functions $v(t)$. That is, $u(t)$ and $v(t)$ are vectors of dimension n whose components are real functions which depends on the time variable t

$$u^T(t) = \begin{bmatrix} u_1(t) & u_2(t) & \dots & u_n(t) \end{bmatrix}. \tag{7.1}$$

In the time domain, DtN-maps are defined by the convolution integral

$$v(t) = H*u = \int_0^t H(t-\tau)u(\tau)d\tau. \tag{7.2}$$

147

The DtN-map kernel $H(t)$ is a square, $n \times n$ matrix function whose components are real distributions defined in R_+, that is,

$$H(t) = \begin{bmatrix} h_{11}(t) & \dots & h_{1n}(t) \\ \dots & \dots & \dots \\ h_{n1}(t) & \dots & h_{nn}(t) \end{bmatrix} \tag{7.3}$$

Each component of the output vector is given by a sum of n convolution integrals

$$v_j(t) = \int_0^t h_{j1}(t-\tau)u_1(\tau)d\tau + \int_0^t h_{j2}(t-\tau)u_2(\tau)d\tau + \dots + \int_0^t h_{jn}(t-\tau)u_n(\tau)d\tau. \tag{7.4}$$

A DtN-map given by the convolution integral (7.2) is linear and time-invariant.

The unilateral Laplace transform maps a DtN-map defined by (7.2) to a system of complex algebraic equations being defined in the open complex plane

$$v(s) = \begin{bmatrix} v_1(s) \\ \dots \\ v_2(s) \end{bmatrix} = H(s)u(s) = \begin{bmatrix} h_{11}(s) & \dots & h_{1n}(s) \\ \dots & \dots & \dots \\ h_{n1}(s) & \dots & h_{nn}(s) \end{bmatrix} \begin{bmatrix} u_1(s) \\ \dots \\ u_2(s) \end{bmatrix} \tag{7.5}$$

Correspondingly, the Fourier transform of a DtN-map defined by (7.2) is given by the system of complex algebraic equations being defined on the imaginary axis iR:

$$v(\omega) = \begin{bmatrix} v_1(\omega) \\ \dots \\ v_2(\omega) \end{bmatrix} = H(\omega)u(\omega) = \begin{bmatrix} h_{11}(\omega) & \dots & h_{1n}(\omega) \\ \dots & \dots & \dots \\ h_{n1}(\omega) & \dots & h_{nn}(\omega) \end{bmatrix} \begin{bmatrix} u_1(\omega) \\ \dots \\ u_2(\omega) \end{bmatrix} \tag{7.6}$$

In this work, we assume that the DtN-map kernels are symmetric. That is, $H^T(t) = H(t)$ or equivalently $h_{jk}(t) = h_{kj}(t)$ for every index $i, j \leq n$ and for any time t in the domain of definition of $H(t)$. Because $H(t)$ is a real matrix function we have $H(-\omega) = \bar{H}(\omega)$. Considering that $H(\omega)$ is symmetric yields $H(-\omega) = H^*(\omega)$. Similarly, the Laplace transform of $H(t)$ satisfy $H(\bar{s}) = H^*(s)$. However, the theory developed in this chapter applies to nonsymmetric DtN-map kernels as well.

7.1.1 Norms of vector- and matrix-valued functions

Analogously to scalar input-output systems, the analysis of vector-valued input-output systems are performed in adequate normed spaces. These are normed spaces of vector and matrix-valued functions. Usually, it will be clear from the context if we refer to a space of scalar or vector-valued functions. Therefore, the symbols used so far to refer to normed spaces of scalar functions will be also used to denote normed spaces of vector- and matrix-valued functions.

First, we consider three normed spaces defined in the time domain. Of particular interest are those defined on the open domain R_+. With $L_1(R_+)$ we refer to the space of real vector-valued functions which are defined in R_+ and which are absolutely integrable. The norm is defined by

$$\|u\|_{L_1(R_+)} = \int_0^\infty \left(\sum_{k=0}^n |u_k(t)| \right) dt. \tag{7.7}$$

Next we introduce the space $L_2(R_+)$ of real vector-valued functions which are defined on R_+ and which are square integrable. The norm is defined by

$$\|u\|_{L_2(R_+)} = \left(\int_0^\infty u^T(t)u(t)dt\right)^{1/2} = \left(\int_0^\infty \left(\sum_{k=0}^n |u_k(t)|^2\right)dt\right)^{1/2}. \tag{7.8}$$

$L_2(R_+)$ is a Hilbert space and its norm is induced by the inner product

$$\langle u, v\rangle = \int_0^\infty u^T(t)v(t)dt = \int_0^\infty \left(\sum_{k=0}^n u_k(t)v_k(t)\right)dt. \tag{7.9}$$

The last normed space of real vector-valued bounded functions defined in R_+ is $L_\infty(R_+)$. The norm of $L_\infty(R_+)$ is given by

$$\|u\|_{L_\infty(R_+)} = \max_k \sup_t |u_k(t)| \tag{7.10}$$

where the supremum is taken in the domain R_+ and the maximum over the index set $k \in [1\ldots n]$.

Equivalently, we may define the Hilbert space $L_2(iR)$ of complex vector-valued functions defined on the imaginary axis iR. Its norm is given by

$$\|u\|_{L_2(iR)} = \left(\int_{-\infty}^\infty u^*(\omega)u(\omega)dt\right)^{1/2} = \left(\int_{-\infty}^\infty \left(\sum_{k=0}^n |u_k(\omega)|^2\right)dt\right)^{1/2}, \tag{7.11}$$

which is induced by the inner product

$$\langle u, v\rangle = \int_{-\infty}^\infty u^*(\omega)v(\omega)dt = \int_{-\infty}^\infty \left(\sum_{k=0}^n u_k^*(\omega)v_k(\omega)\right)dt. \tag{7.12}$$

So far we have introduced normed spaces of vector-valued functions. However, for the analysis of matrix-valued DtN-map kernels we need normed spaces of matrix-valued functions. An important class of norms, the so-called induced or operator norms, are defined by way of vector norms. These are the most natural norms when considering mappings from one normed space of vector-valued functions to another. Usually, these norms are defined by the relation

$$\|A\| = \sup_{x \neq 0} \frac{\|Ax\|}{\|x\|} \tag{7.13}$$

where $\|\cdot\|$ is a vector norm. In this work, we will consider the normed spaces of vector-valued functions $L_1(R_+)$ and $L_\infty(iR)$. $L_1(R_+)$ is induced by studying the mapping of vector-valued functions from $L_\infty(R_+)$ to $L_\infty(R_+)$. Consider the convolution integral (7.2). Any component of the output vector $v(t)$ is given by (7.4). Therefore, we have

$$|v_j(t)| = \left|\sum_{k=1}^n \left(\int_0^t h_{jk}(t-\tau)u_k(\tau)d\tau\right)\right| \leq \sum_{k=1}^n \left(\int_0^t |h_{jk}(t-\tau)||u_1(\tau)|d\tau\right). \tag{7.14}$$

Taking the supremum over the interval $(0, t)$ we obtain

$$\sup_t |v_j(t)| \leq \sup_t \sum_{k=0}^n \left(\int_0^t |h_{jk}(t-\tau)||u_1(\tau)|d\tau\right) \leq \sum_{k=0}^n \left(\int_0^t |h_{jk}(t-\tau)|d\tau\right) \cdot \max_k \sup_t |u_k(t)| \tag{7.15}$$

Therefore, letting $t \to \infty$, we obtain

$$\sup_t |v_j(t)| \le \sum_{k=0}^n \left(\int_0^\infty |h_{jk}(t)| d\tau \right) \|u\|_{L_\infty(R_+)}.$$ (7.16)

Now, we pick out the component of $v(t)$ with the largest supremum. This is the norm of $v(t)$ in $L_\infty(R_+)$

$$\|v\|_{L_\infty(R_+)} = \max_j \sup_t |v_j(t)| \le \max_j \left(\sum_{k=0}^n \left(\int_0^\infty |h_{jk}(t)| d\tau \right) \right) \|u\|_{L_\infty(R_+)}.$$ (7.17)

We define $L_1(R_+)$ as the normed space of matrix-valued functions such that

$$\|H\|_{L_1(R_+)} = \max_j \left(\sum_{k=0}^n \left(\int_0^\infty |h_{jk}(t)| d\tau \right) \right),$$ (7.18)

is finite. It is easy to show that (7.18) is a norm. Clearly, $H(t)$ belongs to $L_1(R_+)$ if and only if all components of $H(t)$ are in $L_1(R_+)$, that is, $h_{jk}(t) \in L_1(R_+)$.

As already pointed out in the previous chapter, $L_1(R_+)$ is the natural space of our approximation problem. In fact, from (7.17) we have

$$\|v - \tilde{v}\|_{L_\infty(R_+)} \le \|H - \tilde{H}\|_{L_1(R_+)} \|u\|_{L_\infty(R_+)},$$ (7.19)

where $\tilde{H}(t)$ refers to the approximated DtN-map and $\tilde{v}(t) = \tilde{H}(t)*u(t)$. The inequality (7.19) states that given an input function $u(t)$ the largest error of the output function $v(t)$ is determined by how close the approximation $\tilde{H}(t)$ is to $H(t)$ with respect to $L_1(R_+)$.

The linear space of matrix-valued functions $L_\infty(iR)$ is obtained if we consider mappings from the space of vector-valued functions $L_2(iR)$ to itself. That is, the space of matrix-valued functions which map finite-energy inputs $u(t)$ to finite-energy outputs $v(t)$. Combining (7.6) with the definition of the norm of $L_2(iR)$ given in (7.11) we obtain

$$\|v\|_{L_2(iR)}^2 = \langle v*H^*, Hv \rangle = \int_{-\infty}^\infty v^*(\omega) H^*(\omega) H(\omega) v(\omega) dt$$ (7.20)

With the singular value decomposition we may represent $H(\omega)$ as

$$H(\omega) = U(\omega) \Sigma(\omega) V^*(\omega).$$ (7.21)

Inserting the last equation into (7.20) yields

$$\|v\|_{L_2(iR)}^2 = \int_{-\infty}^\infty v^*(\omega) V(\omega) \Sigma^2(\omega) V^*(\omega) v(\omega) dt \le \bar{\sigma}^2 \int_{-\infty}^\infty v^*(\omega) v(\omega) dt = \bar{\sigma} \|u\|_{L_2(iR)}^2,$$ (7.22)

where $\bar{\sigma} = \sup_\omega \sigma_1(H(\omega))$, that is, the supremum of the function constructed by picking the largest singular value of $H(\omega)$ in the domain of definition iR. We now define $L_\infty(iR)$ to be the linear normed space of matrix-valued function defined on iR with norm

$$\|H\|_{L_\infty(iR)} = \sup_\omega \sigma_1(H(\omega)).$$ (7.23)

Clearly, $\|H\|_{L_\infty(iR)} \ge 0$. $\|H\|_{L_\infty(iR)} = 0$ if and only if $H(\omega) \equiv 0$. Moreover, because $\sigma_1(H)$ is the square root of the largest eigenvalue of $H(\omega)^*H(\omega)$, we have $\sigma_1(cH) = |c|\sigma_1(H)$, such that $\|cH\|_{L_\infty(iR)} = |c|\|H\|_{L_\infty(iR)}$. The triangle inequality $\|H+G\|_{L_\infty(iR)} \leq \|H\|_{L_\infty(iR)} + \|G\|_{L_\infty(iR)}$ follows from the fact that $\sigma_1(H+G) \leq \sigma_1(H) + \sigma_1(G)$. Therefore, (7.23) is a norm. Because of the equality $\|u\|_{L_2(R.)} = \|u\|_{L_2(iR)}$, the output error induced by the approximation of DtN-map kernels is estimated as

$$\|v - \tilde{v}\|_{L_2(R.)} \leq \|H - \tilde{H}\|_{L_\infty(iR)}\|u\|_{L_2(R.)} \qquad (7.24)$$

Contrary to the estimate (7.19), the last equation doesn't provide us with a bound of the largest error of the output $v(t)$ but with a bound of the largest error of its energy.

Similarly as for the space $L_\infty(iR)$, we may define the linear normed space $L_\infty(T)$ of matrix-valued functions defined on the unit circle T. The norm is given by

$$\|H\|_{L_\infty(T)} = \sup_z \sigma_1(H(z)) \ , \qquad (7.25)$$

where $z = e^{i\theta}$ and $\sigma_1(H(z))$ is the largest singular value of the matrix $H(z)$. Note that uniform convergence of $H_-^n(z)$ to $H(z)$ is equivalent to $\|H - H_-^n\|_{L_\infty(T)} \to 0$ for $n \to \infty$. The norm (7.25) is induced by the mapping from the Hilbert space of vector-valued square integrable functions $u(z)$ defined on the unit circle T to itself.

Remarks:

- For vector and matrix-valued functions of dimension $n = 1$ we obtain the function spaces discussed in chapter 4.

- The inequality (7.19) gives us a mean to estimate the error induced by neglecting the coupling of the propagation modes in matrix-valued DtN-maps. From (7.18) we deduce that if $H(t)$ is diagonally dominant, that is

$$\int_0^\infty |h_{jj}(t)|d\tau \gg \sum_{\substack{k=0 \\ k \neq j}}^n \left(\int_0^\infty |h_{jk}(t)|d\tau\right), \qquad (7.26)$$

then the error of the output with respect to $L_1(R_+)$ will be small.

- The norm with respect to $L_\infty(iR)$ of continuous-time systems given by the state-space realization $\{A, B, C, D\}$ may be computed finding the smallest value γ such that the matrix

$$V = \begin{bmatrix} A + BR^{-1}D^TC & BR^{-1}B^T \\ -C^T(I + DR^{-1}D^T)C & (A + BR^{-1}D^TC)^T \end{bmatrix}, \qquad (7.27)$$

where $R = \gamma^2 I - D^TD$, has no eigenvalues on the imaginary axis [BBK89]. In [BBK89] a simple bisection algorithm is used to approximate numerically the smallest value γ. In [BS90] a more refined and faster algorithm has been proposed.

7.1.2 Causality and stability

Causality and stability of matrix-valued DtN-maps are straightforwardly deduced using the results obtained in chapter 4 for scalar DtN-maps. In fact, referring to the convolution integral representations of DtN-maps given in (7.2), a DtN-map is a causal system if every component $h_{ij}(t)$ of the DtN-map kernel $H(t)$ is causal, that is, if $h_{jk}(t) = 0$ for $t < 0$. Analogously, the regularized DtN-map kernel $H(t)$ is BIBO stable if and only if every component $h_{ij}(t)$ belongs to $L_1(R_+)$. Because of (7.18) this is equivalent to state that $H(t)$ is BIBO stable if and only if $H(t)$ belongs to $L_1(R_+)$.

The characterization of the causality of regularized DtN-maps via the Fourier transform is provided by Titchmarsh's theorem. Hence, a regularized matrix-valued DtN-map kernel $H(\omega)$ is causal if and only if there exists a matrix-valued function $H(s)$ analytic in C_+ with components $h_{ij}(s)$ belonging to $H_2(C_+)$ such that $H(s) \to H(\omega)$ for $\mathrm{Re}(s) \to 0$. In addition, the matrix-valued function $H(s)$ is the Laplace transform of $H(t)$.

7.2 The first approximation step

7.2.1 Approximation on the unit circle

Similarly as for scalar DtN-maps, we subdivide the approximation process in two steps. First, we construct a Laurent series or trigonometric Fourier series expansion of the DtN-map kernel $H(\omega)$. The last is obtained by mapping $H(\omega)$ from the imaginary axis iR to the unit circle T via the Möbius transformation

$$z = \frac{a+s}{a-s}. \tag{7.28}$$

Then, we expand $H(z)$, where $z = e^{i\theta t}$, in the trigonometric Fourier series

$$H(z) = \sum_{k=-\infty}^{\infty} H_k z^{-k}, \tag{7.29}$$

Next, we define the truncated Laurent series $H_-(z)$ analytic in S_∞

$$H_-(z) = \sum_{k=0}^{\infty} H_k z^{-k}. \tag{7.30}$$

Afterward, we construct a linear finite-dimensional discrete-time system considering the truncated Laurent series of order n

$$H_-^n(z) = \sum_{k=0}^{n} H_k z^{-k}. \tag{7.31}$$

Finally, the system induced by $H_-^n(z)$ will be approximated by a reduced-order model. This may be achieved in the context of discrete-time systems or, after having constructed the related continuous-time system, of continuous-time systems.

Using the results developed for scalar systems we may draw the following conclusions:

1. Because $H(t)$ is real we have $H(-\theta) = \overline{H}(\theta)$.
2. $H(\omega)$ is causal if and only if $H_k = 0$ for $k < 0$, where H_k are the Fourier coefficient of (7.29).
3. If $H(\omega)$ is causal, $H_-^n(z)$ converge uniformly to $H(z)$ on T if all components $h_{jk}(z)$ of $H(z)$ are continuous and of bounded variation on T. Consequently, $H_-^n(\omega)$ converge uniformly to $H(\omega)$, that is $\|H - H_-^n\|_{L_-(iR)} \to 0$ for $n \to \infty$.

7.2.2 State-space description of truncated Laurent series

Let us assume that the DtN-map kernel has been approximated by a truncated Laurent series of order n, that is

$$H(z) = \sum_{k=0}^{n} H_k z^{-k} = H_0 + \sum_{k=1}^{n} \frac{H_k z^{n-k}}{z^n}. \tag{7.32}$$

Similar as for scalar systems, we interpret $H(z)$ as a transfer function of a linear finite-dimensional discrete-time system. A realization of (7.32) is given by the block controller form

$$A = \begin{bmatrix} 0 & 0 & \dots & 0 \\ I & 0 & \dots & 0 \\ 0 & I & \dots & 0 \\ \dots & \dots & \dots & \dots \\ 0 & 0 & I & 0 \end{bmatrix} \quad B = \begin{bmatrix} I \\ 0 \\ 0 \\ \dots \\ 0 \end{bmatrix} \quad C^T = \begin{bmatrix} H_1 \\ H_2 \\ H_3 \\ \dots \\ H_n \end{bmatrix} \quad D = H_0 \tag{7.33}$$

If the transfer function is of dimension $m \times m$, the matrix A is $nm \times nm$, B is $nm \times m$, C is $m \times nm$ and D is $m \times m$. Because the transfer matrix has finitely many terms, the system (7.33) is BIBO stable. Moreover, the system matrix A is nilpotent of order n, that is, $A^k = 0$ for $k \geq n$. As a consequence, all the eigenvalues of the matrix A are zero. Computing the partial controllability matrix we see that the realization (7.33) is controllable. In fact the partial controllability matrix is the identity matrix:

$$M_c = [B, AB, \dots, A^{n-1}B] = \begin{bmatrix} I & 0 & \dots & 0 \\ 0 & I & \dots & 0 \\ \dots & \dots & \dots & \dots \\ 0 & \dots & \dots & I \end{bmatrix} \tag{7.34}$$

In addition, the partial observability matrix is equal to the block Hankel matrix of the system

$$M_o = \Gamma_H = \begin{bmatrix} C \\ CA \\ \dots \\ CA^{n-1} \end{bmatrix} = \begin{bmatrix} H_1 & H_2 & \dots & H_n \\ H_2 & H_3 & \dots & 0 \\ \dots & \dots & \dots & \dots \\ H_n & 0 & \dots & 0 \end{bmatrix} \tag{7.35}$$

Hence, the observability of the system depends on the rank of this Hankel matrix. Now, it is easily seen that the Hankel matrix (7.35) has full rank if and only if the Markov parameter H_n has full rank. Under this condition, the linear system (7.33) is a minimal realization of the transfer function (7.32).

Remarks:
• Instead of the controllability form (7.33) we can use the observer form

$$A = \begin{bmatrix} 0 & I & 0 & \dots & 0 \\ 0 & 0 & I & \dots & 0 \\ \dots & \dots & \dots & \dots & I \\ 0 & 0 & 0 & \dots & 0 \end{bmatrix} \quad B = \begin{bmatrix} H_1 \\ H_2 \\ H_3 \\ \dots \\ H_n \end{bmatrix} \quad C^T = \begin{bmatrix} I \\ 0 \\ 0 \\ \dots \\ 0 \end{bmatrix} \quad D = H_0 \tag{7.36}$$

to describe the system defined by the transfer function (7.32). Both forms are completely equivalent provided the Markov parameter H_n has full rank. With the controllability form the observability matrix is the identity matrix and the controllability matrix is the Hankel matrix.

- Because the matrix A is nilpotent of order n we have $A^k B = 0$ and $CA^k = 0$ for $k \geq n$. Therefore, the infinite controllability and observability matrices are equal to their related partial controllability and observability matrices.

7.3   Truncated balanced realization methods

7.3.1   The balanced truncation method

The most appealing fact about the truncated balanced realization method is that it preserves the conceptual simplicity even if matrix-valued transfer functions are considered. In fact, the approximation procedure developed for scalar DtN-maps applies essentially unchanged in the context of matrix-valued DtN-maps. Therefore, much of this section is a mere repetition of section 6.3.1 so that we can keep it very short. However, we will develop the theory using continuous-time systems obtained by mapping discrete-time systems defined by (7.33) using the Möbius transformation (7.28). The system matrices of such a mapped continuous-time system is

$$
A = 2a
\begin{bmatrix}
-\dfrac{I}{2} & 0 & \cdots & 0 & 0 & 0 & 0 \\[2mm]
I & -\dfrac{I}{2} & \cdots & 0 & 0 & 0 & 0 \\[2mm]
-I & I & \cdots & 0 & 0 & 0 & 0 \\[2mm]
I & I & \cdots & -\dfrac{I}{2} & 0 & 0 & 0 \\[2mm]
\cdots & \cdots & \cdots & \cdots & \cdots & \cdots & \cdots \\[2mm]
(-1)^{n-2}I & (-1)^{n-3}I & \cdots & -I & I & -\dfrac{I}{2} & 0 \\[2mm]
(-1)^{n-1}I & (-1)^{n-2}I & \cdots & I & -I & I & -\dfrac{I}{2}
\end{bmatrix}
\tag{7.37}
$$

$$
B = \sqrt{2a}
\begin{bmatrix}
I \\
-I \\
I \\
-I \\
\cdots \\
(-1)^{n-2}I \\
(-1)^{n-1}I
\end{bmatrix}
\qquad
C^T = \sqrt{2a}
\begin{bmatrix}
H_1^T - C_2^T \\
H_2^T - C_3^T \\
H_3^T - C_4^T \\
\cdots \\
H_{n-2}^T - C_{n-1}^T \\
H_{n-1}^T - C_n^T \\
H_n^T
\end{bmatrix}
$$

$$
D = H_0 - H_1 + H_2 - \cdots + (-1)^{n-2}H_{n-2} + (-1)^{n-1}H_{n-1} + (-1)^n H_n
$$

Observe that the matrix A is a Toeplitz matrix whose nonzero, nondiagonal blocks are given by $A_{ij} = 2a(-1)^{i-j}I$. The eigenvalues of this matrix are all $-a$. The blocks of the matrix C are computed recursively beginning with the last block $C_n = H_n^T$. Because of the map (7.28) the controllability and observability grammians of the system (7.37) are equal to those of the system

(7.33) so that

$$W_c = M_c M_c^T = I \qquad\qquad (7.38)$$

$$W_o = M_o^T M_o = \Gamma_H^T \Gamma_H \qquad\qquad (7.39)$$

Because the controllability grammian (7.33) is equal to the identity matrix we have to transform the observability grammian W_o in the diagonal form. Since the Hankel matrix Γ_H has full rank, W_o is positive definite and may be represented in the spectral form $W_o = V\Sigma^2 V^T$, where Σ are the singular values of the Hankel matrix Γ_H and V is a unitary matrix which corresponds to the right unitary matrix of the SVD of Γ_H. Defining the new set of state variables $x = Vy$ we obtain the input-normal realization

$$\bar{A} = V^T A V \qquad \bar{B} = V^T B \qquad \bar{C} = CV \qquad \bar{D} = D. \qquad\qquad (7.40)$$

With $x = V\Sigma^{-1/2}y$ we obtain the internally balanced realization

$$\bar{A} = \Sigma^{1/2} V^T A V \Sigma^{-1/2} \qquad \bar{B} = \Sigma^{1/2} V^T B \qquad \bar{C} = CV\Sigma^{-1/2} \qquad \bar{D} = D. \qquad (7.41)$$

Let us assume that the system $\{A, B, C\}$ is in balanced realization form. We partition the controllability and observability grammians conforming to

$$\Sigma = \begin{bmatrix} \Sigma_1 & 0 \\ 0 & \Sigma_2 \end{bmatrix}, \qquad\qquad (7.42)$$

with

$$\Sigma_1 = \begin{bmatrix} \sigma_1 I_{r_1} & \cdots & 0 \\ \cdots & \cdots & \cdots \\ 0 & \cdots & \sigma_k I_{r_k} \end{bmatrix} \qquad \Sigma_1 = \begin{bmatrix} \sigma_{k+1} I_{r_{k+1}} & \cdots & 0 \\ \cdots & \cdots & \cdots \\ 0 & \cdots & \sigma_p I_{r_p} \end{bmatrix} \qquad (7.43)$$

where $\sigma_k > \sigma_{k+1}$. If the system matrices are partitioned in conformity to (7.42), that is,

$$A = \begin{bmatrix} A_{11} & A_{12} \\ A_{21} & A_{22} \end{bmatrix} \qquad B = \begin{bmatrix} B_1 \\ B_2 \end{bmatrix} \qquad C = \begin{bmatrix} C_1 & C_2 \end{bmatrix}, \qquad (7.44)$$

we define the system $\{A_{11}, B_1, C_1, D\}$ as the truncated balanced realization of $\{A, B, C\}$.

The system $\{A_{11}, B_1, C_1, D\}$ fulfill the Lyapunov equations

$$A_{11}\Sigma_1 A_{11}^T - \Sigma_1 = -B_1 B_1^T$$
$$A_{11}^T \Sigma_1 A_{11} - \Sigma_1 = -C_1^T C_1 \qquad\qquad (7.45)$$

Because Σ_1 is positive definite, the realization is minimal and has McMillan degree $n = r_1 + \ldots + r_k$. Furthermore, the system is stable but not necessarily BIBO stable. However, as was shown in [PS82], any balanced truncation of a BIBO stable system is also BIBO stable.

The approximation error with respect to $L_\infty(iR)$ is bounded by twice the sum of the singular values associated with the deleted states. That is

$$\|H(\omega) - \hat{H}(\omega)\|_{L_\infty(iR)} \le 2(\sigma_{k+1} + \ldots + \sigma_p). \qquad\qquad (7.46)$$

Moreover, if only one state with the associated singular value σ is deleted the approximation error bound (7.46) becomes an equality $\|H(\omega) - \hat{H}(\omega)\|_{L_-(iR)} = 2\sigma$. Furthermore, the maximum error occurs at $\omega = 0$. Because the system matrix D is not involved in the truncation procedure the reduced-order model has perfect matching at infinity, that is $H(\infty) = \hat{H}(\infty) = D$. Therefore, the reduced-order model have good high-frequency fidelity. Therefore, if all singular values are distinct, we may expect that the largest error occurs at low frequencies.

Next we propose an algorithm to compute reduced-order models of the system (7.37) via the truncated balanced realization method:

1. Choose an upper bound for the approximation error, say ε.
2. Setup the observability matrix $W_o = \Gamma_H^T \Gamma_H$ and compute their eigenvalues Λ and eigenvectors V. Sort the eigenvalues and eigenvectors such that $\lambda_1 \geq \lambda_2 \geq \ldots \geq \lambda_n$.
3. Find the index k such that $2(\sigma_k + \ldots + \sigma_p) > \varepsilon \geq 2(\sigma_{k+1} + \ldots + \sigma_p)$. This is the index of the last state variable that will not be deleted.
4. Compute the truncated state-space realization of McMillan degree k by

$$\bar{A}_{11} = \Sigma_k^{1/2} V_k^T A V_k \Sigma_k^{-1/2} \qquad \bar{B}_1 = \Sigma_k^{1/2} V_k^T B \qquad \bar{C}_1 = C V_k \Sigma_k^{-1/2} \qquad \bar{D} = D, \quad (7.47)$$

where the $n \times k$ matrix V_k contains the k-first eigenvectors and Σ_k contains the k-first square roots of the eigenvalues of the observability matrix. The system matrices A, B, C and D are given by (7.37).

Remarks:
- Observe that because of the special triangular structure of the block Hankel matrix the observability grammian W_o is quickly computed block by block using the relations

$$Q_{nk} = H_n^T H_k \qquad k = 1 \ldots n$$
$$Q_{jk} = H_j^T H_k + Q_{j+1, k+1} \qquad j = 1 \ldots n-1, k = 1 \ldots n-j+1 \qquad (7.48)$$

Moreover, because W_o is symmetric we have only to compute $n(n+1)/2$ blocks.
- We can always find a truncated balanced realization of McMillan degree m_ε so that given an error bound ε we have $\|H(\omega) - H_m(\omega)\|_{L_-(iR)} \leq \varepsilon$. That is, the truncated balanced realization allows us to approximate matrix-valued DtN-maps with any accuracy. The proof is similar to those given in section 6.3.1.

7.3.2 The balanced singular perturbation method

In the last section, we pointed out that the maximum error of the truncated balanced realization method will occur at low or intermediate frequencies. However, in many applications it may be of concern to model with high accuracy the long time behavior of a system. This implies a reduced-order model with perfect matching at $s = 0$. The balanced singular perturbation method, [FN82], provides a very accurate low frequency approximation of DtN-map kernels.

Suppose we wish to approximate a balanced continuous-time system $\{A, B, C, D\}$ with a reduced-order model which is accurate at low frequencies. This is obtained by an easy modification of the balanced truncation algorithm discussed in the previous section. The idea is to introduce the Möbius transformation

$$s' = \frac{1}{s}. \qquad (7.49)$$

This map is bijective and onto. Points in the right (left) half plane are mapped to points in the right (left) half plane and points on the imaginary axis are mapped to points on the imaginary axis. The point at infinity is mapped to the origin. Therefore, if $\{A, B, C, D\}$ is BIBO stable the mapped related system is BIBO stable too. Applying the equations (5.146) with the parameter set $\alpha = 0, \beta = 1, \gamma = 1, \delta = 0$ we obtain the related system

$$\tilde{A} = A^{-1}$$
$$\tilde{B} = -A^{-1}B$$
$$\tilde{C} = CA^{-1}$$
$$\tilde{D} = D - CA^{-1}B$$
$$\quad (7.50)$$

This system exists if and only if the system matrix A is invertible. That is, $s = 0$ is not an eigenvalue of A. To this system we apply the balanced truncation procedure described in the last section and obtain the reduced-order model $\{\tilde{A}_r, \tilde{B}_r, \tilde{C}_r, \tilde{D}\}$. Finally, applying the transformation (7.50) to the system $\{\tilde{A}_r, \tilde{B}_r, \tilde{C}_r, \tilde{D}\}$ yields the reduced-order model $\{A_r, B_r, C_r, D\}$. Because the map (7.49) preserves the balanced realization form the reduced-order model is also in the balanced realization form. Furthermore, the reduced-order model has perfect matching at $\omega = 0$, that is, $H(0) = \hat{H}(0)$, and the error bound is $\|H(\omega) - \hat{H}(\omega)\|_{L_\infty(iR)} \leq 2(\sigma_{k+1} + \dots + \sigma_p)$.

These results suggests to use the following algorithm to compute the balanced singular approximation of the system (7.33):

1. Choose an upper bound for the approximation error, say ε.
2. Setup the observability matrix $W_o = \Gamma_H^T \Gamma_H$ and compute their eigenvalues Λ and eigenvectors V. Sort the eigenvalues and eigenvectors such that $\lambda_1 \geq \lambda_2 \geq \dots \geq \lambda_n$. Find the index k such that $2(\sigma_k + \dots + \sigma_p) > \varepsilon \geq 2(\sigma_{k+1} + \dots + \sigma_p)$. This is the index of the last state variable which will not be deleted.
3. Compute the balanced realization

$$\overline{A} = \Sigma^{1/2} V^T A V \Sigma^{-1/2} \qquad \overline{B} = \Sigma^{1/2} V^T B \qquad \overline{C} = C V \Sigma^{-1/2} \qquad \overline{D} = D. \quad (7.51)$$

4. Compute the related continuous-time system using the map

$$\tilde{A} = -\frac{1}{a}(I - A)^{-1}(A + I) = -\frac{1}{a}(A + I)(I - A)^{-1}$$
$$\tilde{B} = \sqrt{\frac{2}{a}}(I - A)^{-1}B$$
$$\tilde{C} = -\sqrt{\frac{2}{a}}C(I - A)^{-1}$$
$$\tilde{D} = D + C(I - A)^{-1}B$$
$$\quad (7.52)$$

This mapping is derived from the Möbius transformation

$$z = \frac{a + 1/s}{a - 1/s} = \frac{as + 1}{as - 1}, \quad (7.53)$$

which results combining (5.133) with (7.49).

5. Partition the system matrices according to

$$\tilde{A} = \begin{bmatrix} \tilde{A}_{11} & \tilde{A}_{12} \\ \tilde{A}_{21} & \tilde{A}_{22} \end{bmatrix} \qquad \tilde{B} = \begin{bmatrix} \tilde{B}_1 \\ \tilde{B}_2 \end{bmatrix} \qquad \tilde{C} = \begin{bmatrix} \tilde{C}_1 & \tilde{C}_2 \end{bmatrix}. \quad (7.54)$$

The reduced-order model of McMillan degree k is defined to be $\{\tilde{A}_{11}, \tilde{B}_1, \tilde{C}_1, \tilde{D}\}$.

6. Finally, map the reduced-order model $\{\tilde{A}_{11}, \tilde{B}_1, \tilde{C}_1, \tilde{D}\}$ via (7.50) to the balanced singularly perturbed system $\{A_{11}, B_1, C_1, D\}$.

Remark:

• As can be shown, the continuous-time system (7.52) is defined by the state-space realization

$$A = -\frac{2}{a}\begin{bmatrix} I/2 & 0 & \dots & 0 & 0 & 0 & 0 \\ I & I/2 & \dots & 0 & 0 & 0 & 0 \\ I & I & \dots & 0 & 0 & 0 & 0 \\ I & I & \dots & I/2 & 0 & 0 & 0 \\ \dots & \dots & \dots & \dots & \dots & \dots & \dots \\ I & I & \dots & I & I & I/2 & 0 \\ I & I & \dots & I & I & I & I/2 \end{bmatrix}$$

$$B = \sqrt{\frac{2}{a}}\begin{bmatrix} I \\ I \\ I \\ I \\ \dots \\ I \\ I \end{bmatrix} \qquad C^T = -\sqrt{\frac{2}{a}}\begin{bmatrix} H_1^T + C_2^T \\ H_2^T + C_3^T \\ H_3^T + C_4^T \\ \dots \\ H_{n-2}^T + C_{n-1}^T \\ H_{n-1}^T + C_n^T \\ H_n^T \end{bmatrix} \qquad (7.55)$$

$$D = H_0 + H_1 + H_2 + \dots + H_{n-2} + H_{n-1} + H_n$$

7.4   Some supplementary results of linear systems theory

7.4.1   Combining linear systems

In order to obtain some deeper insight into the structure of the Hankel-norm approximation methods we need several supplementary results from linear systems theory. Consider two finite-dimensional linear systems with matrix-valued transfer functions $G_1(s)$ and $G_2(s)$ which are defined by the state-space realizations $\{A_1, B_1, C_1, D_1\}$ and $\{A_2, B_2, C_2, D_2\}$, respectively. We assume that both transfer matrices have equal dimension.

The parallel connection of two linear systems is given by the system whose transfer function is defined to be the sum of the transfer functions $G_1(s)$ and $G_2(s)$, that is

$$G(s) = G_1(s) + G_2(s). \qquad (7.56)$$

The state-space realization of $G(s)$ is then given by the matrices

$$A = \begin{bmatrix} A_1 & 0 \\ 0 & A_2 \end{bmatrix} \qquad B = \begin{bmatrix} B_1 \\ B_2 \end{bmatrix} \qquad C = \begin{bmatrix} C_1 & C_2 \end{bmatrix} \qquad D = D_1 + D_2. \qquad (7.57)$$

In fact, we have

$$G(s) = C(sI - A)^{-1}B + D = C_1(sI - A_1)^{-1}B_1 + D_1 + C_2(sI - A_2)^{-1}B_2 + D_2$$
$$= G_1(s) + G_2(s) \qquad (7.58)$$

so that (7.56) is fulfilled.

A little more involved is the state-space realization of the cascade or series connection of two systems. In a series connection the output of the first system is the input of the second system. That is, the transfer function of the compound system results from the matrix multiplication of the transfer functions $G_1(s)$ and $G_2(s)$:

$$G(s) = G_1(s)G_2(s). \tag{7.59}$$

The state-space realization of $G(s)$ is then given by the system matrices

$$A = \begin{bmatrix} A_1 & B_1C_2 \\ 0 & A_2 \end{bmatrix} \qquad B = \begin{bmatrix} B_1D_2 \\ B_2 \end{bmatrix} \qquad C = \begin{bmatrix} C_1 & D_1C_2 \end{bmatrix} \qquad D = D_1D_2 \tag{7.60}$$

or

$$A = \begin{bmatrix} A_1 & 0 \\ B_1C_2 & A_2 \end{bmatrix} \qquad B = \begin{bmatrix} B_2 \\ B_1D_2 \end{bmatrix} \qquad C = \begin{bmatrix} D_1C_2 & C_1 \end{bmatrix} \qquad D = D_1D_2 \tag{7.61}$$

Considering that the inverses of block triangular matrices M are given by

$$M = \begin{bmatrix} M_{11} & M_{12} \\ 0 & M_{22} \end{bmatrix} \qquad M^{-1} = \begin{bmatrix} M_{11}^{-1} & -M_{11}^{-1}M_{12}M_{22}^{-1} \\ 0 & M_{22}^{-1} \end{bmatrix} \tag{7.62}$$

we may verify (7.59) by computing $G(s)$ with the system matrices (7.61). This yields

$$\begin{aligned} G(s) &= C(sI - A)^{-1}B + D \\ &= \begin{bmatrix} C_1 & D_1C_2 \end{bmatrix} \begin{bmatrix} (sI - A_1)^{-1} & -(sI - A_1)^{-1}B_1C_2(sI - A_2)^{-1} \\ 0 & (sI - A_2)^{-1} \end{bmatrix} \begin{bmatrix} B_1D_2 \\ B_2 \end{bmatrix} + D_1D_2 \\ &= (C_1(sI - A_1)^{-1}B_1 + D_1)(C_2(sI - A_2)^{-1}B_2 + D_2) \\ &= G_1(s)G_2(s) \end{aligned} \tag{7.63}$$

Similarly we can show that (7.61) is a state-space realization of the compound system (7.59).

Remark:
• The state-space realizations (7.57), (7.60) and (7.61) applies for discrete-time systems as well.

7.4.2 Adjoint, inverse and all-pass systems

The results of the previous section are useful to derive state-space realizations of special systems. The first special system we will study is the so-called adjoint or conjugate system. Consider a continuous-time system $G(s)$ defined by the state-space realization $\{A, B, C, D\}$. Its adjoint or conjugate system is defined to be the system $\tilde{G}(s)$ belonging to $L_2(iR)$ and having the following property

$$\langle w, Gu \rangle = \langle \tilde{G}w, u \rangle. \tag{7.64}$$

The definition of the inner product in $L_2(iR)$ yields

$$\langle w, Gu \rangle = \int\limits_{-\infty}^{\infty} w^*(i\omega)G(i\omega)u(i\omega)d\omega = \int\limits_{-\infty}^{\infty} (G^*(i\omega)w(i\omega))^*u(i\omega)d\omega = \langle \tilde{G}w, u \rangle. \quad (7.65)$$

By definition, we obtain

$$\tilde{G}(i\omega) = G^*(i\omega). \quad (7.66)$$

Considering that $G(s)$ is defined by way of the state-space realization $\{A, B, C, D\}$, we obtain $G^*(i\omega) = -B^*(i\omega I - (-A^*))^{-1}C^* + D^*$. Because the state-space matrices $\{A, B, C, D\}$ are real the state-space realization of the adjoint or conjugate system $\tilde{G}(s)$ is given by the set

$$\{-A^T, -C^T, B^T, D^T\}. \quad (7.67)$$

From (7.66) it follows that $\tilde{G}(s) = G^T(-s)$. If $G(s)$ is symmetric we obtain $\tilde{G}(s) = G(-s)$.

Remarks:

- Observe that because of (7.67) the adjoint system of a stable system is necessarily unstable. In fact, the eigenvalues of $-A^T$ are $\lambda = -r \pm is$, where λ is any eigenvalue of A with $r < 0$. As a consequence, if $G(s)$ belongs to $L_\infty(iR)$ then its adjoint belongs to $L_\infty(iR)$ too. Moreover, it can be shown that $\|\tilde{G}\|_{L_\infty(iR)} = \|G\|_{L_\infty(iR)}$.

- Similarly as for continuous-time systems we may define the adjoint system of a discrete-time system $G(z)$ by requiring that $\langle w, Gu \rangle = \langle \tilde{G}w, u \rangle$ holds in $L_2(T)$. This yields the relation

$$\tilde{G}(z) = G^T\left(\frac{1}{z}\right). \quad (7.68)$$

Given a state-space description $\{A, B, C, D\}$ for $G(z)$, we find that the adjoint system $\tilde{G}(z)$ has the description $\{A^{-T}, -(CA^{-1})^T, (A^{-1}B)^T, D^T - (CA^{-1}B)^T\}$.

Adjoint systems are used to construct so-called allpass systems. Allpass systems have the property that the norm of the output function is equal to the norm of the input function. Therefore, given an allpass system $G(s)$, then, with respect to the $L_2(iR)$ norm, we obtain

$$\|Gu\|_{L_2(iR)} = \|u\|_{L_2(iR)}, \quad (7.69)$$

where $u(s)$ is any input function belonging to $L_2(iR)$. From the definition (7.69) and the polarization identity of Hilbert spaces it follows that $G(s)$ is an allpass system if and only if

$$\langle Gw, Gu \rangle = \langle w, u \rangle \quad (7.70)$$

for all $u(s)$ and $w(s)$ belonging to $L_2(iR)$. That is, an allpass system preserves the inner product. Consequently, by the use of the adjoint $\tilde{G}(s)$, a system $G(s)$ is allpass if and only if

$$\tilde{G}G = I. \quad (7.71)$$

Now because we consider only input-output symmetric systems, it follows that $G(s)$ is allpass if and only if

$$\tilde{G}(s) = G^{-1}(s), \quad (7.72)$$

provided the inverse system $G^{-1}(s)$ exists. The inverse exists if and only if the system matrix D is nonsingular. That is, if D^{-1} exists. Clearly, if $G^{-1}(s)$ exists then $\tilde{G}G = G\tilde{G} = I$.

The inverse $G^{-1}(s)$ of a system $G(s)$ with state-space realization $\{A, B, C, D\}$ is defined by the state-space realization

$$\{A - BD^{-1}C, -BD^{-1}, D^{-1}C, D^{-1}\}. \tag{7.73}$$

In fact, by (7.60) the state-space realization of the composition GG^{-1} is

$$\hat{A} = \begin{bmatrix} A & BD^{-1}C \\ 0 & A - BD^{-1}C \end{bmatrix} \qquad \hat{B} = \begin{bmatrix} BD^{-1} \\ -BD^{-1} \end{bmatrix} \qquad \hat{C} = \begin{bmatrix} C & C \end{bmatrix} \qquad \hat{D} = I. \tag{7.74}$$

Performing a similarity transformation with

$$T = \begin{bmatrix} I & I \\ 0 & I \end{bmatrix} \tag{7.75}$$

yields

$$\hat{A} = \begin{bmatrix} A & 0 \\ 0 & A - BD^{-1}C \end{bmatrix} \qquad \hat{B} = \begin{bmatrix} 0 \\ -BD^{-1} \end{bmatrix} \qquad \hat{C} = \begin{bmatrix} C & 0 \end{bmatrix} \qquad \hat{D} = I \tag{7.76}$$

This system has the transfer function $GG^{-1} = I$ so that (7.73) is the state-space realization of $G^{-1}(s)$. Observe that because of (7.71) the state-space realizations (7.73) and (7.67) are associated via a similarity transformation.

With (7.67) and (7.71) we are able to provide a characterization of allpass systems. The product $\tilde{G}G(s)$ can be written as

$$\begin{aligned} \tilde{G}G(s) &= (B^T(-sI - A^T)^{-1}C^T + D^T)(C(sI - A)^{-1}B + D) \\ &= B^T(-sI - A^T)^{-1}C^TD + D^TC(sI - A)^{-1}B \\ &\quad + B^T(-sI - A^T)^{-1}C^TC(sI - A)^{-1}B + D^TD \\ &= B^T(-sI - A^T)^{-1}C^TD + D^TC(sI - A)^{-1}B \\ &\quad + B^T(-sI - A^T)^{-1}(Q(sI - A) + (-sI - A^T)Q)(sI - A)^{-1}B + D^TD \\ &= B^T(-sI - A^T)^{-1}(C^TD + QB) + (D^TC + B^TQ)(sI - A)^{-1}B + D^TD \end{aligned} \tag{7.77}$$

Therefore, the relation $\tilde{G}G(s) = I$ implies the existence of a symmetric matrix Q such that

$$\begin{aligned} QA + A^TQ + C^TC &= 0 \\ D^TC + B^TQ &= 0 \end{aligned} \tag{7.78}$$

Moreover, the system matrix D must be unitary

$$D^TD = I. \tag{7.79}$$

Because the system G is input-output symmetric we may define the adjoint $\tilde{G}(s)$ according to $G\tilde{G}(s) = I$. Proceeding similarly as in (7.77), we find that $G(s)$ is allpass if there exists a symmetric matrix P such that

$$\begin{aligned} PA^T + AP + BB^T &= 0 \\ BD^T + PC^T &= 0 \end{aligned} \tag{7.80}$$

and $DD^T = I$. From the very definition of allpass systems (7.69) and the induced norm (7.13) we find that

$$\|G\|_{L_\infty(iR)} = \sup_{\|u\|_{L_2(iR)} \neq 0} \frac{\|Gu\|_{L_2(iR)}}{\|u\|_{L_2(iR)}} = 1. \tag{7.81}$$

Hence, the norm of any allpass system with respect to $L_\infty(iR)$ is exactly unity.

Example: To become familiar with these concepts we consider a simple one dimensional model problem. Let us assume that the transfer function is given by

$$g(s) = \frac{bc}{s-a} + d, \tag{7.82}$$

where all system parameters are real. A state-space realization of this transfer function is $g = \{a, b, c, d\}$. By (7.67) its adjoint system is $\tilde{g} = \{-a, b, -c, d\}$. This yields the transfer function

$$\tilde{g}(s) = -\frac{bc}{s+a} + d = g(-s). \tag{7.83}$$

If $a > 0$ the systems $g(s)$ is stable. Then, by (7.83), $\tilde{g}(s)$ must be unstable. In order to be allpass the parameters of $g(s)$ must fulfill (7.78) and (7.79). That is

$$2qa + c^2 = 0 \qquad dc + bq = 0 \qquad d^2 = 1 \tag{7.84}$$

Clearly, $d = 1$ and $q = -c/b$ such that $a = cb/2$. In other words, $g(s)$ is allpass if it is given by

$$g(s) = \frac{2a}{s-a} + 1. \tag{7.85}$$

In fact, on the imaginary axis we have

$$g(i\omega) = \frac{2a}{i\omega - a} + 1 = \frac{i\omega + a}{i\omega - a} = \frac{a^2 - \omega^2}{\omega^2 + a^2} + i\frac{2a\omega}{\omega^2 + a^2}. \tag{7.86}$$

Taking the absolute value of (7.86) yields $|g(i\omega)| = 1$ for any ω so that $\|G\|_{L_\infty(iR)} = 1$. Note that $g(i\omega)$ is not a constant function and that $\tilde{g}(i\omega)$ is allpass. According to (7.73) the inverse system is given by

$$g^{-1}(s) = -\frac{bc/d^2}{s - (a - bc)} + \frac{1}{d}. \tag{7.87}$$

Note that $g^{-1}(s)$ is stable only if $a - bc > 0$. That is, the inverse of a stable system needs not to be stable. Note that with $a = cb/2$ and $d = 1$ $g^{-1}(s)$ is equal to $\tilde{g}(s)$.

Remarks:

- Note that the first equations in (7.78) and (7.80) are structurally equivalent to the Lyapunov equations. However P and Q need not to be positive definite. In fact, they may be indefinite or even singular.
- All pass systems have no poles on the imaginary axis iR.
- If the inverse system $G^{-1}(s)$ exists, comparing (7.67) and (7.73) we conclude that $G(s)$ is allpass if $A + A^T = C^T C \qquad D^T C = B^T \qquad D^T D = I$

7.5 Hankel-norm methods

7.5.1 Hankel operators and Hankel norm of continuous-time systems

Hankel operators are defined by a map from $L_2(R_+)$ to $L_2(R_+)$ given by

$$v(t) = (\Gamma_G u)(t) = \int_0^\infty G(t + \tau)u(\tau)d\tau, \qquad (7.88)$$

where the input function $u(t)$ as well as the output function $v(t)$ belongs to $L_2(R_+)$. This implies that the kernel $G(t)$ belongs to $L_1(R_+)$. The link to linear systems theory is set up if we consider the change of variable $s = -\tau$ in (7.88). After defining $\hat{u}(s) = u(-\tau)$ we obtain

$$v(t) = (\Gamma_G u)(t) = \int_{-\infty}^0 G(t - s)\hat{u}(s)ds, \qquad (7.89)$$

where $G(t)$ is causal and $\hat{u}(s) = 0$ for $t > 0$. The form (7.89) can be interpreted in the context of linear system theory. In fact, $v(t)$ is the output of a linear system in the interval $t > 0$ driven by an input $\hat{u}(t)$ acting in the interval $t < 0$. Hence the Hankel integral operator maps past input to future output. If the linear system is given by a state-space realization then (7.89) is

$$v(t) = (\Gamma_G u)(t) = \int_{-\infty}^0 Ce^{A(t-s)}B\hat{u}(s)ds. \qquad (7.90)$$

Note that the system matrix D doesn't appear in (7.90) because $D\hat{u}(t) = 0$ for $t > 0$. A remarkable fact is that we may add any anticausal kernel function $F(t)$ to $G(t)$ in (7.90) without influencing the output $v(t)$ for $t > 0$. In fact, because $F(t) = 0$ for $t > 0$ any input $u(t)$ with $u(t) = 0$ for $t > 0$ gives

$$\int_{-\infty}^0 F(t - s)u(s)ds = 0 \text{ for all } t > 0 \qquad (7.91)$$

Hence, we obtain

$$v(t) = (\Gamma_{G-F} u)(t) = \int_{-\infty}^0 (G(t - s) - F(t - s))\hat{u}(s)ds = (\Gamma_G u)(t). \qquad (7.92)$$

so that only the causal part of the kernel is effective for the output.

The Hankel norm is an induced or operator norm. It is defined to be the supremum of the ratio of the future output to the past input

$$\|G\|_\Gamma = \sup_{\|u\|_{L_2(R_-)} \neq 0} \frac{\|\Gamma_G u\|_{L_2(R_-)}}{\|u\|_{L_2(R_-)}} = \sup_{\|\hat{u}\|_{L_2(R_-)} \neq 0} \frac{\|v\|_{L_2(R_-)}}{\|\hat{u}\|_{L_2(R_-)}}. \qquad (7.93)$$

The Hankel norm is bounded by above by the infinity norm because for any input function of unit energy $\|\hat{u}\|_{L_2(R_-)}$, $\|G\|_{L_\infty(iR)}$ is the least upper bound of the total output energy (past and future) whereas the Hankel norm is the least upper bound of the future output energy. Therefore

$$\|G\|_\Gamma \leq \|G\|_{L_\infty(iR)}. \qquad (7.94)$$

Moreover, because an arbitrary anticausal kernel $F(t)$ doesn't influence the future output we have $\|G\|_\Gamma = \|G - F\|_\Gamma$ so that

$$\|G\|_\Gamma \leq \|G - F\|_{L_-(iR)} . \tag{7.95}$$

Similar as for scalar input-output systems, it can be shown that there exists an anticausal transfer function $F(t)$ such that

$$\|G\|_\Gamma = \min\|G - F\|_{L_-(iR)} . \tag{7.96}$$

This is the equivalent of Nehari's theorem for matrix-valued continuous-time systems. It is the corner stone of the optimal Hankel-norm approximation method for matrix-valued systems. Infinite block Hankel matrices were studied in [AAK78]. The theory of Hankel-norm approximation of linear time-invariant continuous-time systems was first developed by Glover [Glo84]. Textbook presentations of this theory are found in [GL96] and [ZDG96].

7.5.2 Suboptimal Hankel-norm approximation

In the last sections, we have provided all instruments which will be necessary to construct the optimal Hankel-norm approximation of a given linear finite-dimensional continuous-time system. That is, given a linear system G of McMillan degree n, find a linear system G_r of McMillan degree $r < n$ such that

$$\|G - G_r\|_\Gamma = \sigma_{r+1}(G), \tag{7.97}$$

where $\sigma_{r+1}(G)$ is the $r + 1$ singular value of the system G. However, before giving the solution of the optimal Hankel-norm approximation problem we consider the so-called suboptimal Hankel-norm approximation problem which instead of (7.97) fulfills the inequality

$$\|G - G_r\|_\Gamma < \gamma , \tag{7.98}$$

where $\sigma_r(G) > \gamma > \sigma_{r+1}(G)$. That is, the limit $\gamma = \sigma_{r+1}(G)$ is not achieved. The method for the construction of G_r is similar to that applied to scalar systems. Suppose we find an anticausal system $F(t)$ such that $(G - G_r - F)/\gamma$ is allpass. Then, because allpass systems has an infinity norm which is unity we obtain

$$\|G - G_r - F\|_{L_-(iR)} = \gamma. \tag{7.99}$$

Hence, we conclude from the inequality (7.94) that $\|G - G_r\|_\Gamma \leq \gamma$.

Let us assume that G is $m \times m$ and has a realization $G = \{A, B, C, D\}$. Now define the augmented system G_a with the realization

$$A_a = A \qquad B_a = \begin{bmatrix} B & 0 \end{bmatrix} \qquad C_a = \begin{bmatrix} C \\ 0 \end{bmatrix} \qquad D_a = \begin{bmatrix} D & 0 \\ 0 & 0 \end{bmatrix}, \tag{7.100}$$

where A_a is $n \times n$, B_a is $n \times 2m$, C_a is $2m \times n$ and D_a is $2m \times 2m$. That is, we imbed G in a larger system G_a. The embedding is necessary to construct the allpass system. Now, define a system $\hat{G}$ of dimension $2m \times 2m$ with the realization $\hat{G} = \{\hat{A}, \hat{B}, \hat{C}, \hat{D}\}$ such that the system

$$E = G_a - \hat{G} \tag{7.101}$$

satisfies

$$\tilde{E}E = E\tilde{E} = \gamma^2 I. \tag{7.102}$$

That is, E/γ an allpass system. Conforming to (7.56), $E = G - \hat{G}$ has the realization

$$A_e = \begin{bmatrix} A_a & 0 \\ 0 & \hat{A} \end{bmatrix} \qquad B_e = \begin{bmatrix} B_a \\ \hat{B} \end{bmatrix} \qquad C_e = \begin{bmatrix} C_a & \hat{C} \end{bmatrix} \qquad D_e = D_a - \hat{D}. \qquad (7.103)$$

According to (7.78) and (7.79) the system defined by (7.103) must fulfill the following equations

$$Q_e A_e + A_e^T Q_e + C_e^T C_e = 0 \qquad (7.104.\text{a})$$

$$D_e^T C_e + B_e^T Q_e = 0 \qquad (7.104.\text{b})$$

$$D_e^T D_e = \gamma^2 I. \qquad (7.104.\text{c})$$

Furthermore, because E is square we impose $E\tilde{E} = \gamma^2 I$. Hence E must even fulfill (7.80)

$$P_e A_e^T + A_e P_e + B_e B_e^T = 0 \qquad (7.105.\text{a})$$

$$B_e D_e^T + P_e C_e^T = 0 \qquad (7.105.\text{b})$$

Now let us analyse the relationship between Q_e and P_e. Multiplying (7.105.a) from the left and the right with Q_e and subtracting (7.104.a) multiplied by $\gamma^2 I$ yields

$$\begin{aligned} &Q_e A_e (\gamma^2 I - P_e Q_e) + (\gamma^2 I - Q_e P_e) A_e^T Q_e + Q_e B_e B_e^T Q_e - \gamma^2 C_e^T C_e \\ &= Q_e A_e (\gamma^2 I - P_e Q_e) + (\gamma^2 I - Q_e P_e) A_e^T Q_e + C_e^T (\gamma^2 I - D_e D_e^T) C_e \qquad (7.106) \\ &= Q_e A_e (\gamma^2 I - P_e Q_e) + (\gamma^2 I - Q_e P_e) A_e^T Q_e = 0 \end{aligned}$$

The last equation is obtained using (7.104.b) and (7.104.c). Because E is bounded in $L_\infty(iR)$, the system matrix A_e has no eigenvalues on the imaginary axis. Therefore, the grammians P_e and Q_e are nonsingular. Hence, because of (7.106) they fulfill

$$P_e Q_e = Q_e P_e = \gamma^2 I. \qquad (7.107)$$

Now let us partition P_e and Q_e according to

$$P_e = \begin{bmatrix} P & P_{12} \\ P_{12}^T & P_{22} \end{bmatrix} \qquad Q_e = \begin{bmatrix} Q & Q_{21}^T \\ Q_{21} & Q_{22} \end{bmatrix}, \qquad (7.108)$$

where each block is $n \times n$. P and Q are the positive definite grammians of the stable system G as is easily shown examining the (1,1)-blocks of (7.104.a) and (7.105.a). From the (1,1)-block of $P_e Q_e - \gamma^2 I = 0$ we obtain $P_{12} Q_{21} = \gamma^2 I - PQ$. Note that because γ is not a singular value of G ($\gamma \notin \sigma(G) = \sqrt{\lambda}(PQ)$) $\gamma^2 I - PQ$ is nonsingular such that $P_{12} Q_{21}$ has rank n. Hence, P_{12} as well as Q_{21} are nonsingular, too. Now we may choose the basis of the system E such that P_{21} is the identity matrix. Consequently, we obtain

$$Q_{12} = \gamma^2 I - PQ = -M. \qquad (7.109)$$

Examining the (1,2)-block and the (2,1)-block of $P_e Q_e - \gamma^2 I = 0$ we obtain the two blocks P_{22} and Q_{22}. Finally, P_e and Q_e are given by

$$P_e = \begin{bmatrix} P & I \\ I & M^{-1}Q \end{bmatrix} \qquad Q_e = \begin{bmatrix} Q & -M \\ -M^T & PM \end{bmatrix}. \tag{7.110}$$

Now we can determine the system matrices of the system E. We begin with the system matrix $\hat{D}$. Conforming to (7.104.c) D_e must be unitary up to a factor γ. The obvious choice is

$$\hat{D} = \begin{bmatrix} D & \gamma I \\ \gamma I & 0 \end{bmatrix}. \tag{7.111}$$

This gives

$$D_e = D_e^T = \begin{bmatrix} 0 & \gamma I \\ \gamma I & 0 \end{bmatrix}, \tag{7.112}$$

which fulfills (7.104.c). From the (1,1)-block of (7.104.b) we found that $\hat{B}$ is

$$\hat{B} = M^{-1}QB_a. \tag{7.113}$$

The (1,1)-block of (7.105.b) gives

$$\hat{C} = C_a P. \tag{7.114}$$

Finally, the (1,2)-block of (7.104.a) and the (2,1)-block of (7.105.a) yield two alternative expressions for the system matrix $\hat{A}$

$$\hat{A} = -(A^T + \hat{B}B_a^T) \qquad \text{and} \qquad \hat{A} = -M^{-1}(A^T M + C_a^T \hat{C}). \tag{7.115}$$

$\hat{A}$ is $n \times n$ and of full rank so that the system $\hat{G}$ is of McMillan degree n. Moreover, it can be shown that $\hat{A}$ has no eigenvalues on the imaginary axis. The most remarkable fact about $\hat{A}$ is that the number of stable poles is equal to the number of negative eigenvalues of $M = QP - \gamma^2 I$. That is, if $\sigma_r(G) > \gamma > \sigma_{r+1}(G)$, where $r = 1, ..., n$, then $\hat{A}$ has r eigenvalues in the open left half-plane C_- and $n - r$ eigenvalues in the open right half-plane C_+. Hence, the stable part of $\hat{G}$ belongs to $H_\infty(C_+)$. However, the size of E and of $\hat{G}$ is $2m \times 2m$ and not $m \times m$ as for the original system G. As can be shown, the reduced-order model G_r is obtained from the stable part of the (m,m)-block of $\hat{G}$, say $\hat{G}_m$. Moreover, it turns out that $\|G - \hat{G}_m\|_\Gamma < \gamma$. Hence, a suboptimal Hankel-norm approximation of G of McMillan degree r, say G_r, is given by the stable part of the system described by the state-space realization

$$\begin{aligned} \hat{A}_m &= -(A^T + M^{-1}QBB^T) \qquad \text{or} \qquad \hat{A}_m = -M^{-1}(A^T M + C^T C P) \\ \hat{B}_m &= M^{-1}QB \qquad \hat{C}_m = CP \qquad \hat{D}_m = D \end{aligned} \tag{7.116}$$

Both expressions defining the matrix $\hat{A}$ turns out to be identical when the Lyapunov equations of the system G are considered. Then the system $\hat{G}_m$ is defined by the system matrices

$$\hat{A}_m = -M^{-1}(QAP + \gamma^2 A^T) \qquad \hat{B}_m = M^{-1}QB \qquad \hat{C}_m = CP \qquad \hat{D}_m = D \tag{7.117}$$

The stable part of this system is the suboptimal Hankel-norm approximation G_r of G of McMillan degree r. If γ is chosen in such a way that $\sigma_r(G) > \gamma > \sigma_{r+1}(G)$ then the error of this system with respect to the Hankel norm is

$$\gamma > \|G - G_r\|_\Gamma > \sigma_{r+1}(G). \tag{7.118}$$

Let us apply the theory given before to construct the suboptimal Hankel-norm approximation G_r of the state-space realization of the truncated Laurent series given in section 7.2.2. The observability grammian P of this discrete-time system is equal to the identity matrix and the controllability grammian is equal to $Q = \Gamma_H^T \Gamma_H$. Now, if the discrete-time system is mapped to its related continuous-time system using the Möbius transformation (7.28) the grammians don't change so that $P = I$ and $Q = \Gamma_H^T \Gamma_H$. Moreover, the system matrices of the continuous-time system are given by (7.37) so that the system matrices (7.117) become

$$
\begin{aligned}
\hat{A}_m &= -M^{-1}(QA + \gamma^2 A^T) \\
\hat{B}_m &= M^{-1}QB \\
\hat{C}_m &= C \\
\hat{D}_m &= D
\end{aligned}
\tag{7.119}
$$

where $M = \Gamma_H^T \Gamma_H - \gamma^2 I$. Observe that the simple structure of A and B allows to compute very quickly the products QA and QB.

Now let us propose an algorithm to compute the suboptimal Hankel-norm approximation of the system (7.33):

1. Choose an upper bound for the approximation error with respect to the Hankel norm, say γ.
2. Compute the observability grammian $W_o = \Gamma_H^T \Gamma_H$.
3. Compute the system matrices $\{\hat{A}_m, \hat{B}_m, \hat{C}_m, \hat{D}_m\}$ conforming to (7.119) using the system matrices (7.37).
4. Perform a similarity transformation of the system $\{\hat{A}_m, \hat{B}_m, \hat{C}_m, \hat{D}_m\}$ in such a way that $\hat{A}_m$ is in the diagonal form and extract the stable part $\{A_r, B_r, C_r, D_r\}$. This is the state-space realization of the suboptimal Hankel-norm approximant G_r.

Remarks:

- Observe that $M = PQ - \gamma^2 I$ is positive definite if $\gamma < \sigma_n$, negative definite if $\gamma > \sigma_1$ and indefinite if $\sigma_1 > \gamma > \sigma_n$, where σ_n and σ_1 are the smallest and largest singular values of the system G.
- The suboptimal Hankel-norm approximation G_r is constructed by transforming (7.116) to the diagonal form and then extracting the stable part. This approach works well only if the system matrix $\hat{A}$ has n distinct eigenvectors. However, this is the rule when working with matrix-valued DtN-maps. Otherwise, the matrix $\hat{A}$ must be transformed in the Schur form.
- There exists a theory of suboptimal Hankel-norm approximation of discrete-time systems too. However, the development of the state-space description of the approximant G_r is not so simple as that of continuous-time systems. This is because the state-space realization of the adjoint system of a discrete-time system is much more involved (see section 7.4.2). For further details of the Hankel-norm approximation of discrete-time systems we refer to [CC92].

7.5.3 Optimal Hankel-norm approximation

The theory developed in the previous section allows to construct the suboptimal Hankel-norm approximation of a finite-dimensional system G. The optimal Hankel-norm approximants G_r of McMillan degree at most r is characterized by the fact that its distance to G with respect to the Hankel norm is $\|G - G_r\|_\Gamma = \sigma_{r+1}(G)$. That is, $\gamma = \sigma_{r+1}(G)$. Assume that $\sigma_{r+1}(G)$ is of multiplicity p. Then $M = PQ - \gamma^2 I$ has rank $n - p$ and is therefore singular. As a consequence, the inverse of M doesn't exist so that the equations in (7.117) are not directly applicable for the computation of optimal Hankel-norm approximations. However, introducing a slight modification, we can still follow the general route described in the foregoing section.

Let us partition the grammians according to

$$P = \begin{bmatrix} P_1 & 0 \\ 0 & \gamma I_p \end{bmatrix} \qquad Q = \begin{bmatrix} Q_1 & 0 \\ 0 & \gamma I_p \end{bmatrix}. \qquad (7.120)$$

$\gamma = \sigma_{r+1}(G)$ and P_1 as well as Q_1 are positive definite and of rank $q = n - p$. Choosing the dimension of $\hat{A}$ to be $q = n - p$, we define the $2n - p \times 2n - p$ grammians P_e and Q_e as

$$P_e = \begin{bmatrix} P_1 & 0 & I_q \\ 0 & \gamma I_p & 0 \\ I_q & 0 & M_q^{-1} \end{bmatrix} \qquad Q_e = \begin{bmatrix} Q_1 & 0 & -M_q \\ 0 & \gamma I_p & 0 \\ -M_q^T & 0 & \Sigma_q^2 M_q \end{bmatrix} \qquad (7.121)$$

where $M_q = Q_1 P_1 - \gamma^2 I_q$. By inspection we find that $P_e Q_e = Q_e P_e = \gamma^2 I$ so that (7.107) is fulfilled.

Similarly as for the suboptimal case, we first imbed the system G in the system G_a of dimension $2m - p \times 2m - p$. The dimension of the system $\hat{G}$ is equal to that of G_a. A state-space realization of the allpass system $E = G_a - \hat{G}$ which is partitioned conforming to (7.121) is

$$A_e = \begin{bmatrix} A_{11} & A_{12} & 0 \\ A_{21} & A_{22} & 0 \\ 0 & 0 & \hat{A} \end{bmatrix} \qquad B_e = \begin{bmatrix} B_1 & 0 \\ B_2 & 0 \\ \hat{B}_1 & \hat{B}_2 \end{bmatrix} \qquad C_e = \begin{bmatrix} C_1 & C_2 & -\hat{C}_1 \\ 0 & 0 & -\hat{C}_2 \end{bmatrix} \qquad D_e = \begin{bmatrix} D_{11} & D_{12} \\ D_{21} & D_{22} \end{bmatrix}. \quad (7.122)$$

Where A_e is of dimension $2n - p \times 2n - p$, B_e is $2n - p \times 2m - p$, C_e is $2m - p \times 2n - p$ and D_e is $2m - p \times 2m - p$ with D_{11} of dimension $2m \times 2m$. Evaluating the blocks similarly as in the previous section, we obtain the following state-space realization of the system $\hat{G}$

$$\hat{A} = -\left(A_{11}^T + \hat{B}\begin{bmatrix} B_1^T \\ 0 \end{bmatrix}\right) \qquad \hat{A} = -M_q^{-1}\left(A_{11}^T M_q + \begin{bmatrix} C_1 & 0 \end{bmatrix}\hat{C}\right)$$

$$\hat{B} = M_q^{-1}\left(Q_1\begin{bmatrix} B_1 & 0 \end{bmatrix} + \begin{bmatrix} C_1^T & 0 \end{bmatrix}D_e\right) \qquad (7.123)$$

$$\hat{C} = \begin{bmatrix} C_1 \\ 0 \end{bmatrix}P_1 + D_e\begin{bmatrix} B_1^T \\ 0 \end{bmatrix}$$

$$\hat{D} = D_a - D_e$$

To complete the state-space realization of $\hat{G}$ we must construct the matrix D_e. Its determination is more involved than for the suboptimal case because evaluating the (1,2)-block of (7.104.b) we obtain a second condition for D_e

$$\begin{bmatrix} C_2^T & 0 \end{bmatrix}D_e + \gamma\begin{bmatrix} B_2 & 0 \end{bmatrix} = 0. \qquad (7.124)$$

Remember that D_e must fulfill $D_e D_e^T = \gamma^2 I$. From the (2,2)-block of (7.104.a) we obtain

$$C_2^T C_2 = B_2 B_2^T \tag{7.125}$$

The last equation implies that there exists an $m \times m$ matrix U such that

$$B_2 = C_2^T U \tag{7.126}$$

Furthermore, if $B_2 B_2^T$ has rank p then U has rank p with singular value decomposition

$$U = \begin{bmatrix} Y_1 & Y_2 \end{bmatrix} \begin{bmatrix} I_p & 0 \\ 0 & 0 \end{bmatrix} \begin{bmatrix} Z_1 \\ Z_2 \end{bmatrix}. \tag{7.127}$$

One such realization of the matrix U is

$$U = C_2 (B_2 B_2^T)^{-1} B_2 \tag{7.128}$$

as can be verified by inspection. Finally, D_e is defined as

$$D_e = -\gamma \begin{bmatrix} U & Y_2 \\ Z_2^T & 0 \end{bmatrix} \tag{7.129}$$

D_e fulfills $D_e D_e^T = \gamma^2 I$ as well as (7.124). Finally, by definition, we obtain

$$\hat{D} = \begin{bmatrix} D + \gamma U & \gamma Y_2 \\ \gamma Z_2^T & 0 \end{bmatrix} \tag{7.130}$$

Now the state-space realization of $\hat{G}$ is complete. The optimal Hankel-norm approximation of G is the stable part of the left-upper (m,m)-block of the system $\hat{G}$. Combining the Lyapunov equations of the system G with the submatrices of (7.123) defining the left-upper (m,m)-block we obtain the state-space realization of $\hat{G}_m$:

$$\begin{aligned}
\hat{A}_m &= -M_q^{-1}(Q_1 A_{11} P_1 + \gamma^2 A_{11}^T + \gamma C_1^T U B_1) \\
\hat{B}_m &= M_q^{-1}(Q_1 B_1 - \gamma C_1^T U) \\
\hat{C}_m &= C_1 P_1 - \gamma U B_1^T \\
\hat{D}_m &= D_{11} + \gamma U
\end{aligned} \tag{7.131}$$

The stable part of this system is the optimal Hankel-norm approximation of G. The equations (7.131) are similar to those defining the suboptimal Hankel-norm approximation (7.117). The most significant difference is that all system matrices have an additional term which is proportional to the singular value $\gamma = \sigma_{r+1}$. Observe that even $\hat{D}_m$ has such a term. Because of this term G_r doesn't coincide with G at infinity so that $\lim_{\omega \to \infty} \| G(\omega) - \hat{G}_m(\omega) \|_{L_\infty(i\mathbb{R})} = \sigma_{r+1}$. However, this perturbation is small if σ_{r+1} is small.

The equations of the state-space matrices are valid even when the system is in input normal or in output normal realization. When computing the matrix U, we have only to take care that (7.125) changes to $\gamma^2 C_2^T C_2 = B_2 B_2^T$ if the system is in input normal and to $C_2^T C_2 = \gamma^2 B_2 B_2^T$ if it is in the output normal form. Because the system associated with the truncated Laurent series is in the input normal form we can use this property to slightly reduce the computational cost of the approximation. Moreover, the singular value σ_{r+1} is usually simple so that the dimension of the matrix $\hat{A}_m$ is $n - 1$. Furthermore, the matrix U is of rank one and is given by

$$U = C_2 B_2 / \gamma (B_2 B_2^T).$$ (7.132)

Nevertheless, the suboptimal Hankel-norm approximation is still numerically more efficient than the optimal Hankel-norm approximation because we are not obliged to isolate the state variable associated with the singular value σ_{r+1}. Furthermore, the suboptimal is virtually as accurate as the optimal Hankel-norm approximation because σ_{r+1} is small.

Let us describe an algorithm to compute the optimal Hankel-norm approximation of the system (7.33):

1. Choose an upper bound for the approximation error with respect to the Hankel norm, say ε.

2. Setup the observability grammian $W_o = \Gamma_H^T \Gamma_H$ and compute their eigenvalues Λ. Sort the eigenvalues and eigenvectors such that $\lambda_1 \geq \lambda_2 \geq \ldots \geq \lambda_n$. Perform a similarity transformation of the system (7.37) such that the observability grammian is diagonal: $W_o = \Lambda$.

3. Choose $\gamma = \sigma_{r+1} = \sqrt{\lambda_r}$ such that $\varepsilon < 2(\sigma_{r+1} + \ldots + \sigma_p)$. This defines the McMillan degree r of the optimal Hankel-norm approximation. Delete from the state-space matrices the state variable associated with the singular value $\gamma = \sigma_{r+1}$.

4. Compute the system matrices $\{\hat{A}_m, \hat{B}_m, \hat{C}_m, \hat{D}_m\}$ using (7.131).

5. Perform a similarity transform of the system $\{\hat{A}_m, \hat{B}_m, \hat{C}_m, \hat{D}_m\}$ in such a way that $\hat{A}_m$ is in the diagonal form and extract the stable part $\{A_r, B_r, C_r, D_r\}$. This is the state-space realization of the suboptimal Hankel-norm approximation G_r.

It can be shown that the error of the optimal Hankel-norm approximation with respect to the infinity norm is bounded by twice the tail of the singular values of the system G. That is, $\|G(\omega) - G_r(\omega)\|_{L_\infty(iR)} \leq 2(\sigma_{r+1} + \ldots + \sigma_p)$, where p is the number of distinct singular values of G. Choosing an appropriate system matrix D, this error can be reduced to the half, that is $\|G(\omega) - G_r(\omega)\|_{L_\infty(iR)} \leq (\sigma_{r+1} + \ldots + \sigma_p)$. However, this is of little interest for our approximation problem because we are more interested to have exact matching at $\omega \to \infty$.

7.6   An example

The aim of this example is to illustrate the approximation methods developed in this chapter. We will apply these methods on a simple mechanical system which can be easily reduced applying the standard modal truncation method.

Consider the two degree of freedom system shown in Figure 7.1 (left). The equations of motion are given by the system of differential equations of second order

$$\begin{bmatrix} 1 & 0 \\ 0 & \mu \end{bmatrix} \begin{bmatrix} \ddot{x}_1 \\ \ddot{x}_2 \end{bmatrix} + \begin{bmatrix} 2\zeta_1 & -2\zeta_1 \\ -2\zeta_1 & 2\zeta_1(1 + \eta\mu\Omega^2) \end{bmatrix} \begin{bmatrix} \dot{x}_1 \\ \dot{x}_2 \end{bmatrix} + \begin{bmatrix} 1 & -1 \\ -1 & 1 + \mu\Omega^2 \end{bmatrix} \begin{bmatrix} x_1 \\ x_2 \end{bmatrix} = \begin{bmatrix} u_1 \\ u_2 \end{bmatrix}$$ (7.133)

The dot refers to a differentiation with respect to the nondimensional time $\tau = \omega_1 t$, where $\omega_1 = \sqrt{k_1/m_1}$. The input variables are defined by $u_1 = f_1/(m_1\omega_1^2)$, $u_2 = f_2/(m_1\omega_1^2)$. The nondimensional parameters are $\mu = m_1/m_2$, $\Omega = \omega_2/\omega_1$, $\eta = \zeta_2/\zeta_1$, where $\omega_2 = \sqrt{k_2/m_2}$, $\zeta_1 = c_1/(2m_1\omega_1)$ and $\zeta_2 = c_2/(2m_2\omega_2)$. As output variable v we choose the velocity vector so that $v^T = \dot{x}^T = [\dot{x}_1, \dot{x}_2]$. Now let us define the matrices B_2 and C_2 as

$$B_2 = \begin{bmatrix} 1 & 0 \\ 0 & 1 \end{bmatrix} \qquad C_2 = \begin{bmatrix} 1 & 0 \\ 0 & 1 \end{bmatrix}.$$ (7.134)

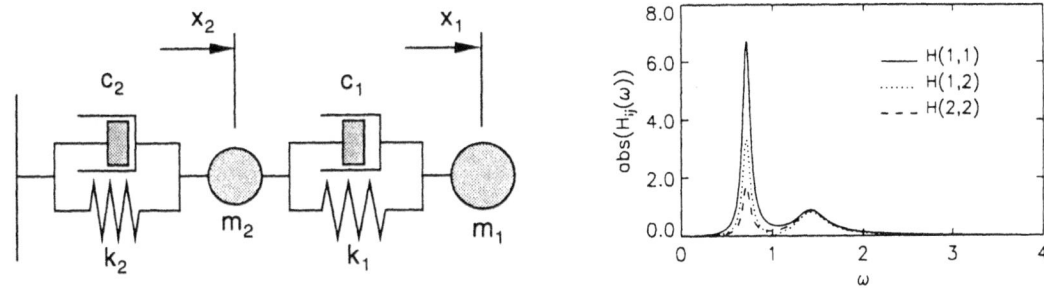

Figure 7.1: Left: Two degrees of freedom system. Right: Components of the transfer function (7.136)

Then, the linear system associating the input vector $u^T = [u_1 \ u_2]$ to the output vector v is

$$M\ddot{x} + C\dot{x} + Kx = Bu$$
$$v = C\dot{x}$$

(7.135)

where the matrices M, C and K are defined by the equations (7.133). With the Laplace transform of (7.135) we obtain the transfer matrix $H(s) = sC(Ms^2 + Cs + K)^{-1}B$. Fixing the parameters to $\mu = 2, \Omega = 1, \eta = 1$ and $\zeta_1 = 1/10$ we obtain

$$H(s) = \frac{s}{2s^4 + s^3 + 5.08s^2 + 0.8s + 2} \begin{bmatrix} 2s^2 + 0.6s + 3 & \dfrac{s+5}{5} \\ \dfrac{s+5}{5} & \dfrac{5s^2 + s + 5}{5} \end{bmatrix}$$

(7.136)

Note that the transfer matrix is symmetric. Furthermore, $H(0) = 0$ and $\lim_{s \to \infty} H(s) = 0$. The absolute values of the components of $H(i\omega)$ are given in Figure 7.1 (right).

The equations (7.135) can be formulated in standard state-space formulation as a system of differential equations of first order. Define the variable y by $\dot{x} = M^{-1}y$. Then, introducing the state vector $z^T = [x^T \ y^T]$ we obtain the system matrices

$$A = \begin{bmatrix} 0 & M^{-1} \\ -K & -CM^{-1} \end{bmatrix} \qquad B = \begin{bmatrix} 0 \\ B_2 \end{bmatrix} \qquad C = [0 \ C_2 M^{-1}] \qquad D = 0$$

(7.137)

This system has four state variables. Clearly, its transfer function is given by (7.136). Moreover, the system (7.137) is observable and controllable.

Now, the dominance of the first peak of the transfer function shown in Figure 7.1 (right) suggests to approximate the system (7.137) with one having only two state variables. First, let us consider the usual modal truncation procedure of structural dynamics. With the eigenvectors of the system matrix A we can define a similarity transformation such that A transforms to

$$\hat{A} = \begin{bmatrix} -0.050 & 0.705 & 0.0 & 0.0 \\ -0.705 & -0.050 & 0.0 & 0.0 \\ 0.0 & 0.0 & -0.2 & 1.4 \\ 0.0 & 0.0 & 1.4 & -0.2 \end{bmatrix}$$

(7.138)

In this form, the system defined by the left upper 2×2 block is independent of that defined by the right lower 2×2 block. The damping coefficient of the last system is $\zeta_2 = 0.2$ and is significantly larger than that of the first system which is $\zeta_1 = 0.05$. Therefore, we delete the system defined by the right lower 2×2 block. This gives the following state-space matrices

$$A_t = \begin{bmatrix} -0.050 & 0.705 \\ -0.705 & -0.050 \end{bmatrix} \qquad B_t = \begin{bmatrix} -0.890 & -0.445 \\ -0.527 & -0.263 \end{bmatrix} \qquad C_t = \begin{bmatrix} -0.531 & -0.368 \\ -0.266 & -0.184 \end{bmatrix} \qquad (7.139)$$

Its transfer matrix is given by

$$H_t(s) = \frac{1}{(s^2 + 0.1s + 0.5)6} \frac{1}{6} \begin{bmatrix} 4s & 2s \\ 2s & s \end{bmatrix} \qquad (7.140)$$

As is shown in Figure 7.2, the reduced-order model captures essentially the dominant mode. The second one is totally neglected.

Next, we apply the balanced truncation procedure. First we have to compute the observability and controllability grammians W_c and W_o. This is performed using the Lyapunov equations. However, for this small BIBO stable system the grammians W_c and W_o can be computed using a much simpler brute force procedure applying the equations (5.97.a) and (5.97.b) and stopping the summation after a finite number of terms. The state matrices of the related discrete-time system are obtained applying (5.147) with the Möbius transformation (5.134). These are given by

$$\begin{aligned} \tilde{A} &= (aI - A)^{-1}(aI + A) \\ \tilde{B} &= \sqrt{2a}(aI - A)^{-1}B \\ \tilde{C} &= \sqrt{2a}C(aI - A)^{-1} \end{aligned} \qquad (7.141)$$

The grammians W_c, W_o of the system (7.137) are

$$W_c = \begin{bmatrix} 5.740 & 2.628 & 0.0 & 0.204 \\ 2.628 & 1.505 & -0.102 & 0.0 \\ 0.0 & -0.102 & 3.092 & 2.204 \\ 0.204 & 0.0 & 2.204 & 3.816 \end{bmatrix} \qquad W_o = \begin{bmatrix} 1.990 & 2.551 & -0.102 & -0.102 \\ 2.551 & 3.520 & 0.306 & 0.102 \\ -0.102 & 0.306 & 3.092 & 1.102 \\ -0.102 & 0.102 & 1.102 & 0.954 \end{bmatrix} \qquad (7.142)$$

The singular values are the square roots of the eigenvalues of the product $W_o W_c$ of the grammians. These are

$$s_1 = 4.165 \qquad s_2 = 4.159 \qquad s_3 = 0.831 \qquad s_4 = 0.825 \qquad (7.143)$$

The similarity matrix T is given by

$$T = \begin{bmatrix} 1.161 & -0.392 & 0.016 & -0.024 \\ 0.572 & 0.417 & -0.017 & -0.011 \\ -0.011 & -0.033 & 0.824 & 0.568 \\ 0.024 & -0.033 & 0.804 & -1.165 \end{bmatrix} \qquad (7.144)$$

transforms the system (7.137) to the balanced realization form. That is, the grammians W_c and W_o are diagonal with components equal to the singular values (7.143). The subsystem of the transformed system (7.137) associated with the two larger singular values are

$$A_b = \begin{bmatrix} 0.0 & 0.707 \\ -0.707 & -0.101 \end{bmatrix} \qquad B_b = \begin{bmatrix} 0.011 & -0.012 \\ 0.824 & 0.402 \end{bmatrix} \qquad C_b = \begin{bmatrix} -0.011 & 0.824 \\ 0.012 & 0.402 \end{bmatrix} \qquad (7.145)$$

and $D_b = 0$. Its associated transfer matrix is given by

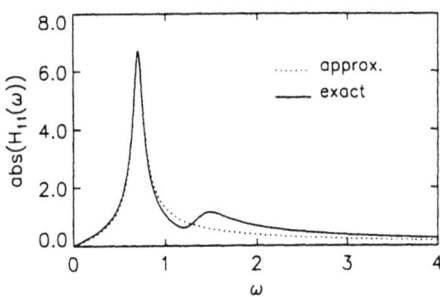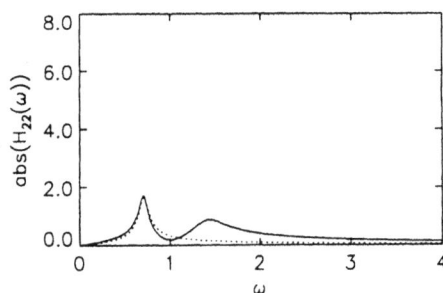

Figure 7.2: Approximation applying the modal truncation method

$$H_b(s) = \frac{1}{s^2 + 0.101s + 0.499}\begin{bmatrix} 0.679s - 0.013 & 0.331s + 0.004 \\ 0.331s + 0.004 & 0.161s + 0.007 \end{bmatrix} \qquad (7.146)$$

It is very similar to that obtained by the modal truncation procedure (see Figure 7.3 for a confirm). The coefficients of the denominator polynomial are equal up to the second significant digit. Those of the numerator polynomials differ beginning with the second digit. However, at $s = 0$ the transfer matrix (7.146) doesn't vanish. This is typical of the balanced truncation method which is exact only for $s \to \infty$.

The singular balanced perturbation method is more appropriate when perfect matching at $s = 0$ is essential. First, the system (7.137) in the balanced realization form is mapped via the formulas (7.50) to its related system. This is in the balanced realization form too and their grammians are given by (7.142). The next step is to delete from this system the two state variables associated with the smallest singular values. This gives the truncated system

$$\tilde{A}_s = \begin{bmatrix} 0.201 & -1.415 \\ 1.415 & 0.000 \end{bmatrix} \qquad \tilde{B}_s = \begin{bmatrix} -1.161 & -0.572 \\ 0.016 & -0.017 \end{bmatrix} \qquad \tilde{C}_s = \begin{bmatrix} -1.161 & 0.016 \\ -0.572 & 0.017 \end{bmatrix} \qquad (7.147)$$

with $\tilde{D}_s = 0$. Finally, applying (7.50) to this system we obtain the approximant via the singular balanced perturbation method. Its state-space realization is

$$A_s = \begin{bmatrix} 0.0 & 0.707 \\ -0.707 & -0.100 \end{bmatrix} \qquad B_s = \begin{bmatrix} -0.011 & -0.012 \\ -0.819 & -0.406 \end{bmatrix} \qquad C_s = \begin{bmatrix} 0.011 & -0.819 \\ -0.012 & -0.406 \end{bmatrix}$$

$$D_s = \begin{bmatrix} -0.027 & -0.007 \\ -0.007 & 0.014 \end{bmatrix} \qquad (7.148)$$

Note that the matrix D_s doesn't vanish. The transfer function is given by

$$H_s(s) = \frac{1}{s^2 + 0.100s + 0.499}\begin{bmatrix} 0.670s - 0.013 & 0.332s + 0.004 \\ 0.332s + 0.004 & 0.164s + 0.007 \end{bmatrix} + \begin{bmatrix} -0.027 & -0.007 \\ -0.007 & 0.014 \end{bmatrix} \qquad (7.149)$$

and is very similar to that obtained using the balanced truncation method. However, this approximation vanish at $s = 0$.

The suboptimal Hankel-norm approximation is computed with the equations (7.117) which define the subsystem of $\hat{G}$ containing the approximation of McMillan degree two. Let $\gamma = 0.9$. Then, with the grammians W_c and W_o given in (7.142) we obtain the system matrix $\hat{A}$ which can be transformed to

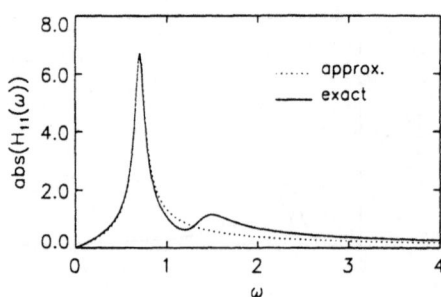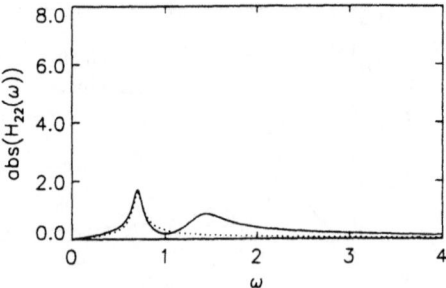

Figure 7.3: Approximation applying the balanced truncation approximation method

$$
\hat{A} = \begin{bmatrix} 4.540 & 0 & 0 & 0 \\ 0 & -0.054 & 0.703 & 0 \\ 0 & -0.703 & -0.054 & 0 \\ 0 & 0 & 0 & 0.455 \end{bmatrix}
\tag{7.150}
$$

In conformity with the theory, this matrix has two instable state variables: the first and the fourth state variable. The poles of these state variables are both real and are given by $p_1 = 4.540$ and $p_2 = 0.455$. The second and the third state variables define a stable system. With these state variables the reduced-order system is given by the system matrices

$$
A = \begin{bmatrix} -0.054 & 0.703 \\ -0.703 & -0.054 \end{bmatrix} \qquad B = \begin{bmatrix} -0.058 & 0.041 \\ 0.151 & 0.077 \end{bmatrix} \qquad C = \begin{bmatrix} -2.129 & 3.590 \\ -0.901 & 2.048 \end{bmatrix}
\tag{7.151}
$$

The transfer function of this system is

$$
H_h(s) = \frac{1}{s^2 + 0.109s + 0.497} \begin{bmatrix} 0.667s - 0.043 & 0.363s + 0.008 \\ 0.363s + 0.008 & 0.194s + 0.020 \end{bmatrix}
\tag{7.152}
$$

Figure 7.3 shows that the suboptimal Hankel-norm approximant underestimates the peak value of the component H_{11} and slightly overestimates that of the component H_{22} of the original transfer function. Similar to the system obtained with the balanced truncation procedure, the transfer function (7.152) doesn't vanish at $s = 0$. The error is even larger.

Finally, we shall compute the optimal Hankel-norm approximation of the system (7.137). We choose $\gamma = s_3 = 0.8308$. First, we transform the system (7.137) in the balanced realization form applying the similarity matrix (7.144). Then, the state variable associated with the singular value s_3 is deleted. Finally, the formulas (7.131) is applied to obtain the state matrices of the system $\hat{G}_m$. After an additional similarity transformation, the matrix $\hat{A}_m$ becomes

$$
\hat{A}_m = \begin{bmatrix} 0.223 & 0 & 0 \\ 0 & -0.037 & 0.694 \\ 0 & -0.694 & -0.037 \end{bmatrix}
\tag{7.153}
$$

As predicted, one of the state variables of the system $\hat{G}_m$ – the first one – is unstable because of the real positive pole $p_1 = 0.223$. This state variable has to be deleted from the state matrices $\hat{A}_m, \hat{B}_m, \hat{C}_m$. Next, we have to define the system matrix $\hat{D}_m$. The submatrices B_2 and C_2 of the system (7.137) which is in the balanced realization form and associated with the singular value s_3 are

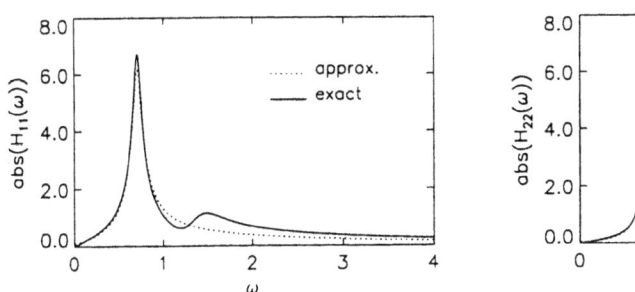

Figure 7.4: Approximation applying the suboptimal Hankel-norm approximation method

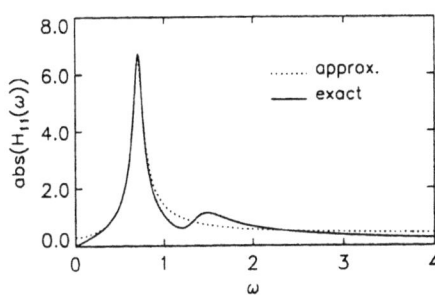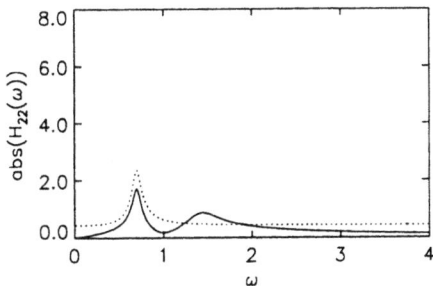

Figure 7.5: Approximation applying the optimal Hankel-norm approximation method

$$B_2 = [0.568, -0.582] \qquad C_2^T = [0.568, -0.582] \tag{7.154}$$

Hence, $B_2 = C_2^T$. The matrix U is of rank one and is defined by $U = C_2 B_2 / (B_2 B_2^T)$. Therefore, we obtain $\hat{D}_m = \gamma U$. Finally, the stable system defining the optimal Hankel-norm approximation are given by the state matrices

$$A_o = \begin{bmatrix} -0.053 & 0.701 \\ -0.701 & -0.053 \end{bmatrix} \qquad B_o = \begin{bmatrix} 0.106 & 0.066 \\ -0.160 & -0.085 \end{bmatrix} \qquad C_o = \begin{bmatrix} 2.394 & -2.452 \\ 1.232 & -1.466 \end{bmatrix}$$

$$D_o = \begin{bmatrix} 0.405 & -0.415 \\ -0.415 & 0.426 \end{bmatrix} \tag{7.155}$$

The transfer function is

$$H_o(s) = \frac{1}{s^2 + 0.107s + 0.484} \begin{bmatrix} 0.647s - 0.052 & 0.365s + 0.010 \\ 0.365s + 0.010 & 0.205s + 0.005 \end{bmatrix} + \begin{bmatrix} 0.405 & -0.415 \\ -0.415 & 0.426 \end{bmatrix} \tag{7.156}$$

Observe that this transfer function is significantly different from those obtained by the four other approximation methods (see Figure 7.5). This is mainly because of the matrix D_o which produces a significant shift of the transfer function. The effect of this shift on the infinity norm is shown in Figure 7.6. The optimal Hankel-norm approximation has an error with respect to the infinity norm which is nearly half in magnitude as those obtained with the other approximation methods. The error of these methods are mainly generated by the truncation of the second mode. Figure 7.7 shows that the largest singular values have essential the same dominant peak at the eigenfrequency of the second mode. In contrast, the error curve of the optimal Hankel-norm approximant is nearly constant except in the neighborhood of the eigenfrequency of the second mode.

approximation method	Hankel norm	infinity norm
modal truncation	0.8333	1.666
balanced truncation	0.8348	1.662
singular balanced truncation	0.8329	1.649
suboptimal Hankel-norm, $\gamma = 0.9$	0.8314	1.658
optimal Hankel-norm, $\gamma = 0.8308$	0.8308	0.8288

Figure 7.6: Errors with respect to the Hankel and infinity norms

The error with respect to the infinity norm of the other approximation methods can be improved by adding adequate state matrices D to the transfer functions. In fact, by adding the matrix $D = 0.833 I_2$ to the transfer function obtained with the modal truncation method reduces its error with respect to the infinity norm to 0.833. This error is not significantly larger than that obtained with the optimal Hankel-norm approximation method.

The effect of the shift is overemphasized in this example because the magnitude of the singular value s_3 is relatively large compared to the magnitude of the infinity norm of the transfer function of the original system ($\|H\|_{L_-(iR)} = 8.324$). When infinite-dimensional systems are approximated, the singular value σ_{r+1} defining the reduced-order models are orders of magnitude smaller than the infinity norm of the infinite-dimensional system so that the shift induced by $\sigma_{r+1} U$ is insignificant. Observe, that the error with respect to the Hankel norm remains unchanged by this modification. Note that the approximant via singular balanced truncation is the second best accurate with respect to the infinity norm. Furthermore, observe that the errors with respect to the infinity norm are much smaller than the estimate $\varepsilon = 2(s_3 + s_4) = 3.312$.

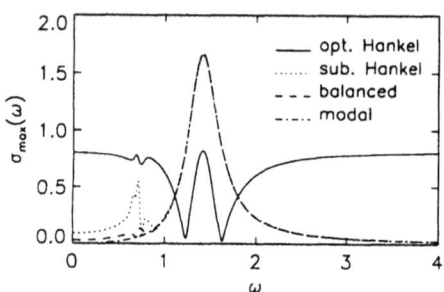

Figure 7.7: Errors of the approximations with respect to the infinity norm

Obviously, the optimal Hankel-norm approximant is the best approximation with respect to the Hankel norm. It is followed by the approximant obtained via the suboptimal Hankel-norm approximation and that via the singular balanced truncation method. The approximant obtained by the balanced truncation method is the less accurate with respect to the Hankel norm. It is even less accurate than that obtained via modal truncation approximation method. However, the errors differ only in the third significant digit.

All methods described in this chapter are accurate and robust when applied to the approximation of DtN-map kernels The numerically less effective method is the optimal Hankel-norm approximation method because it requires the solution of two eigenvalue problems. The suboptimal Hankel-norm approximation method requires the computation of an unsymmetric eigenvalue problem to extract the stable part. In contrast, the balanced and singular balanced truncation methods needs the computation of a more agreeable symmetric and positive-definite eigenvalue problem. This is undoubtedly a great advantage. The singular balanced truncation method is recommended when high accuracy at low frequencies is important. The balanced truncation method is suggested for accurate approximations at high frequencies. It is the cheapest numerical method and gives very accurate results in most practical applications.

Chapter 8
Symmetric realizations

In the previous chapters we developed methods to approximate DtN-maps with scalar and matrix-valued rational functions. These can be associated with systems of linear ordinary differential equations (ODE) of first order. In contrast, the equations of motion of a near field are usually described by systems of ODE of second order. In addition, the standard numerical time integration algorithms of structural dynamics are based on such systems of ODE of second order. Therefore, the systems of ordinary differential equations of first order describing absorbing boundary conditions can't be integrated in this scheme. Using a staggered solution algorithm these two systems of ODE could be solved separately using two different numerical integration algorithms. However, a much more elegant solution would be to represent an absorbing boundary condition by a system of ODE of second order instead of one of first order. In addition, if this system could be made symmetric it could be coupled with that describing the near-field so that the coupled system could finally be solved with standard time integration algorithms.

In this chapter, we develop a procedure to describe absorbing boundary conditions by symmetric systems of ODE of first and second order (symmetric realization). First, the procedure is introduced applying it to some absorbing boundaries of the restrained rod. We shall see that such symmetric realizations always exist and that they are generally not unique. That is, there are more than one symmetric realization. Then we develop a systematic theory. Finally, this technique is applied to construct so-called absorbing boundary elements.

8.1 Absorbing boundary conditions of the restrained rod

The motion of the thin restrained rod is given by a symmetric hyperbolic PDE of second order and a set of initial and boundary conditions

$$u_{,tt} - c^2 u_{,xx} + 2\varepsilon c u_{,t} + \kappa^2 u = f(t, x)$$
$$u(x, 0) = 0$$
$$u_{,t}(x, 0) = 0 \tag{8.1}$$
$$u_{,x}(t, 0) + h_L(t) = 0$$
$$u_{,x}(t, L) + B_t(u) = 0$$

The semi-discretisation of the initial boundary value problem (IBVP) (8.1) with isoparametric finite elements leads to an initial value problem (IVP) described by a n-dimensional system of ordinary differential equations (ODE), where n is the number of degrees of freedom $v(t)$ of the finite element model. That is

$$M\ddot{v} + C\dot{v} + Kv + Bv = F + e_{(n)}\sigma_n - e_{(1)}\sigma_1, \tag{8.2}$$

$$\sigma_1 + h_L = 0, \tag{8.3.a}$$

$$\sigma_n + B_t(u_n) = 0. \tag{8.3.b}$$

$h_L(t)$ is a loading function which acts at the left boundary, e_m is the mth basis vector and $B_t(u_n)$ is the boundary operator associated with the absorbing boundary condition at the right boundary. It acts on the variable u_n. We assume that the reader is familiar with the standard techniques of the finite-element method so that we can skip the details of the discretisation of the IBVP. The matrices M, C and K are called mass, damping and stiffness matrices, respectively. The vector $F(t)$ denote the external loadings acting on the restrained rod. The matrix B is the additional stiffness matrix associated with the elastic restrain. The matrices M, C and B are positive definite and the matrix K is positive semi-definite. The sum of the matrices K and B is a positive definite matrix. First, we implement the absorbing boundary conditions of the nondissipative restrained rod. The implementation of the absorbing boundary conditions of the dissipative restrained rod is described in section 8.1.4.

8.1.1   The asymptotic boundary condition of first order

The boundary operator of the asymptotic boundary condition of first order is

$$B_t(u_n) = \frac{1}{c}\dot{u}_n + \frac{c\kappa^2}{2}\int_0^t u_n(\tau)d\tau. \tag{8.4}$$

The first term on the right hand side is the viscous dashpot. It gives an additional contribution to the global damping matrix:

$$C_B = C + (1/c)E_{(n,n)}, \tag{8.5}$$

where the matrix $E_{(p,q)}$ is unity at the index pair (p, q) and zero elsewhere. The second term

$$\Delta\sigma_n(t) = -\frac{c\kappa^2}{2}\int_0^t u_n(\tau)d\tau \tag{8.6}$$

doesn't fit into the structure of the system of ODE given in (8.2). However, defining an auxiliary variable $\phi(t)$ by

$$\phi(t) = \alpha\int_0^t u_n(\tau)d\tau \tag{8.7}$$

we can formulate (8.6) as a differential equation. The constant α is inserted to enforce the symmetry of the equations of motion. Differentiating the last equation with respect to the time yields

$$\dot{\phi}(t) = \alpha u_n(\tau). \tag{8.8}$$

This is an additional ODE which has to be added to (8.2). Hence, the complete set of equations describing the absorbing boundary condition becomes

$$\sigma_n = -\frac{1}{c}\dot{u}_n - \frac{c\kappa^2}{2\alpha}\phi,\tag{8.9.a}$$

$$\dot{\phi}(t) = \alpha u_n(t).\tag{8.9.b}$$

Now we adjust the constant α so that the coefficients in front of $\phi(t)$ and $u_n(t)$ on the right hand side of the equations (8.9.a) and (8.9.b) are equal. This is achieved with

$$\alpha = \sqrt{\frac{c}{2}}\kappa.\tag{8.10}$$

However, the equations (8.9.a) and (8.9.b) with α given by (8.10) are still not symmetric because the signs in front of ϕ and u_n are different. We can enforce the symmetry multiplying (8.9.b) with -1:

$$\sigma_n(t) = -\frac{1}{c}\dot{u}_n(t) - \alpha\phi(t) \qquad -\dot{\phi}(t) = -\alpha u_n(t)\tag{8.11}$$

Obviously, the last operation doesn't change the mathematical content of the absorbing boundary condition. Now, (8.2) and (8.11) form a symmetric system of ODE. This can be solved by standard numerical time integration algorithms. Equation (8.11) is called a symmetric realization of the absorbing boundary condition.

Now let us define the new $n+1$-dimensional generalized displacement vector w. It consists of the n-dimensional displacement vector u and the auxiliary variable ϕ

$$w^T = [u^T, \phi].\tag{8.12}$$

Hence, the ODE system (8.2) transforms to

$$M_B\ddot{w} + C_B\dot{w} + K_B w = F_B(t),\tag{8.13}$$

with system matrices and right hand side vector

$$M_B = \begin{bmatrix} M & 0 \\ 0 & 0 \end{bmatrix} \qquad C_B = \begin{bmatrix} C + \frac{1}{c}E_{(n,n)} & 0 \\ 0 & -1 \end{bmatrix} \qquad K_B = \begin{bmatrix} K + B & \alpha e_{(n)} \\ \alpha e_{(n)}^T & 0 \end{bmatrix} \qquad F_B = \begin{bmatrix} F \\ 0 \end{bmatrix}\tag{8.14}$$

We call (8.13) the enlarged system. Observe that the matrices of the enlarged system are not positive definite.

The absorbing boundary conditions of Engquist and Halpern [EH88] can be cast in the same form. These are given in the general form

$$\sigma_n = -\frac{1}{c}\dot{u}_n - \delta u_n - \beta c\kappa^2 \int_0^t e^{-\gamma(t-\tau)} u_n(\tau)d\tau\tag{8.15}$$

Performing a Laplace transform of (8.15) the construction a symmetric realization is reduced to a nonlinear algebraic problem which is usually easy to solve. The Laplace transform of the absorbing boundary condition (8.15) has the form

$$\sigma_n = -\frac{s}{c}u_n - \delta u_n - \beta c \kappa^2 \frac{\alpha}{(s+\gamma)}u_n \tag{8.16}$$

As before, we introduce the new auxiliary variable ϕ:

$$\phi = \frac{\alpha}{(s+\gamma)}u_n . \tag{8.17}$$

After a few straightforward manipulations we find the symmetric realization

$$\sigma_n = -\frac{1}{c}\dot{u}_n - \delta u_n - \alpha\phi , \tag{8.18.a}$$

$$-\dot{\phi} - \gamma\phi = -\alpha u_n , \tag{8.18.b}$$

with

$$\alpha = \sqrt{c\beta}\kappa . \tag{8.19}$$

Observe, that α is real only if β is positive. The associated enlarged system has the matrices

$$M_B = \begin{bmatrix} M & 0 \\ 0 & 0 \end{bmatrix} \quad C_B = \begin{bmatrix} \frac{1}{c}E_{(n,\,n)} & 0 \\ 0 & -1 \end{bmatrix} \quad K_B = \begin{bmatrix} K + B + \delta E_{(n,\,n)} & \alpha e_{(n)} \\ \alpha e_{(n)}^T & -\gamma \end{bmatrix} , \tag{8.20}$$

where the parameter α is given by (8.19).

Remarks:

- Observe that the boundary conditions (8.4) and (8.16) are nonlocal in time. Introducing the auxiliary variable $\phi(t)$ we obtain a representation which is local in time.
- The initial condition of the auxiliary variable $\phi(t)$ is $\phi(0) = 0$. This follows directly from (8.6) and the convolution integral in (8.15).
- The symmetry condition may generate minus signs in the diagonal components of enlarged systems. Note that (8.20) is equivalent to a system defined by the matrices

$$\hat{M}_B = \begin{bmatrix} M & 0 \\ 0 & 0 \end{bmatrix} \quad \hat{C}_B = \begin{bmatrix} \frac{1}{c}E_{(n,\,n)} & 0 \\ 0 & 1 \end{bmatrix} \quad \hat{K}_B = \begin{bmatrix} K + B + \delta E_{(n,\,n)} & \alpha e_{(n)} \\ -\alpha e_{(n)}^T & \gamma \end{bmatrix} , \tag{8.21}$$

where $\hat{M}_B$ and $\hat{C}_B$ are symmetric and positive semi-definite. However, $\hat{K}_B$ is nonsymmetric. We emphasize that the form (8.21) is equivalent to the form (8.20). In particular, the eigenvalues of a system defined by (8.21) are the same of a system defined by (8.20) because both have exactly the same characteristic equation. Systems like (8.20) and (8.21) are called nonclassical mechanical systems.

8.1.2   The asymptotic boundary condition of second order

The Laplace transform of the asymptotic boundary condition of second order is

$$\sigma_n + \frac{s}{c}u_n + \frac{2\kappa^2 c s}{c^2\kappa^2 + 4s^2}u_n = 0 . \tag{8.22}$$

The second term represents a dashpot and is readily included into the equation of motion (8.2).

Encouraged by the success of the auxiliary variable approach, we define the auxiliary variable $\phi(t)$ by

$$\phi = \sqrt{2}c\kappa\frac{s}{c^2\kappa^2+4s^2}u_n.\tag{8.23}$$

This yields the couple of equations

$$\sigma_n+\sqrt{2}c\kappa\phi+u_n = 0\tag{8.24.a}$$

$$4\ddot\phi+(c\kappa)^2\phi = \sqrt{2}c\kappa\dot u_n.\tag{8.24.b}$$

Unfortunately, we can't construct a symmetric realization because the first equation contains $\phi(t)$ whereas the second contains the time derivative of $u_n(t)$. However, the time derivative in the right hand side of (8.24.b) suggests a straightforward generalization of the previous approach. Consider the new definition of the auxiliary variable $\phi(t)$ by the couple of equations

$$\sigma_n+\alpha\phi+\beta\dot\phi+\gamma u_n+\frac{1}{c}\dot u_n = 0,\tag{8.25.a}$$

$$4\ddot\phi+c^2\kappa^2\phi = \alpha u_n+\beta\dot u_n.\tag{8.25.b}$$

In these equations, the formal symmetry between $u_n(t)$ and $\phi(t)$ is included by definition. The exigency of the term γu_n in (8.25.a) will be evident later. Now, we compute the set of real parameters α, β and δ so that (8.25.a) and (8.25.b) are equivalent to (8.22). We perform the Laplace transform of the equations (8.25.a) and (8.25.b) and combine these such that $\phi(s)$ can be eliminated. This yields

$$\sigma_n+2\kappa^2\frac{\alpha^2+2\alpha\beta s+\beta^2 s^2}{c^2\kappa^2+4s^2}u_n+\gamma u_n+\frac{s}{c}u_n = 0.\tag{8.26}$$

Note that the rational function in (8.26) isn't proper. Hence, before proceeding to the identification of the parameters α, β and γ, we divide the numerator by the denominator of the rational function in (8.26) and extract the term $\beta^2 s^2$. The strictly proper rational function obtained by this procedure is then assumed to be identical to the strictly proper rational function in (8.23). That is,

$$\frac{\alpha^2+2\alpha\beta s+\beta^2 s^2}{c^2\kappa^2+4s^2}-\frac{\beta^2}{4} = \frac{(\alpha^2-(c\kappa\beta)^2/4)+2\alpha\beta s}{c^2\kappa^2+4s^2} = \frac{s}{c^2\kappa^2+4s^2}.\tag{8.27}$$

Comparing the corresponding coefficients of the nominator of both rational functions we find a set of three nonlinear algebraic equations which define α, β and γ:

$$\alpha = \pm c\kappa\beta/2,\tag{8.28.a}$$

$$\alpha\beta = c\kappa^2,\tag{8.28.b}$$

$$\gamma = -\beta^2/4.\tag{8.28.c}$$

Solving (8.28.b) for β and inserting the resulting expression into (8.28.a) yields

$$\alpha^2 = \pm\frac{c^2\kappa^3}{2}.\tag{8.29}$$

This equation has two solutions. First we consider the positive solution

$$\alpha = \sqrt{\frac{\kappa}{2}}c\kappa. \tag{8.30}$$

Combining the last equation with (8.28.b) and (8.28.c) yields

$$\beta = \sqrt{2\kappa}, \tag{8.31}$$

$$\gamma = -\kappa/2. \tag{8.32}$$

Since c and κ are both real and positive α, β and γ are real and positive too so that the symmetric realization is complete. As can be verified by inspection, the parameter set $\{-\alpha, -\beta, \gamma\}$ is also a solution of (8.28.a) to (8.28.c). The matrices of the enlarged system are

$$M_B = \begin{bmatrix} M & 0 \\ 0 & -4 \end{bmatrix} \qquad C_B = \begin{bmatrix} \frac{1}{c}E_{(n)} & \sqrt{2\kappa}e_{(n)} \\ \sqrt{2\kappa}e_{(n)}^T & 0 \end{bmatrix} \qquad K_B = \begin{bmatrix} K + B - \frac{\kappa}{2}E_{(n)} & c\kappa\sqrt{\frac{\kappa}{2}}e_{(n)} \\ c\kappa\sqrt{\frac{\kappa}{2}}e_{(n)}^T & -c^2\kappa^2 \end{bmatrix} \tag{8.33}$$

Now let us consider the second solution of (8.29). We obtain the parameter set

$$\alpha = i\sqrt{\frac{\kappa}{2}}c\kappa \qquad \beta = -i\sqrt{2\kappa} \qquad \gamma = \kappa/2. \tag{8.34}$$

Unfortunately, the parameters α and β are not real. As a consequence, with this set we can't formulate a real enlarged symmetric system so that we discard this set. We shall see later, that a simple scaling transformation allows to recover the parameter set (8.34).

8.1.3   A scaling transformation

In this section, we introduce a scaling transformation of the enlarged system. We will show that the eigenvalues of the enlarged system as well as the solution of the discretized IBVP are invariant with respect to this transformation. Furthermore, it turns out, that this scaling transformation allows to recover the purely imaginary symmetric realization (8.34).

Consider the scaling transformation of an enlarged system

$$\tilde{M}_B = \begin{bmatrix} M & 0 \\ 0 & -m\eta^2 \end{bmatrix}, \tag{8.35.a}$$

$$\tilde{C}_B = \begin{bmatrix} C + C' & \beta\eta \\ \beta\eta & -c\eta^2 \end{bmatrix}, \tag{8.35.b}$$

$$\tilde{K}_B = \begin{bmatrix} K + B + K' & a\eta \\ a\eta & -k\eta^2 \end{bmatrix}. \tag{8.35.c}$$

The off-diagonal components associated with the new variable ϕ are multiplied with η and the diagonal one with η^2. This transformation results when the row and the column associated with the new variable ϕ are multiplied by the scaling parameter η. Setting $\eta = 1$ we recover the original system matrices.

First we show that the eigenvalues of the enlarged system are invariant with respect to the scaling transformation. Consider the eigenvalue problem specified by

$$(\lambda^2 \tilde{M}_B + \lambda \tilde{C}_B + \tilde{K}_B)\tilde{w} = 0 \tag{8.36}$$

The eigenvalues of the system (8.36) are defined by its characteristic polynomial

$$det(\lambda^2 \tilde{M}_B + \lambda \tilde{C}_B + \tilde{K}_B) = 0 . \tag{8.37}$$

Because in (8.37) a row and a column has been multiplied by the scaling parameter η we obtain

$$det(\lambda^2 \tilde{M}_B + \lambda \tilde{C}_B + \tilde{K}_B) = \eta^2 det(\lambda^2 M_B + \lambda C_B + K_B) = 0 . \tag{8.38}$$

If $\eta \neq 0$ (8.37) is equivalent to

$$det(\lambda^2 M_B + \lambda C_B + K_B) = 0 \tag{8.39}$$

Hence, the eigenvalues are invariant with respect to the scaling transformation. In particular, the stability of enlarged systems is invariant with respect to scaling transformations.

To establish the invariance of the solution of the IBVP we must show that the subvector $u(t)$ is a solution of both, the scaled and the unscaled enlarged systems. Assume that we have a solution of the unscaled enlarged system $w(t)$ given by $w^T(t) = [u^T(t), \phi(t)]$ so that

$$M_B \ddot{w} + C_B \dot{w} + K_B w = F_B(t) \tag{8.40}$$

holds. Let us define a new vector $v(t)$ given by $v^T(t) = [u^T(t), \eta^{-1}\phi(t)]$. The relation between both vectors $w(t)$ and $v(t)$ is given by the matrix equation

$$v(t) = Aw(t) \tag{8.41}$$

with the matrix

$$A = \begin{bmatrix} I & 0 \\ 0 & \eta^{-1} \end{bmatrix} . \tag{8.42}$$

Since A is invertible we express the solution vector $w(t)$ by $v(t)$ and insert it into the equation of motion (8.40). The equation of motion for the new vector $v(t)$ becomes

$$M_B A^{-1} \ddot{v} + C_B A^{-1} \dot{v} + K_B A^{-1} v = F_B(t) . \tag{8.43}$$

The inverse of the matrix A is

$$A^{-1} = \begin{bmatrix} I & 0 \\ 0 & \eta \end{bmatrix} . \tag{8.44}$$

The system (8.43) is obviously not symmetric unless it is multiplied from the left by $A^{-T} = A^{-1}$. The system is then transformed to

$$A^{-T} M_B A^{-1} \ddot{v} + A^{-T} C_B A^{-1} \dot{v} + A^{-T} K_B A^{-1} v = A^{-T} F_B(t) = F_B(t) \tag{8.45}$$

The matrices of the transformed system are exactly the matrices (8.35.a) to (8.35.c). Hence, $v(t)$

is the solution of the scaled system. Because $u(t)$ is contained in both solution vectors $w(t)$ and $v(t)$ both solutions are equivalent with respect to the IBVP. The scaling of the matrices induces a scaling of the variable $\phi(t)$ described by $\bar{\phi}(t) = \eta\phi(t)$.

The scaling transformation can easily be extended to systems with more than one auxiliary variables. It is defined by

$$M \rightarrow A^T M A , \tag{8.46}$$

where the matrix A is given by

$$A = \begin{bmatrix} I & 0 \\ 0 & \Theta \end{bmatrix} \tag{8.47}$$

Θ contains the scaling parameters η_k

$$\Theta = \begin{bmatrix} \eta_1 & 0 & \dots & 0 \\ 0 & \eta_2 & \dots & 0 \\ \dots & \dots & \dots & \dots \\ 0 & 0 & \dots & \eta_q \end{bmatrix} \tag{8.48}$$

A is a diagonal matrix of dimension $p + q$, where p is the number of degrees of freedom of the finite element model and q the number of auxiliary variables associated with the absorbing boundary condition.

The scaling transformation is particularly interesting if $\eta = i$. Then, the diagonal components change sign because they are multiplied by $i^2 = -1$ and the off-diagonal components change from real to purely imaginary numbers. However, this scaling transformation gives us the opportunity to transform purely imaginary off-diagonal components to real numbers while the diagonal components remain real. This is exactly the transformation we need to overcome the restrictions found for the second set of parameters of the asymptotic boundary condition of second order given in (8.34). With this parameter set we obtain the system matrices

$$M_B = \begin{bmatrix} M & 0 \\ 0 & 4 \end{bmatrix} \qquad C_B = \begin{bmatrix} C + \frac{1}{c}E_{(n)} & \sqrt{2}\kappa e_{(n)} \\ \sqrt{2}\kappa e_{(n)}^T & 0 \end{bmatrix} \qquad K_B = \begin{bmatrix} K + B + \frac{\kappa}{2}E_{(n)} & c\kappa\sqrt{\frac{\kappa}{2}}e_{(n)} \\ c\kappa\sqrt{\frac{\kappa}{2}}e_{(n)}^T & c^2\kappa^2 \end{bmatrix} \tag{8.49}$$

Observe that this symmetric realization has positive diagonal components. Hence, because of (8.33) and (8.49) we have two different but equivalent symmetric realizations of the asymptotic absorbing boundary condition of second order. We denote realizations with negative diagonal components as symmetric realizations of the first kind and those with positive diagonal components as symmetric realizations of the second kind.

Remarks:

- The scaling transformation with $\eta = i$ is equivalent to a change of the sign in the right hand side of (8.24.b).

- The scaling transformation (8.47) does not change the eigenvalues of enlarged systems and the solutions of the associated IVP.

8.1.4   The asymptotic boundary conditions of the dissipative rod

The asymptotic boundary condition of first order of the dissipative restrained rod is

$$\sigma_n + \frac{s}{c}u_n + \varepsilon u_n + \frac{c(\kappa^2 - \varepsilon^2)}{2s}u_n = 0. \tag{8.50}$$

The first two terms on the right-hand side are a dashpot and a linear spring. These are included into the global damping and stiffness matrices of the near field. The third term is included as shown in section 8.1.1. The parameter α is

$$\alpha = \sqrt{\frac{c(\kappa^2 - \varepsilon^2)}{2}}. \tag{8.51}$$

Note that α is real provided $\kappa > \varepsilon$. For $\kappa < \varepsilon$ α is purely imaginary so that we have to apply the scaling transformation $\eta = i$ to obtain a real symmetric realization with parameter α given by

$$\alpha = \sqrt{\frac{c(\varepsilon^2 - \kappa^2)}{2}}. \tag{8.52}$$

The asymptotic boundary condition of second order of the dissipative rod is

$$\sigma_n + \frac{2c(\kappa^2 - \varepsilon^2)s}{c^2(\kappa^2 - \varepsilon^2) + 4c\varepsilon s + 4s^2}u_n + \varepsilon u_n + \frac{s}{c}u_n = 0. \tag{8.53}$$

The parameters α, β and δ are defined by three nonlinear equations

$$\alpha = \pm c\sqrt{(\kappa^2 - \varepsilon^2)}\beta/2, \tag{8.54.a}$$

$$2\alpha\beta - c\varepsilon\beta^2 = 2\rho c^2(\kappa^2 - \varepsilon^2), \tag{8.54.b}$$

$$\beta^2 = -4\delta. \tag{8.54.c}$$

First, we assume that $\kappa \geq \varepsilon$ and insert the positive solution of (8.54.a) into (8.54.b) and solve for β. This yields the real parameter set α, β and γ

$$\alpha = \frac{1}{\sqrt{2}}\frac{(\kappa^2 - \varepsilon^2)c}{\sqrt{\sqrt{\kappa^2 - \varepsilon^2} - \varepsilon}} \qquad \beta = \sqrt{\frac{2(\kappa^2 - \varepsilon^2)}{\sqrt{\kappa^2 - \varepsilon^2} - \varepsilon}} \qquad \gamma = -\frac{1}{2}\frac{\kappa^2 - \varepsilon^2}{\sqrt{\kappa^2 - \varepsilon^2} - \varepsilon} \tag{8.55}$$

Note that the set $\{-\alpha, -\beta, \gamma\}$ is also a solution. The parameters α and β are real if the denominator $\sqrt{\kappa^2 - \varepsilon^2} - \varepsilon$ is positive. This holds provided that $\varepsilon < \kappa/2$. If $\varepsilon > \kappa/2$ both parameters become imaginary. Applying the scaling transformation with $\eta = i$, the parameters α and β become real. Hence, the application domain of the symmetric realization can be extended to $\kappa \geq \varepsilon$. However, for $\varepsilon = \kappa/\sqrt{2}$ the denominator in (8.55) vanish so that the parameters α and β are infinite. That is, the realization is singular at $\varepsilon = \kappa/\sqrt{2}$.

This causes serious problems so that we try to find a second symmetric realization inserting the negative solution of (8.54.a) into (8.54.b). This yields the parameter set

$$\alpha = i\sqrt{\frac{2}{\sqrt{\kappa^2 - \varepsilon^2} + \varepsilon}}(\kappa^2 - \varepsilon^2)c \qquad \beta = -i\sqrt{\frac{2(\kappa^2 - \varepsilon^2)}{\sqrt{\kappa^2 - \varepsilon^2} + \varepsilon}} \qquad \delta = \frac{1}{2}\frac{\kappa^2 - \varepsilon^2}{\sqrt{\kappa^2 - \varepsilon^2} + \varepsilon} \tag{8.56}$$

The parameters α and β can be made real by applying a scaling transformation with $\eta = -i$. This set of parameters can be applied if $\epsilon \leq \kappa$. This condition is equal to that of the parameter set (8.55). Remarkably, the parameter set (8.56) doesn't exhibit a singularity at $\epsilon = \kappa/\sqrt{2}$.

Now we come back to the restriction given by the condition $\epsilon \leq \kappa$. This restriction has a sound physical background and is associated with a qualitative change of the solution of the equation of motion of the dissipative restrained rod (see chapter 10). For the asymptotic boundary condition of first order we can recover a set of real parameters applying the scaling transformation $\eta = -i$. For that of the second order we can't break up the restriction by a scaling transformation because the parameters α and β are complex with non-vanishing real part. In this case, no scaling parameter η exists which simultaneously transforms α and β to real numbers.

The source of the problem is associated with the assumption made in (8.25.a) and (8.25.b) regarding the form of the symmetric realization. There, we assumed that only one auxiliary variable is needed to realize the boundary condition. This assumption is admissible if the spectrum of the differential operator on the left-hand side of (8.25.b) is a pair of conjugate complex numbers. In this case, the associated rational function given in (8.27) cannot be split up into two separate partial fractions with real coefficients. However, if the spectrum is a pair of real numbers the associated rational function can be split up in two separate rational functions with real coefficients. Let us have a closer look to the differential equation operator of the asymptotic boundary condition of second order. It is given by

$$4\ddot{\phi} + 4c\epsilon\dot{\phi} + c^2(\kappa^2 - \epsilon^2)\phi \; . \tag{8.57}$$

Its spectrum is

$$\lambda = -\frac{c\epsilon}{2} \pm c\frac{\sqrt{2\epsilon^2 - \kappa^2}}{2} \; . \tag{8.58}$$

For $\kappa > \sqrt{2}\epsilon$ the spectrum is a pair of conjugate complex numbers. If $\kappa = \sqrt{2}\epsilon$ the term in the square-root vanish and we have only one negative real number. Observe that $\kappa = \sqrt{2}\epsilon$ corresponds to the singular point of the parameter set (8.55). Hence, the singularity of this parameter set is associated with the singularity of the differential operator. For $\kappa < \sqrt{2}\epsilon$ the spectrum has two real negative numbers.

If the spectrum is composed by two negative real numbers, then we can try a symmetric realization with two variables $\phi_1(t)$ and $\phi_2(t)$ given by the set of equations

$$\sigma_n + \alpha_1\phi_1 + \alpha_2\phi_2 + \epsilon u_n + \frac{1}{c}\dot{u}_n = 0, \tag{8.59.a}$$

$$\dot{\phi}_1(t) + \lambda_1\phi_1 = \alpha_1 u_n, \tag{8.59.b}$$

$$\dot{\phi}_2(t) + \lambda_2\phi_2 = \alpha_2 u_n, \tag{8.59.c}$$

where λ_1 and λ_2 are the two spectral values

$$\lambda_1 = -\frac{c\epsilon}{2} + c\frac{\sqrt{2\epsilon^2 - \kappa^2}}{2}, \tag{8.60.a}$$

$$\lambda_2 = -\frac{c\epsilon}{2} - c\frac{\sqrt{2\epsilon^2 - \kappa^2}}{2} \; . \tag{8.60.b}$$

Performing the Laplace-transform of the previous equations and successively eliminating the functions $\phi_1(s)$ and $\phi_2(s)$ yields

$$\sigma_n + \frac{\alpha_1^2}{s+\lambda_1} u_n + \frac{\alpha_2^2}{s+\lambda_2} u_n + \varepsilon u_n + \frac{s}{c} u_n = 0. \tag{8.61}$$

If the rational function (8.53) is brought in the same form we can identify the parameters α_1 and α_2. These are

$$\alpha_1 = \frac{1}{2}\sqrt{\frac{(\kappa^2-\varepsilon^2)(\sqrt{2\varepsilon^2-\kappa^2}-\varepsilon)}{\sqrt{2\varepsilon^2-\kappa^2}}}, \tag{8.62.a}$$

$$\alpha_2 = \frac{1}{2}\sqrt{\frac{(\kappa^2-\varepsilon^2)(\sqrt{2\varepsilon^2-\kappa^2}+\varepsilon)}{\sqrt{2\varepsilon^2-\kappa^2}}}. \tag{8.62.b}$$

The terms under the roots are negative for $\varepsilon > \kappa$. Therefore the realization parameters α_1 and α_2 are purely imaginary. Applying a scaling transformation with the matrix

$$\eta = \begin{bmatrix} -i & 0 \\ 0 & -i \end{bmatrix} \tag{8.63}$$

a real symmetric realization is obtained. However, we can extend the last parameter set even in the domain $\kappa \geq \varepsilon > \kappa/\sqrt{2}$. There, the term under the root of (8.62.a) is positive whereas that of (8.62.b) remains negative. A real symmetric realization is obtained applying the scaling transformation

$$\eta = \begin{bmatrix} 1 & 0 \\ 0 & -i \end{bmatrix} \tag{8.64}$$

The enlarged ODE-system given by the parameters set (8.62.a) and (8.62.b) is defined by

$$M_B = \begin{bmatrix} M & 0 & 0 \\ 0 & 0 & 0 \\ 0 & 0 & 0 \end{bmatrix} \quad C_B = \begin{bmatrix} C+\frac{1}{c}E_{(n)} & 0 & 0 \\ 0 & 1 & 0 \\ 0 & 0 & 1 \end{bmatrix} \quad K_B = \begin{bmatrix} K+B+\varepsilon E_{(n)} & |\alpha_2|e_{(n)} & |\alpha_2|e_{(n)} \\ |\alpha_1|e_{(n)}^T & \lambda_1 & 0 \\ |\alpha_2|e_{(n)}^T & 0 & \lambda_2 \end{bmatrix}$$

$$\tag{8.65}$$

With the last symmetric realization we have covered the full range of the parameters ε and κ. Observe that in the parameter domain $\kappa \geq \varepsilon > \kappa/\sqrt{2}$ there exist three different symmetric realizations. This emphasize again, that symmetric realizations are generally not unique.

8.2 Symmetric realizations of scalar transfer functions

So far, we have seen that symmetric realizations of rational functions generally exist and that these may be not unique. In this section, we systematically analyze the symmetric realization problem of general rational functions. We assume that the analytic part of an absorbing boundary condition can be expanded in a finite sum of proper rational functions of first and second order, that is,

$$h(s) = \sum_{j=1}^{n_1} \frac{p_{j0} + p_{j1}s}{q_{j0} + q_{j1}s + q_{j2}s^2} + \sum_{k=1}^{n_2} \frac{p_{k0}}{q_{k0} + q_{k1}s}. \tag{8.66}$$

This is always possible if the system matrix A of the linear system can be diagonalized. That is, if the matrix A of dimension n has n independent eigenvectors. Each term of (8.66), we call it rational function of first or second order, can be analyzed independently from the others. The assembly of the complete system of symmetric realizations will be treated in section 8.4.

8.2.1 General rational functions of second order

Consider the rational function of second order

$$h(s) = \frac{p_0 + p_1 s}{q_0 + q_1 s + q_2 s^2}. \tag{8.67}$$

Since the differential operator associated with the denominator is stable the coefficients q_0, q_1 and q_2 are positive. The rational function is normalized imposing the condition $q_2 = 1$. Now we search real or purely imaginary solutions of the parameter set $\{\alpha, \beta, \gamma\}$ defined by

$$\begin{aligned} -h(s)u(s) + \alpha\phi(s) + s\beta\phi(s) + \gamma u(s) &= 0 \\ -(q_0 + q_1 s + s^2)\phi(s) + \alpha u(s) + s\beta u(s) &= 0 \end{aligned} \tag{8.68}$$

Combining these equations we can eliminate $\phi(s)$ and obtain

$$-h(s)u(s) + \frac{(\alpha + s\beta)^2}{q_0 + q_1 s + s^2}u(s) + \gamma u(s) = 0. \tag{8.69}$$

Inserting (8.67) into the last equation, converting the resulting one into the factored normal form and equating the coefficients of the numerator polynomial to zero, we obtain three nonlinear equations which define the set $\{\alpha, \beta, \gamma\}$

$$\beta^2 + \gamma = 0, \tag{8.70.a}$$

$$2\alpha\beta - p_1 + \gamma q_1 = 0, \tag{8.70.b}$$

$$\alpha^2 - p_0 + \gamma q_0 = 0. \tag{8.70.c}$$

γ can be eliminated combining the first equation with the last two. This yields two nonlinear equations defining the parameters α and β:

$$2\alpha\beta - p_1 - q_1\beta^2 = 0, \tag{8.71.a}$$

$$\alpha^2 - p_0 - q_0\beta^2 = 0. \tag{8.71.b}$$

Combining (8.71.a) and (8.71.b) we eliminate α and obtain a nonlinear equation for β

$$(q_1^2 - 4q_0)\beta^4 + 2(p_1 q_1 - 2p_0)\beta^2 + p_1^2 = 0. \tag{8.72}$$

If we assume that $q_1^2 - 4q_0 \neq 0$ the last equation has two independent solutions $\{\beta_1, \beta_2\}$ which are given by

$$\beta_1 = \sqrt{\frac{2p_0 - p_1 q_1 + 2\sqrt{p_0^2 - q_1 p_0 p_1 + q_0 p_1^2}}{q_1^2 - 4q_0}}, \tag{8.73.a}$$

$$\beta_2 = \sqrt{\frac{2p_0 - p_1 q_1 - 2\sqrt{p_0^2 - q_1 p_0 p_1 + q_0 p_1^2}}{q_1^2 - 4q_0}}. \tag{8.73.b}$$

These equations have real or purely imaginary solutions only if $p_0^2 - q_1 p_0 p_1 + q_0 p_1^2 \geq 0$. Observe that if $p_0^2 - q_1 p_0 p_1 + q_0 p_1^2 = 0$ $\beta_1 = \beta_2$ so that we obtain only one independent solution. With the solutions β_j the associated parameters α_j and γ_j become

$$\alpha_j = \frac{p_1 + q_1 \beta_j^2}{2\beta_j} \qquad \gamma_j = -\beta_j^2, \tag{8.74}$$

Let us analyze the conditions

$$q_1^2 - 4q_0 \neq 0 \quad \text{and} \quad p_0^2 - q_1 p_0 p_1 + q_0 p_1^2 \geq 0. \tag{8.75}$$

Since the denominator of the rational function (8.67) is quadratic we can convert (8.67) in the partial fraction form. We assume that the denominator of the rational function (8.67) has two distinct complex roots λ and $\bar{\lambda}$. The partial fraction form is then given by

$$\frac{p_0 + p_1 s}{q_0 + q_1 s + s^2} = \frac{a}{s - \lambda} + \frac{\bar{a}}{s - \bar{\lambda}}. \tag{8.76}$$

Let us express the parameters p_0, p_1, q_0, q_1 with the parameters $a, \bar{a}, \lambda, \bar{\lambda}$

$$q_1 = -(\lambda + \bar{\lambda}) = 2\text{Re}(\lambda) \qquad q_0 = \lambda \bar{\lambda} = |\lambda|^2$$
$$p_0 = -\bar{\lambda} a - \lambda \bar{a} \qquad p_1 = a + \bar{a} \tag{8.77}$$

With (8.77) the conditions (8.75) transforms to

$$q_1^2 - 4q_0 = (\lambda - \bar{\lambda})^2 = -4(Im(\lambda))^2 \neq 0, \tag{8.78.a}$$

$$p_0^2 - q_1 p_0 p_1 + q_0 p_1^2 = -a\bar{a}(\lambda - \bar{\lambda})^2 = 4|a|^2(Im(\lambda))^2 \geq 0. \tag{8.78.b}$$

Both equations are fulfilled if $Im(\lambda) \neq 0$. Hence, we obtain the remarkable result that if the roots of the denominator are complex conjugate then there exist two independent symmetric realizations. Furthermore, if $p_1 \neq 0$ one realization is of the first kind and the other is of the second kind. In fact, according to (8.78.a) we have $q_1^2 - 4q_0 < 0$. Therefore $\beta_1^2 \beta_2^2 = p_1^2 / (q_1^2 - 4q_0) < 0$ and this result follows immediately.

Using the partial fraction form (8.76) we obtain a much simpler representation of the parameter set $\{\alpha, \beta, \gamma\}$. Replacing p_0, p_1, q_0, q_1 with $a, \bar{a}, \lambda, \bar{\lambda}$ in (8.73.a) and (8.73.b) we obtain the realization of the first kind

$$\alpha_1 = \frac{\sqrt{a}\bar{\lambda}}{\sqrt{\lambda - \bar{\lambda}}} + \frac{\sqrt{\bar{a}}\lambda}{\sqrt{\bar{\lambda} - \lambda}} \qquad \beta_1 = -\frac{\sqrt{a}}{\sqrt{\lambda - \bar{\lambda}}} - \frac{\sqrt{\bar{a}}}{\sqrt{\bar{\lambda} - \lambda}} \qquad \gamma_1 = -\beta_1^2, \tag{8.79}$$

and that of the second kind

$$\alpha_2 = \frac{\sqrt{a}\bar{\lambda}}{\sqrt{\bar{\lambda}-\lambda}} + \frac{\sqrt{\bar{a}}\lambda}{\sqrt{\lambda-\bar{\lambda}}} \qquad \beta_2 = -\frac{\sqrt{a}}{\sqrt{\bar{\lambda}-\lambda}} - \frac{\sqrt{\bar{a}}}{\sqrt{\lambda-\bar{\lambda}}} \qquad \gamma_2 = \beta_2^2. \tag{8.80}$$

Observe, that the realization of the second kind results from that of the first kind as a consequence of the identity $\sqrt{\lambda - \bar{\lambda}} = i\sqrt{\bar{\lambda} - \lambda}$.

Example: Consider the linear system with two degrees of freedom given by

$$\begin{bmatrix} 1 & 0 \\ 0 & 2 \end{bmatrix} \begin{bmatrix} \ddot{x}_1 \\ \ddot{x}_2 \end{bmatrix} + \begin{bmatrix} 1/4 & -1/4 \\ -1/4 & 1/2 \end{bmatrix} \begin{bmatrix} \dot{x}_1 \\ \dot{x}_2 \end{bmatrix} + \begin{bmatrix} 2 & -1 \\ -1 & 4 \end{bmatrix} \begin{bmatrix} x_1 \\ x_2 \end{bmatrix} = \begin{bmatrix} f_1 \\ 0 \end{bmatrix} \tag{8.81}$$

We assume that the initial conditions of the variable x_2 vanish: $x_2(0) = \dot{x}_2(0) = 0$. The initial conditions of the variable x_1 are combined with the loading function on the right hand side and denoted by f_1. Assume that we are mainly interested in the variable x_1. Applying the Laplace transform to (8.81) we can use the second equation to eliminate the variable x_2 in the first equation. The resulting equation which defines the variable x_1 in terms of the loading function f_1 is

$$\left(s^2 + \frac{1}{4}s + 2\right)x_1(s) - \frac{1}{8}\frac{(s+4)^2}{4s^2 + s + 8}x_1(s) = f_1(s) \tag{8.82}$$

Observe that in the time domain (8.82) is a Volterra integro-differential equation because the term with a rational function transforms to a convolution integral.

Now we try to reconstruct a system of differential equations from (8.82). For this goal, we apply the equations (8.79) and (8.80) to the rational term in (8.82). First, this term has to be expanded in a sum of a proper rational function and a constant:

$$\frac{1}{8}\frac{(s+4)^2}{4s^2 + s + 8} = \frac{1}{32}\frac{31s + 56}{4s^2 + s + 8} + 1/32 \tag{8.83}$$

The parameters of the partial fraction expansion of the proper rational function are

$$\lambda = -\frac{1}{8} + \frac{\sqrt{127}}{8} \qquad a = -\frac{31}{256} + i\frac{417}{32512}\sqrt{127} \tag{8.84}$$

Combining the last equations with (8.79) we obtain a first set of parameters of the symmetric realization

$$\alpha_1 = 0.1884 \qquad \beta_1 = -0.4863 \qquad \gamma_1 = -0.2365 \tag{8.85}$$

With this parameter set and using the scaling transformation with $\eta = \sqrt{2}$ which scales the lower diagonal elements of the system matrices so that their absolute values are equal to those of (8.81) we obtain the system of differential equations

$$\begin{bmatrix} 1 & 0 \\ 0 & -2 \end{bmatrix} \begin{bmatrix} \ddot{x}_1 \\ \ddot{\phi} \end{bmatrix} + \begin{bmatrix} 1/4 & -0.6877 \\ -0.6877 & -1/2 \end{bmatrix} \begin{bmatrix} \dot{x}_1 \\ \dot{\phi} \end{bmatrix} + \begin{bmatrix} 1.7323 & -0.2662 \\ -0.2662 & -4 \end{bmatrix} \begin{bmatrix} x_1 \\ \phi \end{bmatrix} = \begin{bmatrix} f_1 \\ 0 \end{bmatrix} \tag{8.86}$$

The second set of parameters is obtained combining (8.84) with (8.80). This yields

$$\alpha_1 = \frac{\sqrt{2}}{2} \qquad \beta_1 = \frac{\sqrt{2}}{8} \qquad \gamma_1 = \frac{1}{32} \tag{8.87}$$

If this set of parameters is combined with a scaling transformation with $\eta = \sqrt{2}$ we obtain

$$\begin{bmatrix} 1 & 0 \\ 0 & 2 \end{bmatrix} \begin{bmatrix} \ddot{x}_1 \\ \ddot{\phi} \end{bmatrix} + \begin{bmatrix} 1/4 & -1/4 \\ -1/4 & 1/2 \end{bmatrix} \begin{bmatrix} \dot{x}_1 \\ \dot{\phi} \end{bmatrix} + \begin{bmatrix} 2 & -1 \\ -1 & 4 \end{bmatrix} \begin{bmatrix} x_1 \\ \phi \end{bmatrix} = \begin{bmatrix} f_1 \\ 0 \end{bmatrix} \qquad (8.88)$$

Note that the system matrices of (8.88) are equal to those of (8.81) so that the variables ϕ and x_2 are equivalent. On the other hand, the variable ϕ of the system (8.86) is not equivalent to x_2. However, both systems (8.86) and (8.88) define exactly the same relationship between the variable x_1 and the loading f_1 because (8.82) can be derived from (8.86) and (8.88). Note that the realization of the second kind yields a system with positive definite matrices whereas the realization of the first kind yields a system with symmetric but indefinite ones.

Remark:
- If the parameter set $\{\alpha, \beta\}$ is a solution of the equations (8.71.a) and (8.71.b) then $\{-\alpha, -\beta\}$ is also a solution. Furthermore, if α is purely imaginary then β is purely imaginary too.

8.2.2 Special rational functions of second order

If a is purely imaginary $p_1 = a + \bar{a} = 0$ so that (8.72) reduces to

$$(q_1^2 - 4q_0)\beta^4 - 4p_0\beta^2 = 0 \qquad (8.89)$$

with a unique nontrivial independent solution

$$\beta_1 = \sqrt{\frac{-4p_0}{q_1^2 - 4q_0}}. \qquad (8.90)$$

If $p_0 > 0$ the realization is of the first kind and if $p_0 < 0$ it is of the second kind. Hence, in this case, there exists only one nontrivial realization. This is

$$\alpha_1 = \frac{q_1}{\sqrt{-(q_1^2 - 4q_0)p_0}} = \frac{(\lambda + \bar{\lambda})}{\sqrt{-(\lambda - \bar{\lambda})^2(a\bar{\lambda} + \bar{a}\lambda)}}$$

$$\beta_1 = \sqrt{\frac{-4p_0}{q_1^2 - 4q_0}} = 2\sqrt{\frac{a\bar{\lambda} + \bar{a}\lambda}{-(\lambda - \bar{\lambda})^2}} \qquad (8.91)$$

$$\gamma_1 = -\beta_1^2$$

If we assume that $p_1 = a + \bar{a} = 0$ and $q_1 = 0$ the last equations yields

$$\alpha_1 = 0 \qquad \beta_1 = \sqrt{\frac{p_0}{q_0}} = \sqrt{\frac{a\bar{\lambda} + \bar{a}\lambda}{\lambda\bar{\lambda}}} \qquad \gamma_1 = -\beta_1^2. \qquad (8.92)$$

As before, if $p_0 > 0$ the symmetric realization is of the first kind and if $p_0 < 0$ it is of the second kind.

If $p_0 = 0$ we can apply the formulas for the symmetric realizations of the general case. The special case with $p_0 = 0$ and $q_1 = 0$ leads to much simpler formulas. Assuming that $p_1 = a + \bar{a} > 0$, the realization of the first kind is

$$\alpha_1 = \sqrt{\frac{p_1\sqrt{q_0}}{2}} = \sqrt{\frac{(a + \bar{a})\lambda\bar{\lambda}}{2}} \qquad \beta_1 = \sqrt{\frac{p_1}{2\sqrt{q_0}}} = \sqrt{\frac{a + \bar{a}}{\lambda\bar{\lambda}}} \qquad \gamma_1 = -\beta_1^2, \qquad (8.93)$$

and that of the second kind is

$$\alpha_2 = \sqrt{\frac{p_1\sqrt{q_0}}{2}} = \sqrt{\frac{(a+\bar{a})\lambda\bar{\lambda}}{2}} \qquad \beta_2 = \sqrt{\frac{p_1}{2\sqrt{q_0}}} = \sqrt{\frac{a+\bar{a}}{\lambda\bar{\lambda}}} \qquad \gamma_2 = \beta_2^2. \qquad (8.94)$$

Notice that $\alpha_1 = \alpha_2$ and $\beta_1 = \beta_2$. If $p_1 = a + \bar{a} < 0$ the realization of first kind is given by (8.94) and that of the second kind by (8.93).

Finally, let us consider the special case of a denominator with two real roots. The partial fraction form is then given by

$$\frac{p_0 + p_1 s}{q_0 + q_1 s + s^2} = \frac{a_1}{s - \lambda_1} + \frac{a_2}{s - \lambda_2}, \qquad (8.95)$$

where $a_1, a_2, \lambda_1, \lambda_2$ are real numbers. Then, the condition (8.78.b) yields

$$p_0^2 - q_1 p_0 p_1 + q_0 p_1^2 = -a_1 a_2 (\lambda_1 - \lambda_2)^2 \ge 0. \qquad (8.96)$$

Hence, real symmetric realizations exist if and only if $\text{sgn} a_1 = -\text{sgn} a_2$. This is a rather strong restriction. Therefore, for general applications we must split up the rational function of second order in two rational functions of first order.

8.2.3   General rational functions of first order

Consider the rational function

$$h(s) = \frac{p_0}{q_0 + s}, \qquad (8.97)$$

where q_0 is real and positive and p_0 is real. A first symmetric realization is obtained considering the couple of equations

$$\begin{aligned} -h(s)u(s) + \alpha\phi(s) &= 0 \\ -(q_0 + s)\phi(s) + \alpha u(s) &= 0 \end{aligned}. \qquad (8.98)$$

Combining these two equations we can eliminate $\phi(s)$ and obtain

$$-h(s) + \frac{\alpha}{q_0 + s} = 0. \qquad (8.99)$$

The last equation yields a unique equation which defines α

$$\alpha^2 = p_0. \qquad (8.100)$$

Note that α as well as $-\alpha$ are solutions. The realization is

$$\alpha = \sqrt{p_0}. \qquad (8.101)$$

This yields a realization of first kind if $p_0 > 0$ and one of second kind if $p_0 < 0$. Hence a realization of (8.98) exists for every p_0.

A second symmetric realization is given by

$$-h(s)u(s) + \alpha\phi(s) + s\beta\phi(s) + \gamma u(s) = 0$$
$$-(q_0 s + s^2)\phi(s) + \alpha u(s) + s\beta u(s) = 0 \tag{8.102}$$

Combining the two equations we can eliminate $\phi(s)$ and obtain

$$-h(s) + \frac{(\alpha + s\beta)^2}{q_0 s + s^2} + \gamma = 0 \tag{8.103}$$

Inserting (8.97) into the last equation, converting the resulting one into the factored normal form and equating the coefficients of the numerator polynomial to zero we obtain three nonlinear equations defining the parameter set $\{\alpha, \beta, \gamma\}$

$$\beta^2 + \gamma = 0, \tag{8.104.a}$$

$$2\alpha\beta - p_0 + \gamma q_0 = 0, \tag{8.104.b}$$

$$\alpha^2 = 0. \tag{8.104.c}$$

The last equation gives $\alpha = 0$. Combining the first two equations and considering that α vanish yields

$$q_0 \beta^2 + p_0 = 0. \tag{8.105.a}$$

This gives the realization

$$\alpha = 0 \qquad \beta = \sqrt{\frac{-p_0}{q_0}} \qquad \gamma = -\beta^2. \tag{8.106}$$

We obtain a realization of first kind if $p_0 < 0$ and one of second kind if $p_0 > 0$. Hence a realization (8.98) exists for every p_0.

Remark:
• Note that the symmetric realization (8.98) have a dashpot and a spring element but no mass element. On the other hand, the symmetric realization (8.102) have a mass and a dashpot element but no spring element.

8.2.4 Perturbed rational functions of first order

An enlarged system with zero stiffness or zero mass may produce trouble. In fact, a zero stiffness impedes the computation of static states and a zero mass impedes the application of an explicit time integration algorithm. Therefore, a symmetric realization of first order elements analogous to that of second order is generally preferable. A way to obtain such a symmetric realization is to perturb the rational function of first order so that it is transformed to one of second order. Consider the perturbation

$$h_\varepsilon(s) = \frac{p_0}{2}\frac{1 + i\varepsilon}{s + (q_0 + i\varepsilon)} + \frac{p_0}{2}\frac{1 - i\varepsilon}{s + (q_0 - i\varepsilon)} = \frac{p_0(q_0 + \varepsilon^2) + p_0 s}{s^2 + 2q_0 s + (q_0^2 + \varepsilon^2)}. \tag{8.107}$$

The unique pole of the rational function is divided into a pair of conjugate complex poles as shown in Figure 8.1 (left). Expanding the last expression in a Taylor series with respect to ε we find that the error due to the perturbation is of the second order, that is, $|(\sigma - \sigma_\varepsilon)| = O(\varepsilon^2)$. Symmetric realizations of the rational function (8.107) are obtained with the equations (8.79) or (8.80), respectively. The choice of the perturbation parameter ε depends on several considerations. According to the realizations (8.79) or (8.80) the parameters α and β are inverse proportional to $\sqrt{\lambda - \bar{\lambda}} = \sqrt{2i\varepsilon}$. Hence, α and β are of the order $1/\sqrt{\varepsilon}$ and γ is of the order $1/\varepsilon$. That is, a small ε produces large off-diagonal components in the system matrices. However, as will be shown, large off-diagonal components may lead to unstable systems.

A judicious choice of ε is derived when we consider that in the time domain the perturbation (8.97) is given by

$$h_\varepsilon(t) = p_0 e^{-q_0 t} \quad \rightarrow \quad \sigma_p(t) = p_0 e^{-q_0 t}(\cos(\varepsilon t) + \varepsilon \sin(\varepsilon t)). \qquad (8.108)$$

The decrease of the exponential function is controlled by the magnitude of the parameter q_0 and may be characterized by the reference time $T_d = 1/q_0$. Since the perturbation parameter ε is identical to the circular frequency ω of the oscillation, its period is given by $T_\omega = 2\pi/\varepsilon$. Hence, to keep the perturbation error small we have to require that $T_\omega \gg T_d$. Therefore, this perturbation approach is particularly suitable if the reference time T_d is short (large q_0). An example which shows that ε may be defined to be sufficiently large without inducing a considerable perturbation error is shown in Figure 8.1. The parameters are $p_0 = 1$, $q_0 = 1$ and $\varepsilon = 0.25$.

8.3   Symmetric realizations of matrix-valued transfer functions

Similarly as for scalar absorbing boundaries, we assume that matrix-valued absorbing boundaries are defined by a finite number of rational matrix functions of first and second order

$$H(s) = \sum_{j=1}^{n_2} \frac{P_{j0} + P_{j1}s}{q_{j0} + q_{j1}s + q_{j2}s^2} + \sum_{k=1}^{n_1} \frac{P_{k0}}{q_{k0} + q_{k1}s}. \qquad (8.109)$$

The matrices P_{j0}, P_{j1} and P_{k0} are real and symmetric and have the same dimension of $H(s)$. Because the system is assumed to be stable, the parameters of the denominators are positive. For simplicity, we assume that $q_{j2} = 1$ and $q_{k1} = 1$. The matrices P_{j0}, P_{j1} have at most rank two and P_{k0} at most rank one, respectively. Furthermore, we assume that every rational matrix functions of second order can be represented in the partial fraction form

$$H_j(s) = \frac{P_{j0} + P_{j1}s}{q_{j0} + q_{j1}s + q_{j2}s^2} = \frac{a_j a_j^T}{s - \lambda_j} + \frac{\bar{a}_j \bar{a}_j^T}{s - \bar{\lambda}_j}, \qquad (8.110)$$

where λ_j and $\bar{\lambda}_j$ are poles of the rational matrix function and a_j is a complex vector with the same dimension of $H(s)$. Note that an arbitrary stable rational transfer function $H(s)$ can always be decomposed in the form (8.109) if the system matrix A of dimension $n \times n$ has n independent eigenvectors.

8.3.1   General rational matrix functions of second order

Consider the partial fraction form of a rational matrix function of second order

$$H(s) = \frac{P_0 + P_1 s}{q_0 + q_1 s + s^2} = \frac{a a^T}{s - \lambda} + \frac{\bar{a} \bar{a}^T}{s - \bar{\lambda}}. \qquad (8.111)$$

Observe that $q_0 = \lambda \bar{\lambda} = |\lambda|^2$ and $q_1 = -(\lambda + \bar{\lambda}) = -2Re(\lambda)$. We already assumed that the vector a is complex $(a - \bar{a} \neq 0)$. Furthermore, let us assume that $\bar{a}$ is not proportional to a ($\bar{a} \neq ca$), where c is a arbitrary complex number. With these assumptions the matrices $P_0 = -\lambda a a^T - \bar{\lambda} \bar{a} \bar{a}^T$ and $P_1 = a a^T + \bar{a} \bar{a}^T$ are of rank two.

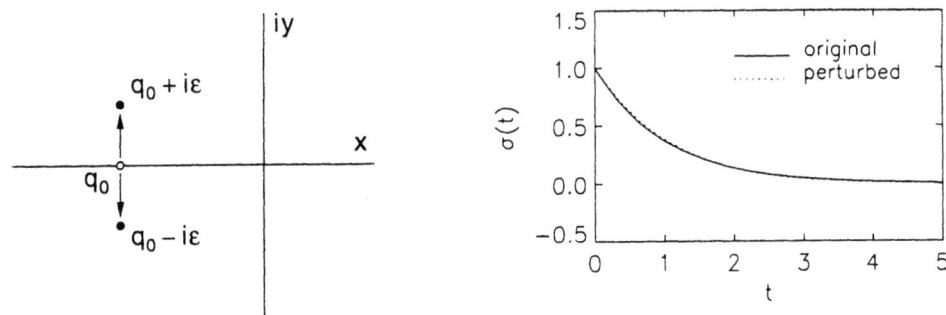

Figure 8.1: Right: Pole splitting due to perturbation. Left: Exact and perturbed responses

A symmetric realization of second order is defined by

$$-H(s)u(s) + f\phi(s) + sg\phi(s) + Eu(s) = 0$$
$$-(q_0 + q_1 s + s^2)\phi(s) + f^T u(s) + sg^T u(s) = 0 \quad ,$$

(8.112)

where f and g are real vectors of the same dimension of a and E is a real symmetric matrix of the same dimension of $H(s)$. $\phi(s)$ is the new auxiliary scalar variable. It can be eliminated by combining the two equations (8.112). This yields an equation defining f, g and E

$$-H(s) + \frac{(f + sg)(f^T + sg^T)}{q_0 + q_1 s + s^2} + E = 0.$$

(8.113)

From the last equation we can derive the following set of equations

$$E + gg^T = 0,$$

(8.114.a)

$$f f^T - \lambda\bar{\lambda} gg^T = -\bar{\lambda}aa^T - \lambda\bar{a}\bar{a}^T,$$

(8.114.b)

$$fg^T + gf^T + (\lambda + \bar{\lambda})gg^T = aa^T + \bar{a}\bar{a}^T.$$

(8.114.c)

The structure of the last two equations suggests that the vectors f and g can be represented as a linear combination of the vectors a and $\bar{a}$. That is, $f = \alpha a + \bar{\alpha}\bar{a}$ and $g = \beta a + \bar{\beta}\bar{a}$. Inserting these equations into (8.114.b) and (8.114.c) and equating all products of the vectors a and $\bar{a}$ to zero yields the set of equations

$$\alpha\bar{\alpha} - \lambda\bar{\lambda}\beta\bar{\beta} = 0,$$

(8.115.a)

$$\alpha\bar{\beta} + \bar{\alpha}\beta + (\lambda + \bar{\lambda})\beta\bar{\beta} = 0,$$

(8.115.b)

$$\alpha^2 - \lambda\bar{\lambda}\beta^2 + \bar{\lambda} = 0,$$

(8.115.c)

$$2\alpha\beta + (\lambda + \bar{\lambda})\beta^2 - 1 = 0.$$

(8.115.d)

The first two equations suggest that α is proportional to β. Assuming $\alpha = -\bar{\lambda}\beta$ we obtain

$$\beta = \pm\frac{1}{\sqrt{\lambda - \bar{\lambda}}}.$$

(8.116)

Now let us check if the fourth equation (8.115.d) holds. Inserting $\alpha = -\bar{\lambda}\beta$ and (8.116) into (8.115.d) we obtain $-2\bar{\lambda}\beta^2 + (\lambda + \bar{\lambda})\beta^2 - 1 = (\lambda - \bar{\lambda})\beta^2 - 1 = 0$. Therefore a solution of the equations (8.114.b) and (8.114.c) is given by the set

$$f_1 = \frac{\bar{\lambda}}{\sqrt{\lambda - \bar{\lambda}}}a + \frac{\lambda}{\sqrt{\bar{\lambda} - \lambda}}\bar{a} \qquad g_1 = -\frac{a}{\sqrt{\lambda - \bar{\lambda}}} - \frac{\bar{a}}{\sqrt{\bar{\lambda} - \lambda}} \qquad E_1 = -g_1 g_1^T \qquad (8.117)$$

Observe that $f_1 = \overline{f_1}$ and $g_1 = \overline{g_1}$. Hence both are real and are realizations of the first kind. The realization of second kind is defined by

$$-H(s) - \frac{(f + sg)(f^T + sg^T)}{q_0 + q_1 s + s^2} + E = 0 \qquad (8.118)$$

Proceeding analogously as with the realization of the first kind yields

$$f_2 = \frac{\bar{\lambda}}{\sqrt{\bar{\lambda} - \lambda}}a + \frac{\lambda}{\sqrt{\lambda - \bar{\lambda}}}\bar{a} \qquad g_2 = -\frac{a}{\sqrt{\bar{\lambda} - \lambda}} - \frac{\bar{a}}{\sqrt{\lambda - \bar{\lambda}}} \qquad E_2 = g_2 g_2^T. \qquad (8.119)$$

Hence, similar as for scalar systems, we obtain two symmetric realizations: one of the first and one of the second kind.

Example: Consider the rational transfer function

$$H(s) = \frac{P_0 + P_1 s}{5 + 4s + s^2} = \frac{\begin{bmatrix} -19 & 12 & -11 \\ 12 & -1 & 3 \\ -11 & 3 & -4 \end{bmatrix} + \begin{bmatrix} -12 & 11 & -9 \\ 11 & -8 & 7 \\ -9 & 7 & 6 \end{bmatrix}s}{5 + 4s + s^2} \qquad (8.120)$$

The poles of (8.120) are $\lambda_1 = \lambda = -2 + i$ and $\lambda_2 = \bar{\lambda} = -2 - i$ so that the partial fraction expansion is

$$H(s) = \frac{H_1}{s - \lambda} + \frac{\bar{H}_1}{s - \bar{\lambda}} \qquad (8.121)$$

with

$$H_1 = \begin{bmatrix} -6 - \frac{5}{2}i & \frac{11}{2} + 5i & -\frac{9}{2} - \frac{7}{2}i \\ \frac{11}{2} + 5i & -4 - \frac{15}{2}i & \frac{7}{2} + \frac{11}{2}i \\ -\frac{9}{2} - \frac{7}{2}i & \frac{7}{2} + \frac{11}{2}i & -3 - 4i \end{bmatrix} \qquad (8.122)$$

H_1 has rank one and can be represented as $H_1 = aa^T$ with

$$a = \frac{1}{2}\begin{bmatrix} 1 - 5i \\ -3 + 5i \\ 2 - 4i \end{bmatrix} \qquad (8.123)$$

Applying the equations (8.117) yields the realization of the first kind

$$f_1 = \begin{bmatrix} 1 \\ 2 \\ -1 \end{bmatrix} \qquad g_1 = \begin{bmatrix} 2 \\ -1 \\ 2 \end{bmatrix} \qquad E_1 = \begin{bmatrix} -4 & 2 & -2 \\ 2 & -1 & 1 \\ -2 & 1 & -1 \end{bmatrix} \qquad (8.124)$$

With the equations (8.119) we obtain the realization of the second kind

$$f_2 = \begin{bmatrix} -8 \\ 9 \\ -7 \end{bmatrix} \qquad g_2 = \begin{bmatrix} -3 \\ 4 \\ 3 \end{bmatrix} \qquad E_2 = \begin{bmatrix} 9 & -12 & 9 \\ -12 & 16 & -12 \\ 9 & -12 & 9 \end{bmatrix} \qquad (8.125)$$

8.3.2 Special rational matrix functions of second order

In the previous section we assumed that $\bar{a}$ is not proportional to a. Here we drop this assumption and require explicitly that $\bar{a} = ca$, where c is a complex number. A detailed analysis shows that this assumption is fulfilled only if a is a real or a purely imaginary vector ($a = ib$), where b is a real vector. If the vector a is real we obtain $P_0 = -(\lambda + \bar{\lambda})aa^T$ and $P_1 = 2aa^T$. Obviously, P_0 as well as P_1 has rank one. Furthermore P_1 can't vanish. We assume that the vectors f and g are proportional to the vector a, that is, $f = \alpha a$ and $g = \beta a$. Clearly, α and β have to be real. The set of equations defining the realization parameters are

$$E + gg^T = 0, \qquad (8.126.\text{a})$$

$$ff^T - \lambda\bar{\lambda}gg^T = -(\lambda + \bar{\lambda})aa^T , \qquad (8.126.\text{b})$$

$$fg^T + gf^T + (\lambda + \bar{\lambda})gg^T = 2aa^T . \qquad (8.126.\text{c})$$

Proceeding as in the previous section we obtain the realization of the first kind

$$f_1 = \left(\frac{\bar{\lambda}}{\sqrt{\lambda - \bar{\lambda}}} + \frac{\lambda}{\sqrt{\bar{\lambda} - \lambda}} \right)a \qquad g_1 = -\left(\frac{1}{\sqrt{\lambda - \bar{\lambda}}} + \frac{1}{\sqrt{\bar{\lambda} - \lambda}} \right)a \qquad E_1 = -gg^T . \qquad (8.127)$$

This solution is also obtained with the equations (8.117). Hence, these equations are still valid if the vector a is real. The realization of the second kind is defined by the set of equations (8.119).

A purely imaginary vector $b = ia$ yields the two matrices $P_0 = (\lambda + \bar{\lambda})aa^T$ and $P_1 = -2aa^T$ of rank one. Hence, defining the equations (8.126.b) and (8.126.c) the right-hand sides change the sign. As a consequence, the realization parameters obtained for a real vector a are also solutions for the case with a purely imaginary vector $a = ib$. The only difference is that the realization of the first kind becomes a realization of the second kind and vice versa.

If the poles are purely imaginary we have $\lambda + \bar{\lambda} = 0$ so that P_0 vanish. In this case, the realization of the first kind is considerably simplified:

$$f_1 = \sqrt{|\lambda|}a \qquad g_1 = \frac{1}{\sqrt{|\lambda|}}a \qquad E_1 = -\frac{1}{|\lambda|}aa^T = -\frac{P_1}{2|\lambda|}. \qquad (8.128)$$

It is easy to show that the realization of the second kind is given by

$$f_2 = \sqrt{|\lambda|}a \qquad g_2 = -\frac{1}{\sqrt{|\lambda|}}a \qquad E_2 = \frac{1}{|\lambda|}aa^T = \frac{P_1}{2|\lambda|}. \qquad (8.129)$$

Hence, similar to the analogous scalar case, there exist always two realizations.

Another special case arises if P_1 vanish. This occurs if $aa^T + \bar{a}\bar{a}^T = 0$. This is equivalent to the identity $a = \sqrt{i}b$, where b is an arbitrary real vector. The matrix P_0 becomes $P_0 = -i(\lambda - \bar{\lambda})bb^T = 2Im(\lambda)bb^T$ which is obviously real. The set of equations defining the realization parameters are given by

$$E + gg^T = 0 , \tag{8.130.a}$$

$$ff^T - \lambda\bar{\lambda}gg^T = -i(\lambda - \bar{\lambda})bb^T , \tag{8.130.b}$$

$$fg^T + gf^T + (\lambda + \bar{\lambda})gg^T = 0 . \tag{8.130.c}$$

Assuming a solution in the form $f = \alpha b$ and $g = \beta b$ yields the set of equations

$$\alpha^2 - \lambda\bar{\lambda}\beta^2 = -i(\lambda - \bar{\lambda}) , \tag{8.131.a}$$

$$2\alpha\beta + (\lambda + \bar{\lambda})\beta^2 = 0 . \tag{8.131.b}$$

One solution is given by the pair

$$\alpha_1 = \sqrt{-i(\lambda - \bar{\lambda})} = \sqrt{2Im(\lambda)} \qquad \beta_1 = 0 . \tag{8.132}$$

and the other by

$$\alpha_2 = \frac{(\lambda + \bar{\lambda})}{\sqrt{i(\lambda - \bar{\lambda})}} = \sqrt{\frac{-2}{Im(\lambda)}}Re(\lambda) \qquad \beta_2 = 2\sqrt{\frac{-1}{Im(\lambda)}} . \tag{8.133}$$

Obviously, these solution generate a realization of the first kind if $Im(\lambda)$ is chosen to be negative and a realization of the second kind $Im(\lambda)$ is chosen to be positive. The realization parameters are given by

$$f_1 = \sqrt{2Im(\lambda)}b \qquad g_1 = 0 \qquad E_1 = 0. \tag{8.134}$$

and

$$f_2 = \sqrt{\frac{2}{Im(\lambda)}}Re(\lambda)b \qquad g_2 = 2\sqrt{\frac{1}{Im(\lambda)}}b \qquad E_1 = -\frac{4}{Im(\lambda)}bb^T . \tag{8.135}$$

These realizations are also obtained when (8.117) and (8.119) are used. Hence these equations covers the symmetric realization of most of the rational matrix functions of second order.

8.3.3   General rational matrix functions of first order

Rational matrix functions of first order are given by

$$H(s) = \frac{P_0}{q_0 + s} , \tag{8.136}$$

where the vector a has the dimension of $H(s)$. In analogy to the scalar system, we construct a symmetric realization in the form

$$\begin{aligned} -H(s)u(s) + f\phi(s) &= 0 \\ -(q_0 + s)\phi(s) + f^T u(s) &= 0 \end{aligned} \tag{8.137}$$

Combining the two equations we eliminate $\phi(s)$ and obtain

$$-H(s) + \frac{ff^T}{q_0 + s} = 0 . \tag{8.138}$$

The last equation yields

$$ff^T = P_0. \tag{8.139}$$

Since P_0 is a real and symmetric matrix of rank one it can be represented in their spectral form

$$P_0 = \mu w w^T, \tag{8.140}$$

where μ is the eigenvalue of P_0 and w is the associated eigenvector. The last equation yields $f = \sqrt{\mu} w$.

If P_0 is positive definite ($\mu > 0$) the realization is of the first kind and if P_0 is negative definite ($\mu < 0$) the realization is of the second kind.

Another symmetric realization is defined by the following equations

$$\begin{aligned} -H(s)u(s) + f\phi(s) + sg\phi(s) + Eu(s) &= 0 \\ -(q_0 s + s^2)\phi(s) + f^T u(s) + s g^T u(s) &= 0 \end{aligned} \tag{8.141}$$

Combining these equations we can eliminate $\phi(s)$ and obtain

$$-H(s) + \frac{(f + sg)(f^T + sg^T)}{q_0 s + s^2} + E = 0. \tag{8.142}$$

Comparing the last equation with (8.127) we obtain three equations defining the set $\{f, g, E\}$

$$E + gg^T = 0, \tag{8.143.a}$$

$$ff^T = 0, \tag{8.143.b}$$

$$fg^T + gf^T - q_0 gg^T = P_0 = \mu w w^T. \tag{8.143.c}$$

The solution is given by

$$f = 0 \qquad g = \sqrt{-\frac{\mu}{q_0}} w \qquad E = -\frac{\mu}{q_0} w w^T = -\frac{P_0}{q_0}. \tag{8.144}$$

Hence, if P_0 is positive definite ($\mu > 0$) the realization is of the second kind and if P_0 is negative definite ($\mu < 0$) the realization is of the first kind.

A last symmetric realization is constructed perturbing the rational matrix function of first order given by (8.136). A perturbed rational matrix function of second order is

$$H_\varepsilon(s) = \frac{1}{2}\frac{P_0(1 + i\varepsilon)}{s + (q_0 + i\varepsilon)} + \frac{1}{2}\frac{P_0(1 - i\varepsilon)}{s + (q_0 - i\varepsilon)} = \frac{P_0(q_0 + \varepsilon^2) + P_0 s}{s^2 + 2q_0 s + (q_0^2 + \varepsilon^2)}. \tag{8.145}$$

As for scalar rational functions, the first perturbation term is of the order $O(\varepsilon^2)$. Its symmetric realization is obtained with the equations (8.117) and (8.119).

Remarks:

- The approach used in this work to construct symmetric realizations is that proposed by Weber [Web94]. However, we gave here a new and detailed derivation and analysis of symmetric realizations. Moreover, many new features are discussed (e.g. nonuniqueness, perturbation etc.) which were not covered before.

- Another approach has been developed by Wolf [Wol91a, Wol91b]. This approach is based on the physical representation of symmetric realizations as lumped parameter models. However, the parameters describing the mass, damping and stiffness of these models are frequently negative so that their physical interpretation is arduous. Furthermore, often the physical interpretation of the symmetric realization is fully disclosed only after having coupled the realization with the rest of the equations (see the example on page 190).

8.4   The matrix representation of symmetric realizations

Assembling all elementary elements of a transfer function generates a matrix representation of the symmetric realization. It is defined by the matrix equations

$$-H(s)u(s) + F\phi(s) + sG\phi(s) + Eu(s) = 0$$
$$-(K + Cs + Ms^2)\phi(s) + F^T u(s) + sG^T u(s) = 0 \quad .$$

$$(8.146)$$

$\phi(s)$ is a column vector containing all auxiliary variables. The dimension of the vector $\phi(s)$ is $n_\phi = n_1 + n_2$, where n_1 corresponds to the number of rational functions of first order and n_2 to those of second order. F and G are matrices of dimension $n_H \times n_\phi$. Their columns are composed by the realization parameters f and g. That is,

$$F = \begin{bmatrix} f_1 & f_2 & \cdots & f_{n_{\phi-1}} & f_{n_\phi} \end{bmatrix} \qquad G = \begin{bmatrix} g_1 & g_2 & \cdots & g_{n_{\phi-1}} & g_{n_\phi} \end{bmatrix}$$

$$(8.147)$$

If the transfer function is a scalar $n_H = 1$ so that F as well as G are row vectors. The $n_H \times n_H$ matrix E is the sum of the corresponding E_j

$$E = E_1 + E_1 + \ldots + E_{n_{\phi-1}} + E_{n_\phi} .$$

$$(8.148)$$

The diagonal matrices K, C and M have dimension $n_\phi \times n_\phi$ and contain the coefficients of the denominators

$$K = \begin{bmatrix} q_{01} & 0 & \cdots & 0 \\ 0 & q_{02} & \cdots & 0 \\ \cdots & \cdots & \cdots & \cdots \\ 0 & 0 & \cdots & q_{0n_\phi} \end{bmatrix} = \begin{bmatrix} \lambda_1 \bar{\lambda}_1 & 0 & \cdots & 0 \\ 0 & \lambda_2 \bar{\lambda}_2 & \cdots & 0 \\ \cdots & \cdots & \cdots & \cdots \\ 0 & 0 & \cdots & \lambda_{n_\phi} \bar{\lambda}_{n_\phi} \end{bmatrix}$$

$$(8.149)$$

$$C = \begin{bmatrix} q_{11} & 0 & \cdots & 0 \\ 0 & q_{12} & \cdots & 0 \\ \cdots & \cdots & \cdots & \cdots \\ 0 & 0 & \cdots & q_{1n_\phi} \end{bmatrix} = \begin{bmatrix} -(\lambda_1 + \bar{\lambda}_1) & 0 & \cdots & 0 \\ 0 & -(\lambda_2 + \bar{\lambda}_2) & \cdots & 0 \\ \cdots & \cdots & \cdots & \cdots \\ 0 & 0 & \cdots & -(\lambda_{n_\phi} + \bar{\lambda}_{n_\phi}) \end{bmatrix}$$

The matrix M corresponds to the unit matrix. Note that these matrices can be scaled using the scaling transformation described in section 8.1.3.

8.5 Absorbing boundary elements

In this section we will use the theory of symmetric realization of matrix-valued rational functions to construct so-called absorbing boundary elements. These are implementations of absorbing boundary conditions which are local in space. They represent rough approximations of exact DtN-maps so that are generally not as accurate as DtN-maps are. Nevertheless, for certain applications (e.g. impulsive loads) absorbing boundary elements may furnish sufficiently accurate solutions with much less numerical effort than using more accurate nonlocal absorbing boundary conditions. In addition, because absorbing boundary elements are local in space they fit well into the framework of the finite element method. In fact, the viscous boundary conditions are so popular because they are easy to implement in a finite element program. In contrast, the little interest of the finite element community for Engquist-Majda boundary conditions is mainly due to the fact that no sufficiently simple finite element implementation exists so far.

First, we shall construct absorbing boundary elements of the Engquist-Majda boundary conditions for the scalar wave equation using their analytical formulations. Later, we derive the same boundary conditions with the technique used in chapter 3 to develop nonlocal DtN-maps of semi-infinite strata. Then, we derive absorbing boundary elements for circular artificial boundaries using the absorbing boundary conditions of Bayliss and Turkel.

8.5.1   Implementation of the Engquist-Majda boundary conditions

In the following, we assume that the domain is two-dimensional and that it is described by an orthogonal Cartesian coordinate system. Furthermore, let us assume that the artificial boundary is defined along a vertical straight line. The Laplace transform of the Engquist-Majda boundary condition of first order is

$$\sigma + \frac{s}{c}u - \frac{c}{2s}u_{,yy} = 0 , \tag{8.150}$$

where σ denotes the generalized stresses and u denotes the displacement field. The weak formulation of the latter equation is

$$\int_{\Omega_e} \left(\sigma v + \frac{s}{c}uv + \frac{c}{2s}u_{,y}v_{,y} \right) dy = 0 . \tag{8.151}$$

We assume that the displacement variable u is discretized using standard isoparametric finite element techniques. The shape functions $N(y)$ are chosen to be consistent with those of the adjacent bulk element. Integration of (8.151) over an element gives

$$f + \frac{s}{c}Pu + \frac{c}{2s}Qu = 0 . \tag{8.152}$$

f is the vector of the nodal forces and u that of the nodal displacements. P and Q are defined by

$$P = \int_h N^T N dy \qquad Q = \int_h N_{,y}^T N_{,y} dy . \tag{8.153}$$

Using isoparametric elements with two and three nodes the element matrices P and Q are

$$P_2 = \frac{h}{6}\begin{bmatrix} 2 & 1 \\ 1 & 2 \end{bmatrix} \qquad P_3 = \frac{h}{30}\begin{bmatrix} 4 & 2 & -1 \\ 2 & 8 & 2 \\ -1 & 2 & 4 \end{bmatrix} \tag{8.154}$$

$$Q_2 = \frac{1}{h}\begin{bmatrix} 1 & -1 \\ -1 & 1 \end{bmatrix} \qquad Q_3 = \frac{1}{3h}\begin{bmatrix} 7 & -8 & 1 \\ -8 & 16 & -8 \\ 1 & -8 & 7 \end{bmatrix} \qquad (8.155)$$

These element matrices P are integrated into the global damping matrix of the near field. P_2 as well as P_3 are positive definite.

To include the last term of (8.152) into the global system of equations of motion we use the symmetric realization procedure discussed in section 8.3. The matrix Q_2 is symmetric and positive semi-definite. It has rank one so that according to the spectral theorem of Hermitian matrices it can be represented by

$$Q_2 = \lambda v v^T = q q^T, \qquad (8.156)$$

where λ is the eigenvalue of Q_2 not equal to zero and v is the associated eigenvector which is normalized with respect to the euclidian norm. A straightforward calculation leads to

$$q = \frac{1}{\sqrt{h}}\begin{bmatrix} 1 \\ -1 \end{bmatrix}. \qquad (8.157)$$

Since the matrix Q_2 has rank one we argue that only one additional degree of freedom is necessary to construct the symmetric realization. In fact, as can be verified by eliminating the auxiliary degrees of freedom ϕ, the pair of equations

$$f + \frac{1}{c}P\dot{u} + \sqrt{\frac{c}{2}}q\phi = 0$$
$$\dot{\phi} = \sqrt{\frac{c}{2}}q^T u \qquad (8.158)$$

is the symmetric realization of (8.152). Hence, each absorbing boundary element augments the number of the degrees of freedom of the global system of ODE by one. The symmetric realization (8.158) can be formulated by way of a damping and a stiffness matrix as

$$C_e = \frac{1}{6c}\begin{bmatrix} 2h & h & 0 \\ h & 2h & 0 \\ 0 & 0 & -6c \end{bmatrix} \qquad K_e = \sqrt{\frac{c}{2h}}\begin{bmatrix} 0 & 0 & 1 \\ 0 & 0 & -1 \\ 1 & -1 & 0 \end{bmatrix} \qquad (8.159)$$

The associated generalized displacement vector is $x^T = [u_1, u_2, \phi]$. It contains two displacement components and the auxiliary variable. Notice that both matrices are indefinite.

The matrix Q_3 is symmetric and positive semi-definite but it is of rank two. Therefore, its symmetric realization needs two auxiliary variables. The symmetric realization is

$$f + \frac{1}{c}P\dot{u} + \sqrt{\frac{c}{2}}q_1\phi_1 + \sqrt{\frac{c}{2}}q_2\phi_2 = 0$$
$$\dot{\phi}_1 = \sqrt{c}q_1^T u \qquad (8.160)$$
$$\dot{\phi}_2 = \sqrt{c}q_2^T u$$

The vectors q_1 and q_2 are computed using the spectral decomposition of Q_3 and are

$$q_1^T = \frac{1}{\sqrt{h}}\begin{bmatrix} -1 & 0 & 1 \end{bmatrix} \qquad q_2^T = \frac{2}{\sqrt{3h}}\begin{bmatrix} 1 & -2 & 1 \end{bmatrix} \tag{8.161}$$

The damping and stiffness matrices of the absorbing boundary elements follows immediately.

Remarks:

- The symmetry of the matrix Q is a necessary prerequisite for a symmetric realization.
- Observe that only the projection of the displacement vector u against q given by (8.157) or the projection of u in the subspace spanned by q_1 and q_2 defined in (8.161) produces a nonzero force vector associated with the term $cQu/2s$ in (8.152).

The symmetric realization of the Engquist-Majda boundary condition of second order is a little more involved. We start with the Laplace transform of its analytic expression

$$\sigma - \frac{c^2}{4s^2}\sigma_{,yy} + \frac{s}{c}u - \frac{3c}{4s}u_{,yy} = 0. \tag{8.162}$$

Now let us define a new variable $\sigma' = \sigma + (s/c)u$. This transforms the last equation to

$$4s^2\sigma' - c^2\sigma'_{,yy} - 2csu_{,yy} = 0. \tag{8.163}$$

Since both variables u and σ' are partially differentiated with respect to y we perform a discretization of u as well as of σ' with the same shape functions $N(y)$. This yields

$$4s^2Ps + c^2Qs + 2csQu = 0. \tag{8.164}$$

Defining $f = Ps$, (8.164) can be written as

$$4s^2f + c^2QP^{-1}f + 2csQu = 0. \tag{8.165}$$

Hence, the nodal forces f are given by

$$f = -2cs(4s^2I + c^2QP^{-1})^{-1}Qu. \tag{8.166}$$

Noting that $f = f + \frac{s}{c}Pu$ we obtain the discretized Engquist-Majda boundary condition

$$f + \frac{s}{c}Pu + 2cs(4s^2I + c^2QP^{-1})^{-1}Qu = 0. \tag{8.167}$$

This can be written in the more "symmetric" form

$$f + \frac{s}{c}Pu + 2csP(4s^2P + c^2Q)^{-1}Qu = 0. \tag{8.168}$$

The discretization of the last term with two-nodes isoparametric elements yields

$$2csP(4s^2P + c^2Q)^{-1}Q = \frac{1}{2}\frac{csh}{h^2s^2 + 3c^2}\begin{bmatrix} 1 & -1 \\ -1 & 1 \end{bmatrix} = \frac{1}{2}\frac{csh}{h^2s^2 + 3c^2}qq^T \tag{8.169}$$

where the vector q is given by

$$q = \begin{bmatrix} 1 \\ -1 \end{bmatrix} \tag{8.170}$$

Hence, only an auxiliary variable is needed to formulate the symmetric realization. That of the first kind is given by the set of equations

$$f + s\frac{P}{c}u + \alpha q\phi + \beta sq\phi + Eu = 0$$

$$h^2 s^2 \phi + 3c^2 = \alpha q^T u + \beta s q^T u$$

(8.171)

where the parameter pair $\{\alpha, \beta\}$ is defined by

$$\alpha = \frac{3^{1/4}}{2}c \qquad \beta = \frac{3^{-1/4}}{2}h$$

(8.172)

and the matrix E is

$$E = -\frac{1}{4\sqrt{3}}qq^T.$$

(8.173)

The parameter pair $\{\alpha, \beta\}$ of the realization of the second kind is given by

$$\alpha = \frac{3^{1/4}}{2}c \qquad \beta = -\frac{3^{-1/4}}{2}h$$

(8.174)

and the matrix E is

$$E = \frac{1}{4\sqrt{3}}qq^T.$$

(8.175)

The symmetric realization for a three-node isoparametric element needs two auxiliary variables. The third term in (8.168) can be decomposed into a sum of two rational functions so that it becomes

$$2csP(4s^2P + c^2Q)^{-1}Q = \frac{1}{2}\frac{csh}{h^2s^2 + 3c^2}\begin{bmatrix} 1 & 0 & -1 \\ 0 & 0 & 0 \\ -1 & 0 & 1 \end{bmatrix} + \frac{2}{3}\frac{csh}{h^2s^2 + 15c^2}\begin{bmatrix} 1 & -2 & 1 \\ -2 & 4 & -2 \\ 1 & -2 & 1 \end{bmatrix}.$$

(8.176)

Both matrices on the right-hand side are positive semi-definite and of rank one. The associated vectors q are given by

$$q_1^T = \begin{bmatrix} 1 & 0 & -1 \end{bmatrix} \qquad q_2^T = \begin{bmatrix} 1 & -2 & 1 \end{bmatrix}$$

(8.177)

The realization parameters associated with the vector q_1 are identical to those given in (8.172) and (8.174). The parameters of the realization of the first kind associated with the vector q_2 are

$$\alpha_1 = \sqrt{1/3}\,15^{1/4}c \qquad \beta_1 = \sqrt{1/3}\,15^{-1/4}h$$

(8.178)

and those of the second kind are

$$\alpha_2 = \sqrt{1/3}\,15^{1/4}c \qquad \beta_2 = -\sqrt{1/3}\,15^{-1/4}h.$$

(8.179)

Hence, the symmetric realization is given by

$$f + \frac{P}{c}\ddot{u} + \alpha_1 q_1 \phi_1 + \beta_1 q_1 \dot{\phi}_1 + \alpha_2 q_2 \phi_2 + \beta_2 q_2 \dot{\phi}_2 + E_1 u + E_2 u = 0$$

$$h^2 \ddot{\phi} + 3c^2 = (\alpha_1 q_1^T u + \beta_1 q_1^T \dot{u}) \quad\quad\quad (8.180)$$

$$h^2 \ddot{\phi} + 15c^2 = (\alpha_2 q_2^T u + \beta_2 q_2^T \dot{u})$$

8.5.2 Absorbing boundary elements using a plane stratum

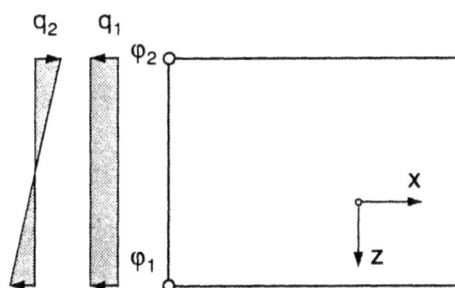

Figure 8.2: Fluid stratum discretized by a two-node finite element in z-direction.

In this section, we develop the local absorbing boundary elements of Engquist and Majda with the semi-analytical method used in chapter 3 to construct accurate DtN-maps. We mainly introduce two simplifications. First, the stratum is reduced to a single element. This guarantees that the absorbing boundary elements are local. Second, the eigenvalue problem is solved asymptotically for $s \to \infty$. We will show that the absorbing boundary elements constructed by this method are equal to those developed in the previous section.

Consider the plane fluid stratum shown in Figure 8.2. We assume that the solution of the equation of motion is given in the form of a plane harmonic wave $\varphi(t, x, z) = \phi(z)e^{-kx+st}$. Here, we use k and s instead of ik and $i\omega$. The finite element modeling of the eigenvalue problem yields

$$(k^2 A_e - G_e - s C_e - s^2 M_e)\Phi = 0, \quad\quad\quad (8.181)$$

where the element matrices of a two-nodes isoparametric element are

$$A_e = \frac{h}{6}\begin{bmatrix} 2 & 1 \\ 1 & 2 \end{bmatrix} \quad\quad G_e = \frac{1}{h}\begin{bmatrix} 1 & -1 \\ -1 & 1 \end{bmatrix} \quad\quad M_e = \frac{h}{6c^2}\begin{bmatrix} 2 & 1 \\ 1 & 2 \end{bmatrix}. \quad\quad (8.182)$$

In order to perform the asymptotic expansion we divide (8.182) with s^2. This yields

$$(\kappa^2 A_e - \varepsilon^2 G_e - M_e)\Phi = 0, \qu\quad\quad\quad (8.183)$$

where $\varepsilon = 1/s$ and $\kappa = k/s$. ε is a small parameter because $s \to \infty$. From section 3.1.3 we know that the eigenvectors are frequency independent so that the matrices in (8.183) are simultaneously diagonalizable. We define the modal basis using the column vectors of the unitary matrix

$$U = \frac{1}{\sqrt{2}}\begin{bmatrix} 1 & 1 \\ 1 & -1 \end{bmatrix}. \quad\quad\quad (8.184)$$

This basis is sketched in Figure 8.2. The congruence transformation of (8.183) with (8.184) yields

$$(\kappa^2 A - \varepsilon^2 G - M)\varphi_e = 0 \qu\quad\quad\quad (8.185)$$

where the matrices are given by

$$A = \frac{h}{6}\begin{bmatrix} 3 & 0 \\ 0 & 1 \end{bmatrix} \quad\quad G = \frac{1}{h}\begin{bmatrix} 0 & 0 \\ 0 & 2 \end{bmatrix} \quad\quad M = \frac{h}{6c^2}\begin{bmatrix} 3 & 0 \\ 0 & 1 \end{bmatrix}. \quad\quad (8.186)$$

Because the eigenvectors are frequency independent we have only to expand the diagonal eigen-value matrix $K = \text{diag}(\kappa_i)$ with respect to the small parameter ε

$$K = K_0 + K_1\varepsilon + K_2\varepsilon^2 + \ldots = \sum_{n=0}^{\infty} K_n\varepsilon^n. \qquad (8.187)$$

Combining the last equation with (8.186) and assembling all terms with equal power of ε yields the infinite set of equations

$$
\begin{aligned}
\varepsilon^0: && AK_0^2 - M &= 0 \\
\varepsilon^1: && 2AK_0K_1 &= 0 \\
\varepsilon^2: && 2AK_0K_2 + AK_1^2 - G &= 0 \\
\varepsilon^3: && K_0K_3 + K_1K_2 &= 0 \\
&& \ldots & \\
\varepsilon^n: && \sum_{i=0}^{n} K_iK_{n-i} &= 0
\end{aligned}
\qquad (8.188)
$$

These equations are solved recursively beginning with the first equation. They exhibit a simple structure because all the matrices are diagonal and the eigenvectors are frequency independent.

Next, we introduce the approximated DtN-map kernel of the order m, denoted by $H_{(m)}$, which, in analogy to (3.31), is defined as

$$H_{(m)} = sUAK_{(m)}U^T = \sum_{n=-1}^{m-1} H_{-n}s^{-n}, \qquad (8.189)$$

where $K_{(m)}$ corresponds to the sum of the asymptotic expansion (8.187) up to the term with the mth power of $\varepsilon = 1/s$:

$$K_{(m)} = \sum_{n=0}^{m} K_n(1/s)^n \qquad (8.190)$$

and $H_{-n} = sU^TAK_nU$ is the coefficient of the term of the order $O(1/s^n)$ of the approximated DtN-map kernel $H_{(m)}$.

The first equations in (8.188) yields

$$K_0 = \left(A^{-1}M\right)^{1/2} = \pm\frac{1}{c}I. \qquad (8.191)$$

We consider only waves travelling in the positive x-direction so that we choose the positive solution. Combining it with (8.189) yields the approximated DtN-map kernel of order zero:

$$H_1 = -sU^TAK_0U = \frac{sh}{6c}\begin{bmatrix} 2 & 1 \\ 1 & 2 \end{bmatrix} \qquad (8.192)$$

This is the finite element implementation of the viscous boundary condition. The second equation in (8.188) yields $K_1 = 0$. Furthermore, because of this all terms K_k with odd index k vanish: $K_{2j+1} = 0$, $j = 0\ldots\infty$. As a consequence, all terms of the approximated DtN-map kernel with even index vanish: $H_{2j} = 0$, $j = 0\ldots\infty$. Hence, the next nonzero term in the asymptotic expansion (8.187) is

$$K_2 = \frac{c}{2s} A^{-1} G.$$ (8.193)

The term H_{-1} becomes

$$H_{-1} = sUAK_2U^T = \frac{c}{2}UGU^T = \frac{c}{2}G_e = \frac{c}{2h}\begin{bmatrix} 1 & -1 \\ -1 & 1 \end{bmatrix}.$$ (8.194)

H_{-1} is positive semi-definite and is identical to $cQ_2/2$ given in (8.152) which was derived using the analytical expression of the Engquist-Majda boundary condition of first order.

The fifth equation in (8.188) gives

$$K_4 = \frac{1}{2}K_0^{-1}K_2^2 = \frac{c}{2}A^{-2}G^2$$ (8.195)

so that

$$H_{-3} = sUAK_4U^T = -\frac{c}{2}UA^{-1}G^2U^T = \frac{3c^3}{2h^3}\begin{bmatrix} -1 & 1 \\ 1 & -1 \end{bmatrix}.$$ (8.196)

H_{-3} is negative semi-definite. Next, we compute the Padé approximation of degree two considering the term $z(H_{-1} + H_{-3}z^2)$, where $z = 1/s$. Replacing z with s in the Padé approximation we obtain

$$Q_{[1,2]} = \frac{shc}{2(h^2s^2 + 3c^2)}\begin{bmatrix} 1 & -1 \\ -1 & 1 \end{bmatrix}.$$ (8.197)

Observe that $Q_{[1,2]}$ is proportional to G_e and therefore semi-definite. The last equation corresponds exactly to the right hand side of (8.169). Therefore, the method described in this section leads to the same finite element implementation of the Engquist-Majda boundary conditions obtained in section 8.5.1.

With the same technique we can construct absorbing boundary elements with three nodes. In this case, it is advantageous to choose a modal basis Ψ normalized according to $\Psi^T A\Psi = I$. With respect to this basis system, the terms of the approximated DtN-map are diagonal. The computation of the Padé approximations is therefore much simpler because the rational approximation contain two rational functions of second order. As can be shown, we obtain exactly the same absorbing boundary elements that have been derived in section 8.5.1.

Proceeding along this path, we can construct absorbing boundary elements of any arbitrary order. However, as we shall see, local absorbing boundary conditions of the order higher than the first are not significantly more effective as those of the first order (at least for a stratum).

Remarks:
- Because U defined by (8.184) diagonalize the eigenvalue problem (8.181) we could first solve the eigenvector problem analytically and then develop the asymptotic expansion. However, the technique presented in this section allows to construct asymptotic expansions of DtN-maps even when the analytical solution of the eigenvalue problem (8.181) is not available.
- The realization procedure developed in this section has the advantage of being conceptually simpler than that of section 8.5.1. As we shall see, it allows a simple physical interpretation of these local absorbing boundary elements.

8.5.3 Implementation of the Bayliss-Turkel boundary conditions

So far we developed absorbing boundary elements of local absorbing boundary conditions formulated with respect to an orthogonal Cartesian coordinate system. However, the same procedure can be also applied with absorbing boundary conditions formulated with respect to a cylindrical coordinate system. Consider, the three absorbing boundary conditions formulated by Bayliss and Turkel [BT80]:

$$\varphi_n + \frac{1}{c}\dot\varphi = 0, \tag{8.198.a}$$

$$\varphi_n + \frac{1}{c}\dot\varphi + \frac{1}{2R}\varphi = 0, \tag{8.198.b}$$

$$\dot\varphi_n + \frac{c}{R}\varphi_n + \frac{1}{c}\ddot\varphi + \frac{3}{2R}\dot\varphi + \frac{3c}{8R^2}\varphi - \frac{c}{2}\varphi_{,yy} = 0. \tag{8.198.c}$$

φ_n is the normal derivative $\varphi_n = \nabla\varphi \cdot n$ at the boundary (n is the unit outer vector normal to the boundary, see Figure 8.3). R is the radius of the circular artificial Γ_B boundary and y is the variable pointing in the circumferential direction (perpendicular to n).

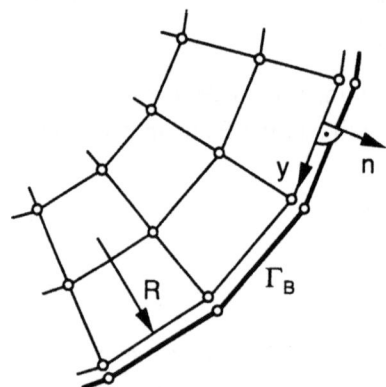

We assume that the discretized artificial boundary is linear between two nodes (see Figure 8.3). The implementation of the absorbing boundary conditions (8.198.a) and (8.198.b) is straightforward and gives

$$f + \frac{h}{6c}\begin{bmatrix} 2 & 1 \\ 1 & 2 \end{bmatrix}\dot\varphi = 0, \tag{8.199.a}$$

$$f + \frac{h}{6c}\begin{bmatrix} 2 & 1 \\ 1 & 2 \end{bmatrix}\dot\varphi + \frac{h}{12R}\begin{bmatrix} 2 & 1 \\ 1 & 2 \end{bmatrix}\varphi = 0. \tag{8.199.b}$$

Figure 8.3: Discretization of the circular artificial boundary

h is the length of the element. Equation (8.199.a) is the viscous boundary condition. In (8.199.b) a spring term is added to the viscous term. The matrix of the spring term is positive definite. Both these boundary conditions are easily introduced into the global damping and stiffness matrices generated by the finite element semi-discretization of the near field.

To implement the third boundary condition described by (8.198.c) we need the realization procedure developed in the last chapter. First we map (8.198.c) via Laplace transform to the algebraic equation

$$\varphi_n + \frac{s}{c}\varphi + \frac{1}{2R}\varphi - \frac{1}{8R^2}\frac{c}{(s + c/R)}\varphi - \frac{c}{2}\frac{\varphi_{,yy}}{(s + c/R)} = 0. \tag{8.200}$$

Next, we perform the semi-discretization and obtain

$$f + \frac{sP}{c}\varphi + \frac{P}{2R}\varphi - \frac{1}{8R^2}\frac{cP}{(s + c/R)}\varphi + \frac{c}{2}\frac{Q}{(s + c/R)}\varphi = 0. \tag{8.201}$$

P and Q are the matrices introduced in section 8.5.1. Because P and Q appears in (8.201) we need two auxiliary degrees of freedom to implement this boundary condition. Moreover, the matrix P has rank two so that it has to be represented as a sum of two matrices of rank one:

$$P = P_1 + P_2 = \frac{h}{12}\begin{bmatrix} 1 & -1 \\ -1 & 1 \end{bmatrix} + \frac{h}{4}\begin{bmatrix} 1 & 1 \\ 1 & 1 \end{bmatrix} \tag{8.202}$$

Because the matrix P_1 is proportional to the matrix Q the component of the second term of (8.201) associated with P_1 can be combined with the last term of (8.201). Using the symmetric realization (8.138) we obtain

$$f_1 = \frac{\sqrt{6}}{24R}\sqrt{\frac{c}{h}}\sqrt{48R^2 - h^2}\begin{bmatrix} -1 \\ 1 \end{bmatrix} \tag{8.203}$$

The component of the second term of (8.201) associated with P_2 gives a symmetric realization with

$$f_2 = \frac{\sqrt{2ch}}{8R}\begin{bmatrix} 1 \\ 1 \end{bmatrix} \tag{8.204}$$

With these results the damping and stiffness matrices of the absorbing boundary element are

$$C_e = \frac{1}{6c}\begin{bmatrix} 2h & h & 0 & 0 \\ h & 2h & 0 & 0 \\ 0 & 0 & 6c & 0 \\ 0 & 0 & 0 & -6c \end{bmatrix} \qquad K_e = \frac{1}{12R}\begin{bmatrix} 2h & h & \alpha & -\beta \\ h & 2h & \alpha & \beta \\ \alpha & \alpha & 12c & 0 \\ -\beta & \beta & 0 & -12c \end{bmatrix} \tag{8.205}$$

where $\alpha = (3/2)\sqrt{ch}$ and $\beta = (\sqrt{6}/2R)\sqrt{c/h}\sqrt{48R^2 - h^2}$.

Remarks:

- Notice that because the term in (8.201) being proportional to P is negative we need a realization of the second kind.

- The factor β is positive provided that $R > h/(4\sqrt{3})$. This condition is easily fulfilled.

- Recently, [KB93] derived time-domain finite element implementations of local absorbing boundary conditions for circular artificial boundaries. The main idea of the implementation was similar to these of this work: auxiliary variables were introduced to symmetrize the implementation. However, the definition of the additional variables was fundamentally different. In fact, the implementation of the Bayliss and Turkel boundary condition (8.198.c) needs four auxiliary variables per local boundary element. The implementation developed in this work needs only two.

8.5.4 Nonlocal and local Engquist-Majda boundary conditions

For semi-infinite strata, the Engquist-Majda boundary conditions can be defined in spatially non-local form. Consider the discretized form of the local Engquist-Majda boundary condition of first order which is given by (8.152):

$$f_e + \frac{s}{c}P_e u + \frac{c}{2s}Q_e u = 0. \tag{8.206}$$

There, the discretization was carried out for a single element. The discretization on a straight artificial boundary with n absorbing boundary elements yields

$$f + \frac{s}{c}Pu + \frac{c}{2s}Qu = 0, \tag{8.207}$$

where the global matrices P and Q are assembled using the element matrices P_e and Q_e. We assumed that the upper and lower Dirichlet boundary conditions are $u_{,y}(-H) = u_{,y}(H) = 0$.

On the other hand, we can apply the method of section 4.1 to construct a DtN-map. The reduced eigenvalue problem defining the modal basis is

$$\lambda^2 Pu - Qu = 0. \tag{8.208}$$

In this equation, the matrices P and Q are the same of (8.207). The eigenvalues of (8.208) are the cut-off frequency of the stratum times the wave speed c. Since the matrix P is symmetric and positive definite and the matrix Q is symmetric and positive semi-definite the eigenvalues λ^2 are real. Furthermore, one is equal to zero and the other are positive. We normalize the real eigenvectors according to $U^T QU = I$, where the matrix U is composed by the set of orthogonal eigenvectors $U = [u_1,...,u_n]$ of (8.208). Then, because of (8.208) we have

$$\Lambda^2 = U^T QU \tag{8.209}$$

with the real diagonal matrix Λ^2 defined by $\Lambda_{ii} = \lambda_i^2$. Next, we define the displacement v with respect to the modal basis $u = Uv$. If (8.206) is multiplied with U^T from the left and u is expressed by the modal displacement vector v we obtain

$$U^T f + \frac{s}{c} Iv + \frac{c}{2s} \Lambda v = 0. \tag{8.210}$$

Let $g = U^T f$ be the vector of modal forces. Then we obtain the uncoupled system of equations

$$g + \frac{s}{c} Iv + \frac{c}{2s} \Lambda^2 v = 0. \tag{8.211}$$

This set of equations expresses the nonlocal Engquist-Majda boundary condition of first order with respect to the modal basis defined by the eigenvectors of the eigenvalue problem (8.208).

Next we show that (8.211) and, as a consequence, (8.206) are equivalent to the approximation of the order $O(1/s)$ of the accurate nonlocal DtN-map. According to section 3.1.6, the DtN-map of a fluid layer with homogeneous boundary conditions and formulated with respect to the modal basis defined by the eigenvectors of (8.208) is

$$g + Kv = 0, \tag{8.212}$$

where K is diagonal and is given by (see section 3.1.3)

$$K = \left(\Lambda^2 + \left(\frac{s}{c} \right)^2 I \right)^{1/2} \tag{8.213}$$

Expanding K asymptotically for $s \to \infty$ up to the order $O(1/s)$ we obtain

$$g + \frac{s}{c} Iv + \frac{c}{2s} \Lambda^2 v = 0. \tag{8.214}$$

This corresponds exactly to (8.211) and establishes the equivalence between (8.206) and the approximation of the order $O(1/s)$ of the nonlocal DtN-map. Hence, the local and nonlocal implementations of the Engquist-Majda boundary condition of first order are equivalent.

The symmetric realization of the nonlocal Engquist-Majda boundary condition of second order is readily obtained from (8.211). In fact, with respect to modal coordinates it is

$$g + \frac{s}{c}Iv + \sqrt{\frac{c}{2}}\Lambda\phi = 0$$

$$s\phi = \sqrt{\frac{c}{2}}\Lambda v$$

(8.215)

where ϕ is a vector. Observing that $U^{-1} = U^T P$ and $U^{-T} = PU$ we obtain the symmetric realization formulated with respect to natural coordinates

$$f + \frac{1}{c}P\dot{u} + \sqrt{\frac{c}{2}}PU\Lambda\phi = 0$$

$$\dot{\phi} = \sqrt{\frac{c}{2}}\Lambda U^T P u$$

(8.216)

Along the same line we can analyze if the local and nonlocal Engquist-Majda boundary conditions of second order are equivalent. First, let us consider the approximation of the order $O(1/s^3)$ of the accurate DtN-map. With respect to the modal basis defined by (8.208) the approximated DtN-map is

$$K_{-3} = \frac{s}{c}I + \frac{c}{2s}\left(1 - \frac{c^2}{4s^2}\Lambda^2\right)\Lambda^2 .$$

(8.217)

Performing the Padé approximation of degree two of the second term on the right hand side of the latter equation we obtain

$$K_{[1,2]} = \frac{s}{c}I + 2cs(4s^2I + c^2\Lambda^2)^{-1}\Lambda^2 .$$

(8.218)

On the other hand, the Padé approximation of the order $O(1/s^3)$ of the one element stratum gives

$$f_e + \left(\frac{s}{c}P_e + \frac{c}{2s}\left(1 - \frac{3c^3}{s^2h^2}\right)Q_e\right)u = 0 .$$

(8.219)

Assembling all boundary elements and formulating the approximated DtN-map with respect to the modal basis yields

$$g + \left(\frac{s}{c}I + \frac{c}{2s}\left(1 - \frac{3c^2}{s^2h^2}\right)\Lambda^2\right)v = 0$$

(8.220)

In order to be equivalent with the nonlocal Engquist-Majda boundary condition of second order, the term in parenthesis should be equal to the right hand side of (8.217). However, it doesn't.

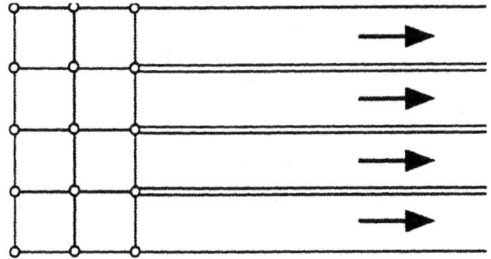

Figure 8.4: Interpretation of absorbing boundary elements as a discontinuous stratum

The construction of the local boundary condition by way of the one element stratum allows a simple interpretation of this fact. The semi-infinite stratum defined by the local absorbing boundary condition is similar to that shown in Figure 8.4. This stratum is composed of a number of strips. The neighboring strips share a common node at the artificial boundary. Because no continuity condition has been imposed between these strips the deformation field of the stratum is discontinuous across the strips. In contrast, the deformation field of the multi-element stratum used to construct nonlocal accurate DtN-maps is continuous. Hence, in the discontinuous stratum no energy exchange occurs transversely to the propagation direction. This fundamentally changes the dynamics of the stratum.

The continuous and the discontinuous stratum show the same behavior for waves with constant amplitude of the field over the height of the stratum because no energy exchange occurs transversely to the propagation direction. This is reflected by the viscous boundary condition. However, all other propagation modes have a different dynamics. The reason why the local and the nonlocal implementation of the Engquist-Majda boundary condition of first order are equivalent is due to the fact that if the element matrices Q_e defined in (8.152) are assembled to a global matrix we obtain the matrix Q defined in (8.208). This relationship doesn't hold if we assemble the element matrices Q_e^2 of the local implementation of the Engquist-Majda boundary condition of second order. Clearly, this assembly doesn't produce the global matrix Q^2 associated with the term Λ^4 in (8.217) so that the local and the global implementations of the Engquist-Majda boundary conditions of second order are different.

Chapter 9

Coupling near and far field

In this chapter we briefly derive the equations of motion of the near field. These equations have a nonstandard structure when the fluid-structure interaction is considered. The equations of motion of the near field are then coupled to those defining the absorbing boundary conditions. This system of equations, we call it the enlarged system, is still symmetric so that it can be integrated with standard numerical time integration algorithms. The last part of this chapter is devoted to the question of well-posedness of these equations of motion. We first analyze the equations of motion of the near field. Finally, we give a sufficient condition for the stability of enlarged systems.

9.1 The finite element model of the near field

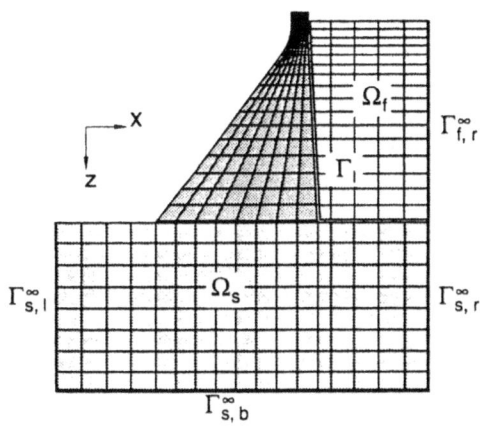

Figure 9.1: Finite element model of the near field with dam, foundation and reservoir.

In this section, we shall formulate the equations of motion of the near field. Consider the two-dimensional finite element model given in Figure 9.1. The near field is composed by the dam and a finite portion of the foundation and reservoir. The state variables are formulated with respect to an orthogonal Cartesian frame of reference with coordinates x and z. The solid medium (dam and foundation) is denoted by Ω_s and the fluid medium by Ω_f. We assume that the absorbing boundary conditions are defined along the artificial boundary denoted by $\Gamma_{f,r}^\infty$, $\Gamma_{s,r}^\infty$, $\Gamma_{s,b}$ and $\Gamma_{s,l}^\infty$. Additional boundary conditions are imposed at the surfaces of the dam, foundation and reservoir. Interface conditions are defined at the solid-fluid interface.

9.1.1 The finite element model of the solid domain

In this section we consider only solid elements with linear elastic materials. The field variables are the displacements u_i with $i \in \{x, z\}$. Consider a solid body in the open domain Ω_s sketched in Figure 9.1. We shall assume that the boundary Γ_s of the solid body is decomposed into four disjoint parts $\Gamma_s = \Gamma_s^u \cup \Gamma_s^\sigma \cup \Gamma_I \cup \Gamma_s^\infty$. At the boundary Γ_s^u we impose essential boundary conditions

$$u_i = q_i, \tag{9.1}$$

where q_i are prescribed displacements. Γ_s^σ refers to the subset of the boundary where natural boundary conditions (stresses) are imposed

$$\sigma_{ij} n_j = g_i. \tag{9.2}$$

σ_{ij} is the symmetric stress tensor, n_j is the unit normal vector pointing outward of the solid and h_i are the prescribed tractions. At the boundary $\Gamma_s^\infty = \Gamma_{s,l}^\infty \cup \Gamma_{s,b}^\infty \cup \Gamma_{s,r}^\infty$ we impose absorbing boundary conditions. These can be expressed as stresses which are proportional to the displacements. That is, we can use (9.2) and replace the right hand side with $h_i(u_k)$. On Γ_I we formulate the interface conditions between the solid and the fluid: continuity of the velocity and tractions components perpendicular to the boundary as well as zero tractions parallel to it (slip condition):

$$v_i^s n_i = v_i^f n_i \qquad \sigma_{ij} n_j = p \qquad \sigma_{ij} t_j = 0. \tag{9.2.a}$$

v_i^s and v_i^f are the components of the velocity vector of the solid and the fluid. p refers to the pressure field in the fluid and t_j are the components of the unit vector tangential to the boundary.

The weak form of the equation of motion of the solid is

$$\int_{\Omega_s} \varepsilon_{ij}(w) \sigma_{ij} d\Omega_s = \int_{\Omega_s} w_i \ddot{u}_i d\Omega_s + \int_{\Omega_s} w_i f_i d\Omega_s +$$
$$\int_{\Gamma_s^\sigma} w_i g_i d\Gamma_s^s + \int_{\Gamma_s^\infty} w_i h_i(u) d\Gamma_s^s + \int_{\Gamma_I} w_i n_i p d\Gamma_I \tag{9.3}$$

where w_i are elements of the appropriate space of trial functions, ε_{ij} is the symmetric strain tensor and f_i are the distributed internal forces. The modeling with isoparametric finite elements yields the system of ODE of second order

$$M\ddot{u} + Ku = f_s + f_\sigma + f_s^\infty + f_s^I, \tag{9.4}$$

where u is the vector of the nodal displacements. The matrices and vectors in (9.4) are assembled using element matrices and vectors defined according to

$$M_e = \int_{\Omega_s^e} \rho_s N_s^T N_s d\Omega_s^e, \tag{9.5.a}$$

$$K_e = \int_{\Omega_s^e} B_s^T D B_s d\Omega_s^e, \tag{9.5.b}$$

$$f_{es} = \int_{\Omega_s^e} N_s^T f d\Omega_s^e, \tag{9.5.c}$$

$$f_{e\sigma} = \int_{\Gamma_s^e} N_s^T g d\Gamma_s^{s,e}, \tag{9.5.d}$$

$$f_{es}^I = \int_{\Gamma_I^e} (n N_{sr})^T p d\Gamma_I^e \tag{9.5.e}$$

N_s is the matrix of the shape functions, B_s is the strain-displacement matrix and D is the elasticity matrix. The force vector associated with the solid-fluid interface f_s^I will be discussed later.

The forces due to the absorbing boundary conditions f_s^∞ are defined by convolution integrals or by linear systems as the result of the rational approximation of DtN-maps. Energy dissipation is included with a Rayleigh-damping matrix $C = \alpha_R M + \beta_R K$ on the left-hand side of (9.4). Finally, (9.4) takes the standard form

$$M\ddot{u} + C\dot{u} + Ku = f_s + f_\sigma + f_s^\infty + f_s^l. \tag{9.6}$$

Remark:
- The equations of motion (9.6) doesn't significantly change if instead of a linear elastic solid we would assume a nonlinear solid. Instead of the term $C\dot{u}$ and Ku we would obtain the term $K(u, \dot{u})$ on the left-hand side of (9.16).

9.1.2   The finite element model of the fluid domain

So far, many finite element models of the fluid has been proposed for the solution of transient fluid-structure interaction problems. These models mainly differ by the choice of the field variables. An overview about this topic with many references may be found in [MO92]. The most natural strategy for fluid-structure interaction problems is to choose the displacements as field variables. This has the advantage to be easily coupled to the solid elements and to produce a symmetric system of ODE. A special interface element is necessary to enforce the slip condition. However, the displacement formulation exhibit so-called circulation or zero energy modes. These may corrupt the numerical solution of fluid-structure interaction problems [HO78]. This anomaly can be controlled with underintegrated stiffness element matrices with an irrotational constraint [WK83] or with underintegrated stiffness element matrices combined with projected mass element matrices [CT90].

Another approach uses the pressure as field variable [ZT78]. It has the advantage to need only one field variables per finite element node (the displacement formulation needs one per dimension). A special interface element is needed to couple fluid and solid elements. However, the coupling produces a nonsymmetric system of ODE. Hence, its numerical solution requires a nonsymmetric solver.

Much more appealing is the formulation with the velocity potential [Eve81, BO85] because the resulting system of ODE is symmetric. Similarly to the pressure formulation, the velocity potential formulation needs one field variable per finite element node. A special interface element couples the fluid elements with the solid elements. However, static problems cannot be solved with the velocity potential formulation.

A finite element formulation which has the advantage to provide a symmetric system of equations for the near field and is also suitable for solving static problems is the mixed ψ-p formulation, where ψ denotes the displacement potential and p the pressure [SG88, PA89]. Unfortunately, the coupling with the far field generates either a nonsymmetric system of equations or a symmetric system of ODE of third or higher order [PA89].

In this work, we use the velocity potential formulation. The equation of motion is

$$\nabla^2 \varphi = \frac{\ddot{\varphi}}{c_f^2} + f. \tag{9.7}$$

where $\varphi(t, x)$ refers to the velocity potential, c_f is the wave speed of the fluid and $f(t, x)$ is an external driving force. The velocity of the fluid $v_f(t, x)$ is recovered forming the gradient of the velocity potential

$$v_f(t, x) = \nabla \varphi(t, x). \tag{9.8}$$

The relation between the pressure and the velocity potential is

$$p(t, x) = -\rho_f \dot\phi(t, x).$$ (9.9)

Assume that the domain Ω_f sketched in Figure 9.1 is filled with a fluid. We shall assume that the boundary Γ of the body is decomposed in four disjoint parts $\Gamma = \Gamma_\varphi \cup \Gamma_n \cup \Gamma_f^\infty \cup \Gamma_I$. At the boundary Γ_φ we impose a velocity potential $\varphi_0(t)$ (essential boundary condition)

$$\varphi = \varphi_0.$$ (9.10)

Natural boundary conditions are imposed on the subset Γ_n of the boundary

$$\frac{\partial \varphi}{\partial n} = \nabla \varphi \cdot n = g,$$ (9.11)

where n is the unit normal vector pointing outward of the fluid and $g(t)$ is the flux. We assume that the absorbing boundary conditions are formulated in a subset of Γ_f^∞. These can be described by (9.11) with the right hand side $h(\varphi)$. At the interface of the fluid with the solid denoted by Γ_I we impose the interface conditions described in (9.3). Formulated in terms of the velocity potential we obtain

$$\frac{\partial \varphi}{\partial n} = n^T \dot u \qquad p n_i = \sigma_{ij} n_j.$$ (9.12)

The weak formulation of the equation of motion of the fluid is

$$\int_{\Omega_f} \nabla \phi \nabla \varphi d\Omega_f = \int_{\Omega_f} c_f^{-2} \phi \ddot\varphi d\Omega_f + \int_{\Omega_f} \phi f_i d\Omega_f + \int_{\Gamma_n} \phi g d\Gamma_v + \int_{\Gamma_f^\infty} \phi h d\Gamma_f^\infty + \int_{\Gamma_I} \phi n^T v^s d\Gamma_I,$$ (9.13)

where ϕ is an arbitrary element of an appropriate space of trial functions. In order to assure the symmetry of the coupled equations, the equations of motion (9.13) is multiplied with the fluid density ρ_f. The modeling with finite element yields the matrix equation

$$M_f \ddot\varphi + K_f \varphi = f_f + f_n + f_f^\infty + f_f^I.$$ (9.14)

The matrices and vectors in (9.4) are assembled using element matrices defined according to

$$M_{ef} = \rho_f c_f^{-2} \int_{\Omega_f^e} N_f^T N_f d\Omega_f^e,$$ (9.15.a)

$$K_{ef} = \rho_f \int_{\Omega_f^e} D_f^T D_f d\Omega_f^e,$$ (9.15.b)

$$f_{ef} = \rho_f \int_{\Omega_f^e} N_f^T f_f d\Omega_f^e,$$ (9.15.c)

$$f_{en} = \rho_f \int_{\Gamma_f^e} N_{fГ}^T g d\Gamma_f^e$$ (9.15.d)

$$f_{ef}^I = \rho_f \int_{\Gamma_f^e} N_{fГ}^T n^T v^s d\Gamma_f^e.$$ (9.15.e)

N_f is the matrix of the shape functions. $D_f = L N_f$, where $L = \mathrm{diag}(\partial/\partial x_j)$ is a diagonal differential operator matrix. The force vector associated with the solid-fluid interface f_f^I will be discussed later. The forces due to the absorbing boundary conditions f_f^∞ are defined by convolution integrals or by linear systems as the result of the rational approximation of DtN-maps. A

Rayleigh-damping term $C = \alpha_R M_f + \beta_R K_f$ provides for the energy dissipation. Finally, (9.14) takes the form

$$M\ddot{\varphi} + C\dot{\varphi} + K\varphi = f_f + f_n + f_f^\infty + f_I. \tag{9.16}$$

9.1.3 The equations of motion of the coupled near field

The coupling matrices are obtained combining (9.5.e) with (9.9)

$$f_{es}^I = -\rho_f \int_{\Gamma_I} (N_s^T n) N_f d\Gamma_I^e \dot{\varphi}_e = -C_{IF}\dot{\varphi}_e. \tag{9.17}$$

In two-dimensions we have

$$N_s^T n = [h_1 n_1, h_1 n_2, h_2 n_1, h_2 n_2, ..., h_N n_1, h_N n_2], \tag{9.18}$$

where h_j are the shape functions and n_j are the Cartesian components of the normal vector. Analogously (9.15.e) becomes

$$f_{ef}^I = \rho_f \int_{\Gamma_I} N_f^T n^T N_s d\Gamma_f^e v^s = C_{IF}^T \dot{u}^s. \tag{9.19}$$

Combining (9.17) and (9.19) with (9.6) and (9.16) yields

$$\begin{bmatrix} M_s & 0 \\ 0 & M_f \end{bmatrix} \begin{bmatrix} \ddot{u} \\ \ddot{\varphi} \end{bmatrix} + \begin{bmatrix} C_s & C_{IF} \\ -C_{IF}^T & C_f \end{bmatrix} \begin{bmatrix} \dot{u} \\ \dot{\varphi} \end{bmatrix} + \begin{bmatrix} K_s & 0 \\ 0 & K_f \end{bmatrix} \begin{bmatrix} u \\ \varphi \end{bmatrix} = \begin{bmatrix} f_s + f_\sigma + f_s^\infty \\ f_f + f_n + f_f^\infty \end{bmatrix}. \tag{9.20}$$

The last system is a so-called damped gyroscopic system. The gyroscopic matrix is the skew symmetric part of the global damping matrix

$$C_I = \begin{bmatrix} 0 & C_{IF} \\ -C_{IF}^T & 0 \end{bmatrix}, \tag{9.21}$$

It generates a nonsymmetric system of equations. We enforce the symmetry multiplying (9.16) and (9.19) with -1 and combining the resulting equations with (9.17) and (9.19)

$$\begin{bmatrix} M_s & 0 \\ 0 & -M_f \end{bmatrix} \begin{bmatrix} \ddot{u} \\ \ddot{\varphi} \end{bmatrix} + \begin{bmatrix} C_s & C_{IF} \\ C_{IF}^T & -C_f \end{bmatrix} \begin{bmatrix} \dot{u} \\ \dot{\varphi} \end{bmatrix} + \begin{bmatrix} K_s & 0 \\ 0 & -K_f \end{bmatrix} \begin{bmatrix} u \\ \varphi \end{bmatrix} = \begin{bmatrix} f_s + f_\sigma + f_s^\infty \\ -f_f - f_n - f_f^\infty \end{bmatrix}. \tag{9.22}$$

The last matrix equation is fundamentally different from the standard matrix equations of structural dynamics because the mass, damping and stiffness matrices are indefinite and not positive definite.

9.2 Coupling near and far field

9.2.1 A time-domain formulation

Now we are ready to combine the equations of motion of the near field formulated in (9.22) with the symmetric realizations of the approximated DtN-maps of the far fields. The time-domain formulation presented in this section generates a fully coupled system of ODE of second order. This can be solved with standard implicit numerical time integration schemes.

According to section 8.4 the most general symmetric realization of a DtN-map is given by

$$f^\infty + f_p^\infty + F_\infty y + G_\infty \dot{y} + K_\infty (w - w_p) + C_\infty (\dot{w} - \dot{w}_p) = 0$$

$$M_y \ddot{y} + C_y \dot{y} + K_y y + F_\infty^T (w - w_p) + G_\infty^T (\dot{w} - \dot{w}_p) = 0 \tag{9.23}$$

With w we refer to those degree of freedom of the near field which are related to nodes located on the artificial boundary Γ^∞. w_p is the particular solution and F_p^∞ is the force vector arising from the loadings appearing in the equations of motion of the far fields. y is the vector of the auxiliary degrees of freedom generated by the process of symmetric realization of the approximated DtN-map kernel. The matrices F_∞, G_∞, K_∞, C_∞ are given by

$$F_\infty = A\Psi F_\Psi \qquad G_\infty = A\Psi G_\Psi \qquad K_\infty = A\Psi K_\Psi \Psi^T A$$

$$C_\infty = A\Psi C_\Psi \Psi^T A \tag{9.24}$$

where Ψ is the matrix containing the basis vectors, A is the matrix which has been used to scale the basis vectors and F_Ψ, G_Ψ, K_Ψ, C_Ψ are the matrices which stems from the symmetric realization of the approximated DtN-map kernel formulated with respect to the basis vectors Ψ. Combining (9.23) with (9.22) yields the symmetric matrix equation

$$\begin{bmatrix} M_{vv} & M_{vw} & 0 \\ M_{vw}^T & M_{ww} & 0 \\ 0 & 0 & M_y \end{bmatrix} \begin{bmatrix} \ddot{v} \\ \ddot{w} \\ \ddot{y} \end{bmatrix} + \begin{bmatrix} C_{vv} & C_{vw} & 0 \\ C_{vw}^T & C_{ww} + C_\infty & G_\infty \\ 0 & G_\infty^T & C_y \end{bmatrix} \begin{bmatrix} \dot{v} \\ \dot{w} \\ \dot{y} \end{bmatrix} + \begin{bmatrix} K_{vv} & K_{vw} & 0 \\ K_{vw}^T & K_{ww} + K_\infty & F_\infty \\ 0 & F_\infty^T & K_y \end{bmatrix} \begin{bmatrix} v \\ w \\ y \end{bmatrix}$$

$$= \begin{bmatrix} f_v \\ f_w + f_p^\infty + K_\infty w_p + C_\infty \dot{w}_p \\ F_\infty^T w_p + G_\infty^T \dot{w}_p \end{bmatrix} \tag{9.25}$$

The vector v refers to those degrees of freedom of the vector $x^T = [u^T, \varphi^T]$ which are not associated with nodes that are on the artificial boundary Γ^∞. Observe that v contains degrees of freedom of the solid as well as of the fluid. Furthermore, $x^T = [v^T, w^T]$. Observe that the global mass, damping and stiffness matrices are generally indefinite. Moreover, the diagonal matrices M_y, C_y and K_y are not necessarily positive definite. We call (9.25) the enlarged system.

9.2.2 Two frequency-domain formulations

The frequency-domain analysis of the fully coupled system is important because it allows to test the accurateness of the approximation of the far field. This is possible because in the frequency domain the coupling of far and near field is straightforward. Using either natural or modal coordinates the DtN-map is

$$f^\infty + H_\infty(\omega)(w - w_p) + i\omega C_\infty(w - w_p) + f_p^\infty = 0. \tag{9.26}$$

$H_\infty(\omega)$ is the bounded and $i\omega C_\infty$ is the singular part of the DtN-map kernel. Observe that if the DtN-map has been computed with respect to a modal basis system, H_∞ and C_∞ are

$$H_\infty(\omega) = A\Psi H_\Psi(\omega)\Psi^T A \qquad C_\infty = A\Psi C_\Psi \Psi^T A. \tag{9.27}$$

Combining (9.26) with the equations of motion of the near field yields

$$\left(-\omega^2 \begin{bmatrix} M_{vv} & M_{vw} \\ M_{vw}^T & M_{ww} \end{bmatrix} + i\omega \begin{bmatrix} C_{vv} & C_{vw} \\ C_{vw}^T & C_{ww} + C_\infty \end{bmatrix} + \begin{bmatrix} K_{vv} & K_{vw} \\ K_{vw}^T & K_{ww} \end{bmatrix} + \begin{bmatrix} 0 & 0 \\ 0 & H_\infty(\omega) \end{bmatrix} \right) \begin{bmatrix} v \\ w \end{bmatrix}$$
$$= \begin{bmatrix} f_v \\ f_w + f_p^\infty + H_\infty w_p + i\omega C_\infty w_p \end{bmatrix} .$$

(9.28)

The first three matrices in the left-hand side of the last equation are frequency independent.

On the other hand, the approximated system (9.25) transforms in the frequency domain to

$$-\omega^2 \begin{bmatrix} M_{vv} & M_{vw} & 0 \\ M_{vw}^T & M_{ww} & 0 \\ 0 & 0 & M_y \end{bmatrix} \begin{bmatrix} \ddot{v} \\ \ddot{w} \\ \ddot{y} \end{bmatrix} + i\omega \begin{bmatrix} C_{vv} & C_{vw} & 0 \\ C_{vw}^T & C_{ww} + C_\infty & G_\infty \\ 0 & G_\infty^T & C_y \end{bmatrix} \begin{bmatrix} \dot{v} \\ \dot{w} \\ \dot{y} \end{bmatrix} + \begin{bmatrix} K_{vv} & K_{vw} & 0 \\ K_{vw}^T & K_{ww} + K_\infty & F_\infty \\ 0 & F_\infty^T & K_y \end{bmatrix} \begin{bmatrix} v \\ w \\ y \end{bmatrix}$$
$$= \begin{bmatrix} f_v \\ f_w + f_p^\infty + K_\infty w_p + C_\infty \dot{w}_p \\ F_\infty^T w_p + G_\infty^T \dot{w}_p \end{bmatrix}$$

(9.29)

Obviously, all matrices of this system are frequency independent. This give us the opportunity to compute the eigenvalues and the eigenvectors of the enlarged system. A task which is much harder to realize with a system like (9.28) because the frequency-dependence of the bounded DtN-map kernel $H_\infty(\omega)$ generates a nonlinear eigenvalue problem.

The system (9.29) is generally the product of two approximation steps. The first approximation stems from the projection of the DtN-map to the basis vectors Ψ if these don't span the complete displacement space. Using the system (9.28) an estimate of the loss of accuracy is quickly obtained comparing the results of test problems with the projected and nonprojected DtN-maps. Finally, an estimate of the approximation error introduced by the rational approximation of DtN-map kernels is obtained computing the test problem with the formulation (9.29). Generally, the error due to the projection of DtN-maps on a modal basis having smaller dimension than the natural basis is much more significant than that due to the rational approximation.

9.3   Well-posedness and stability of enlarged systems

In chapter 2 we discussed well-posedness of IBVP with absorbing boundary conditions in an analytical setting. There, we found that an IBVP is well-posed if the Laplace transform of the solution is an analytic function in the complex half plane $\text{Re}(s) > 0$. A similar condition assures well-posedness of initial value problems (IVP) of linear ODE-systems. The IVP is well-posed, if and only if the solution $x(s)$ of the system $(s^2 M + sC + K)x = f(s)$ is analytic in the complex half-plane $\text{Re}(s) > 0$. This means that the quadratic matrix pencil $R(s) = (s^2 M + sC + K)$ is invertible or nonsingular in the half plane $\text{Re}(s) > 0$ or equivalently that the real part of all eigenvalues of the matrix pencil $\lambda^2 M + \lambda C + K$ are negative or zero, that is, $\text{Re}(\lambda) \le 0$. If $\text{Re}(\lambda) < 0$ the IVP is strongly well-posed. The IVP is weakly well-posed if there exists one eigenvalue with $\text{Re}(\lambda) = 0$. Therefore, analyzing the stability of linear system we find the answer to the question of well-posedness.

Usually, the linear problems of structural dynamics are strongly well-posed. The proof is trivial because the mass, damping and stiffness matrices are symmetric and positive definite. However, this approach fails when it is applied to prove the stability of systems described by matrices which are still symmetric but indefinite. Many stability criterions for linear mechanical systems have been formulated so far. A survey can be found in [Mül77]. However, these criterions are generally not helpful when proving the stability of enlarged systems. Furthermore, the test of all eigenvalues of enlarged systems for $\text{Re}(\lambda) \le 0$ is not applicable in practice because of the tremendous computational costs due to the numerical computation of all eigenvalues.

9.3.1   Stability of the equations of motion of the coupled near field

The equations of motion of the coupled near field given in (9.22) exhibit the unusual property discussed before: all the matrices are indefinite because of the minus sign in front of the mass, damping and stiffness matrices associated with the fluid medium. However, in this case the proof of stability is still easy. Assume that the stiffness matrices K_s, K_f are positive definite. This is usually the case in practical computations. The matrices C_s, C_f vanish or are at least positive semi-definite.

The stability of the undamped coupled near field was proved in [BO85]. However, the proof was established under the assumption that the eigenvectors are real. Similarly, in [OV89] the proof was performed under the assumption that the eigenvalues are purely imaginary. In [Web94] a proof of the damped system is given. However, this was established under the assumption that the eigenvectors are real. In this section, no assumptions are made about the eigenvectors or the eigenvalues of the coupled near field.

Consider the eigenvalue problem

$$\left(\lambda^2 \begin{bmatrix} M_s & 0 \\ 0 & -M_f \end{bmatrix} + \lambda \begin{bmatrix} C_s & C_{IF} \\ C_{IF}^T & -C_f \end{bmatrix} + \begin{bmatrix} K_s & 0 \\ 0 & -K_f \end{bmatrix} \right) \begin{bmatrix} u \\ \varphi \end{bmatrix} = A(\lambda)x = 0 . \tag{9.30}$$

The eigenvalues are defined by the characteristic equation $\det(A(\lambda)) = 0$. Since multiplication of a row or a column of a matrix with -1 changes only the sign of a determinant, the eigenvalues of the symmetric system (9.22) are identical to those of the damped gyroscopic system (9.20). As a consequence, (9.20) is stable if and only if (9.22) is stable. In (9.20) the mass and stiffness matrices of the coupled system are positive definite. Therefore, we try to prove the stability of (9.20) with standard arguments.

First, we analyze the undamped gyroscopic system $C_s = C_f = 0$. Consider the eigenvalue problem

$$\left(\lambda^2 \begin{bmatrix} M_s & 0 \\ 0 & M_f \end{bmatrix} + \lambda \begin{bmatrix} 0 & C_{IF} \\ -C_{IF}^T & 0 \end{bmatrix} + \begin{bmatrix} K_s & 0 \\ 0 & K_f \end{bmatrix} \right) \begin{bmatrix} u \\ \varphi \end{bmatrix} = \lambda^2 M x + \lambda C_g x + K x = 0 . \tag{9.31}$$

Assume that (λ, x) is an eigenvalue-eigenvector pair. Then, because the system matrices are real, $(\bar{\lambda}, \bar{x})$ is also an eigenvalue-eigenvector pair of (9.31). If (9.31) is multiplied from the left with the transposed complex-conjugate of x, x^*, we obtain

$$\lambda^2 x^* M x + \lambda x^* C_g x + x^* K x = \lambda^2 m + \lambda c + k = 0 , \tag{9.32}$$

where $m > 0$ and $k > 0$ because M and K are positive definite. Furthermore, c is purely imaginary because C_g is skew-symmetric. Similarly, we obtain

$$\bar{\lambda}^2 m - \bar{\lambda} c + k = 0 \tag{9.33}$$

with the eigenvalue-eigenvector pair $(\bar{\lambda}, \bar{x})$. Multiplying (9.32) with $\bar{\lambda}$ and (9.33) with λ and adding the resulting equations yields

$$(\lambda + \bar{\lambda})(|\lambda|^2 m + k) = 0. \tag{9.34}$$

Since $|\lambda|^2 m + k > 0$ we obtain $\lambda + \bar{\lambda} = \text{Re}(\lambda) = 0$. Hence, the eigenvalues of (9.31) are purely imaginary. That is, the system (9.31) is conservative. Therefore, the undamped system (9.20) is weakly well-posed and as a consequence also the undamped system (9.22).

Now assume that $\lambda = i\omega$, then (9.32) may be written as $\omega^2 m - \omega i c - k = 0$. The roots of the last equation are

$$\omega_{1,2} = \frac{ic}{2m} \pm \sqrt{\frac{k}{m} + \left(\frac{ic}{2m}\right)^2}. \tag{9.35}$$

But if $c \neq 0$ ω_1 and ω_2 are complex numbers. This contradicts (9.34) so that $c = x^* C x = 0$. Hence, the eigenvectors are real.

Now let us consider the eigenvalue problem of the damped gyroscopic system

$$\lambda^2 M x + \lambda C_g x + \lambda C_D x + K x = 0, \tag{9.36}$$

where C_D is assumed to be positive semi-definite. On physical grounds, we expect that at least some of the eigenvalues of this system are complex with nonvanishing real part. Obviously (λ, x) as well as $(\bar{\lambda}, \bar{x})$ are eigenvalue-eigenvector pairs of (9.36). Proceeding as before we obtain

$$(\lambda + \bar{\lambda})(|\lambda|^2 m + k) + |\lambda|^2 c_D = 0. \tag{9.37}$$

As a consequence, we have

$$(\lambda + \bar{\lambda}) = 2\text{Re}(\lambda) = \frac{-2|\lambda|^2 c_D}{(|\lambda|^2 m + k)}. \tag{9.38}$$

Since $c_D \geq 0$ we obtain $\text{Re}(\lambda) \leq 0$. Therefore, the damped gyroscopic system is stable and, as a consequence, weakly well-posed. If C_D is positive definite the system is asymptotically stable and therefore even strongly well-posed.

A second and very elegant proof of the stability of the system (9.20) is provided by Lyapunov's direct method. The theory of Lyapunov states that if in the neighborhood of a an equilibrium point of a linear or nonlinear system there exists a positive definite function $V(t, x)$ whose total derivative with respect to time $\dot{V}(t, x) = dV(t, x)/dt$ is not positive then the system is stable in a neighborhood of this equilibrium point (linear systems are globally stable). Moreover, if the function $V(t, x)$ is decreasing ($V(x) < V(y)$ for $|x| < |y|$) and its total derivative $V(t, x)$ is strictly negative then the system is asymptotically stable. The function $V(t, x)$ is called the Lyapunov functional. In practical applications, the main difficulty of this method is to find an appropriate Lyapunov functional.

Now let us apply Lyapunov's direct method to the equations of motion of the near field in the form (9.20). Consider the Lyapunov functional

$$V(x, \dot{x}) = \dot{x}^T M \dot{x} + x^T K x, \tag{9.39}$$

where $x^T = [u^T, \varphi^T]$ and M, K are the mass and stiffness matrices of the coupled system (9.20). The quadratic form (9.39) is the sum of the kinetic and potential energy of the system. Since M, K are assumed to be positive definite, $V(x, \dot{x}) > 0$ is positive for any pair of nonvanishing vectors $x, \dot{x}$. According to the stability theory of Lyapunov, the system (9.20) is stable if $\dot{V}(x, \dot{x}) \leq 0$ for any arbitrary pair of vectors $x, \dot{x}$. Differentiating (9.39) with respect to time yields

$$\dot{V}(x, \dot{x}) = 2\dot{x}^T M \ddot{x} + 2x^T K \dot{x}. \tag{9.40}$$

Using the homogeneous differential equation (9.20) we replace $M\ddot{x}$ in the last equation and obtain

$$\dot{V}(x, \dot{x}) = -2\dot{x}^T C_D \dot{x}. \tag{9.41}$$

Now, if C_D is positive definite $\dot{V}(x, \dot{x}) < 0$ so that (9.20) is asymptotically stable. If $C_D = 0$ $\dot{V}(x, \dot{x}) \equiv 0$ and the system (9.20) is stable. Clearly, this system is stable too, if C_D is positive semi-definite.

Another topic of concern when dealing with systems of the form (9.22) is related to the question if they are numerically integrable. In order to be able to perform a numerical time integration, the matrix $T = M + c_c(C_D + C_I) + c_k K$ with $c_c > 0$ and $c_k > 0$ must be invertible. In other words, T^{-1} must exist. Usually, in structural dynamics the mass and the stiffness matrices are positive definite so that T is invertible. However, this argument doesn't work with systems like (9.22) because all matrices are indefinite.

We will prove the invertibility of T by contradiction. Invertibility of T means that the null space of T is the zero vector. That is, $Tx \neq 0$ for any $x \neq 0$. Assume that T is not invertible, e.g. there exists a vector x, not equal to zero, so that $Tx = 0$. Then, for every vector $y \neq 0$ we have $yTx = 0$. Consider $x^T = [u, \varphi]$ and $y^T = [u, -\varphi]$, then $y^T Tx = 0$ is given by

$$u^T M_s u + \varphi^T M_f \varphi + c_c u^T C_{Ds} u + c_c \varphi^T C_{Df} \varphi + c_c u^T C_{IF} \varphi - c_c \varphi^T C_{IF}^T u$$
$$+ c_k u^T K_s u + c_k \varphi^T K_f \varphi = 0 \tag{9.42}$$

Because $c_c u^T C_{IF} \varphi - c_c \varphi^T C_{IF}^T u = 0$ and M_s, M_f, K_s as well as K_f are positive definite the left hand term of (9.42) is always larger than zero. Hence, contrary to the assumption, $Tx \neq 0$ so that T is invertible because x was arbitrary. Therefore, the coefficients of the diagonal matrix D in the factorization $T = L^T D L$ are all nonzero. Clearly, the diagonal matrix D is not positive definite because T is not positive definite.

Remarks:
- Observe that the key arguments to prove the stability of the coupled near field are based on the definiteness of the mass, damping and stiffness matrices. Therefore, a similar procedure fails when applied to the system (9.25).
- Clearly, we could formulate the homogeneous linear system as ODE-system of first order $\dot{z} = Az$ and then solve the Lyapunov equation $AQ + QA = I$. If Q turns out to be positive definite, then the system is asymptotically stable. However, for large systems, the computation of Q is extremely time consuming so that this approach is virtually useless in practice.

9.3.2 Stability of enlarged systems: a sufficient condition

In this section, we shall consider homogeneous systems of ODE of the type given in equation (9.25). Irrespective of the kind of symmetric realization of the far field, the mass, damping and stiffness matrices of these enlarged systems are generally indefinite. Therefore, the techniques used in the foregoing section will fail. The few qualitative theorems for general systems, see [Mül77] for an overview, give rarely more insight into the stability properties because of the difficulty to characterize the coupling matrices F, G of symmetric realizations. As mentioned before, the stability theory of Lyapunov is not a practicable alternative because the numerical solution of the Lyapunov equations of such a large system is definitely too time consuming.

The main difficulty in analyzing the stability of such systems is due to the fact, that when two stable systems are coupled together the resulting system need not to be stable because its stability is determined by the parameters of the coupling. Consider this small symmetric system with two degrees of freedom

$$\begin{bmatrix} m & 0 \\ 0 & m \end{bmatrix} \begin{bmatrix} \ddot{u}_1 \\ \ddot{u}_2 \end{bmatrix} + \begin{bmatrix} k & \alpha \\ \alpha & k \end{bmatrix} \begin{bmatrix} u_1 \\ u_2 \end{bmatrix} = 0 \tag{9.43}$$

with $m > 0$ and $k > 0$. The eigenvalues of (9.43) are

$$s_{1,2} = \pm \sqrt{\frac{(\alpha - k)}{m}} \quad \text{and} \quad s_{3,4} = \pm \sqrt{\frac{-(\alpha + k)}{m}}. \tag{9.44}$$

If $\alpha = 0$ the system uncouples to two stable subsystems. If $|\alpha| < k$ all eigenvalues are distinct and purely imaginary. Hence, the system is stable. If $|\alpha| = k$ two eigenvalues are purely imaginary and two eigenvalues are equal to zero. If $|\alpha| > k$ two eigenvalues are purely imaginary and two eigenvalues are real. However, one of the two real eigenvalues is positive so that the system is unstable. Therefore, the system is stable if and only if $|\alpha| < k$. In this small system, the characterization of stability is very simple because it depends only on one parameter. However, a similar characterization of enlarged systems is much more difficult because of the many parameters determining the coupling. However, a sufficient condition for stability of enlarged systems can be given using the theory of passive systems.

First, we show that the stability of enlarged systems is independent of the kind of symmetric realization. Consider the eigenvalue problem

$$\left(s^2 \begin{bmatrix} M_{vv} & M_{vw} & 0 \\ M_{vw}^T & M_{ww} & 0 \\ 0 & 0 & M_y \end{bmatrix} + s \begin{bmatrix} C_{vv} & C_{vw} & 0 \\ C_{vw}^T & C_{ww} + C_\infty & G_\infty \\ 0 & G_\infty^T & C_y \end{bmatrix} + \begin{bmatrix} K_{vv} & K_{vw} & 0 \\ K_{vw}^T & K_{ww} + K_\infty & F_\infty \\ 0 & F_\infty^T & K_y \end{bmatrix} \right) \begin{bmatrix} v \\ w \\ y \end{bmatrix} = 0 \tag{9.45}$$

Eliminating the vector y we may reduce the last equation to

$$\left(s^2 \begin{bmatrix} M_{vv} & M_{vw} \\ M_{vw}^T & M_{ww} \end{bmatrix} + s \begin{bmatrix} C_{vv} & C_{vw} \\ C_{vw}^T & C_{ww} + C_\infty \end{bmatrix} + \begin{bmatrix} K_{vv} & K_{vw} \\ K_{vw}^T & K_{ww} \end{bmatrix} + \begin{bmatrix} 0 & 0 \\ 0 & H_\infty(s) \end{bmatrix} \right) \begin{bmatrix} v \\ w \end{bmatrix} = 0, \tag{9.46}$$

where

$$H_\infty(s) = -(F_\infty^T + s G_\infty^T)(s^2 M_y + s C_y + K_y)^{-1}(F_\infty + s G_\infty) + K_\infty. \tag{9.47}$$

Observe that $H_\infty(s)$ is nonsingular in $C_+ \cup iR$ because it is analytical in $C_+ \cup iR$. Assume that (9.46) is stable. Then, for all s in C_+ the matrix equation (9.46) has only the trivial solution $x^T = [v, w] = 0$. As a consequence, we obtain from (9.45)

$$(s^2 M_y + s C_y + K_y)y = 0 \tag{9.48}$$

Because the matrix pencil $s^2 M_y + s C_y + K_y$ doesn't vanish in C_+ we obtain $y = 0$ so that $z^T = [v, w, y] = 0$ for all s in C_+. Hence, (9.45) is stable. Conversely, if (9.45) is stable we have only the trivial solution $x^T = [v, w, y] = 0$ in C_+ so that (9.46) is stable too. Furthermore, because $H_\infty(s)$ is independent of the kind of the symmetric realization the stability of the system (9.45) is also independent of the kind of the symmetric realization.

Now let us formulate a sufficient stability criterion. Consider a system given in the form (9.46):

$$M\ddot{x} + C\dot{x} + Kx + H_\infty^s * x = 0. \tag{9.49}$$

M, C and K are the matrices of the coupled near field. H_∞^s is the kernel function of the absorbing boundary condition containing the singular part: $H_\infty^s(s) = H_\infty(s) + s C_\infty$. We assume that the coupled near field is formulated according to (9.31). That is, M and K are positive definite and $C = C_D + C_{IF}$, where C_D is symmetric and C_{IF} is skew-symmetric. As a consequence, H_∞^s is not symmetric. Assume that at the time $t = 0$ we have the initial conditions $x = x_0$ and $\dot{x} = \dot{x}_0$. Now, let us multiply (9.49) from the left with $\dot{x}^T$ and integrate the resulting equation from $t = 0$ to $t = T$. This yields

$$\left(\frac{\dot{x}^T M \dot{x}}{2} + \frac{x^T K x}{2} \right) \Big|_0^T + \int_0^T \dot{x}^T C \dot{x} \, dt + \int_0^T \dot{x}^T H_\infty^s * x = 0 \tag{9.50}$$

The first term is the sum of the kinetic and elastic energy $E(T)$ of the near field. Therefore we can write (9.50) as

$$E(T) = E(0) - \int_0^T \dot{x}^T C \dot{x} \, dt - \int_0^T \dot{x}^T H_\infty^s * x \, dt. \tag{9.51}$$

Because M and K are positive definite $E(T) \geq 0$. Now, if the system (9.49) is stable we obtain $E(T) \leq E(0)$ for all $T \geq 0$. That is, $E(T)$ must be a decreasing function of T or at most a constant. The term $\int_0^T \dot{x}^T C \dot{x} \, dt$ is positive or at least zero. In fact, C_D is positive semi-definite and $\dot{x}^T C_{IF} \dot{x} = 0$ because C_{IF} is skew-symmetric. Now if $W_\infty(T) = \int_0^T \dot{x}^T H_\infty^s * x \, dt$ is positive or at least zero for any time T then we would obtain $E(T) \leq E(0)$. That is, the system (9.49) is stable.

Now let us apply the theory of passive systems. In chapter 4 we defined passive systems by way of the functional

$$W(t) = \int_{-\infty}^t u^T(s)y(s)ds = \int_{-\infty}^t u^T(\tau)(h*u)(\tau)d\tau \tag{9.52}$$

where $u(t)$ was the input function and $y(t)$ was the output function defined by the linear time-invariant convolution integral

$$y(t) = (h*u)(t) = \int_{-\infty}^\infty h(\tau - s)u(s)ds. \tag{9.53}$$

with kernel function $h(t)$. However, the integral $W_\infty(T) = \int_0^T \dot{x}^T H_\infty^s * x \, dt$ has not the symmetric form of (9.52). However, it can be cast in the form (9.52) by defining the kernel $G_\infty^s(t)$ by

$$H_\infty^s(t) = \frac{d}{dt} G_\infty^s(t).$$ (9.54)

This yields

$$W_\infty(T) = \int_0^T \dot{x}^T G_\infty^s * \dot{x} \, dt$$ (9.55)

which exhibits the symmetric form required in (9.52).

One remarkable fact of the theory of passive systems is that the Laplace transform $F(s)$ of kernels $F(t)$ of passive systems are positive-real matrices. These are defined by the following properties [New66]:

a) $F(s)$ is analytic in C_+
b) $F(\bar{s}) = \bar{F}(s)$. That is, $F(t)$ is a real matrix.
c) $F(s) + F^T(\bar{s})$ is positive semi-definite Hermitian in C_+.
Note that because of b) $F(s) + F^T(\bar{s}) = F(s) + F^*(s)$.

A test for positive-real rational matrices is: A $n \times n$ real-rational matrix $F(s)$ is positive real if

a) $F(s)$ is analytic in C_+
b) The poles of $F(s)$ on iR are simple.
c) The residues of $F(s)$ for each pole on iR are positive semi-definite.
d) $F(i\omega) + F^*(i\omega)$ is positive semi-definite for all $\omega \geq 0$ if ω is not a pole.

Let us consider the Laplace transform of the kernel $G_\infty^s(s)$. It is given by

$$G_\infty^s(s) = \frac{H_\infty^s(s)}{s} = \frac{C(sI - A)^{-1}B + D}{s} + C_\infty = \frac{H_\infty(s)}{s} + C_\infty$$ (9.56)

C_∞ is positive definite because it corresponds to the viscous boundary condition. Hence, we have to test $H_\infty(s)/s$. $H_\infty(s)/s$ is analytic in C_+ and has a pole at $s = 0$. The residue of $H_\infty(s)/s$ is $H_\infty(0) = -CA^{-1}B + D$. Hence, because of c) $H_\infty(0)$ have to satisfy

$$H_\infty(0) = -CA^{-1}B + D \geq 0.$$ (9.57)

$H_\infty(0)$ is symmetric because the solid and fluid parts uncouple for $\omega = 0$. Therefore, (9.57) imposes that $H_\infty(0)$ has to be positive semi-definite. This implies that the sum of the stiffness matrix of the near field K with $H_\infty(0)$ is positive semi-definite: $K + H_\infty(0) \geq 0$. Note that a well-posed static problem requires that $K + H_\infty(0) > 0$ so that this condition is stronger.

Condition d) yields

$$H_\infty(i\omega) - H_\infty^*(i\omega) \geq 0 \text{ for } \omega > 0.$$ (9.58)

Therefore, a sufficient condition for the system (9.49) to be stable is that the regular part of the kernel function of the absorbing boundary condition fulfils the conditions (9.57) and (9.58). Note that (9.58) is a generalization of the dissipativity condition formulated in [BBM88] and discussed in chapter 3. In fact, if the transfer function $H_\infty(s)$ is symmetric we obtain

$$\text{Im}(H_\infty(i\omega)) \geq 0 \text{ for } \omega > 0.$$ (9.59)

The physical interpretation of the last condition is that the energy flux across the boundary is positive at any frequency ω. That is, the absorbing boundary condition dissipates energy at any frequency ω.

Example: Now let us consider some absorbing boundary elements. The absorbing boundary element derived from the Engquist-Majda boundary condition of first order are not dissipative. In fact, the residues of $H(s)$ at the pole $\omega = 0$ is indefinite and the imaginary part of $H(i\omega)$ given by

$$\text{Im}(H(i\omega)) = \frac{\omega P}{c} - \frac{cQ}{2\omega} \qquad (9.60)$$

is indefinite at low frequencies because of the second term which is negative and proportional to the positive semi-definite matrix Q. For $\omega > 6c/h$ $\text{Im}(H(\omega))$ is positive definite and therefore dissipative. Observe, that this lower bound increases when the size of the elements decreases. Moreover, only the skew-symmetric mode $\varphi^T = [1, -1]$ is nondissipative. The Engquist-Majda boundary conditions of first order can be made dissipative by the perturbation

$$H(s) = \frac{sP}{c} + \frac{cQ}{2(s+\delta)} . \qquad (9.61)$$

Choosing $\delta > \sqrt{6}c/h$ the perturbed Engquist-Majda boundary condition becomes dissipative so that the enlarged system will be stable. Similarly, the absorbing boundary element derived from the Engquist-Majda boundary conditions of second order is dissipative only for $\omega > \sqrt{9}c/h$.

Finally, let us consider the implementation of the third Bayliss and Turkel boundary condition (8.201). First note that condition (9.58) is fulfilled because P is positive-definite and Q is positive semi-definite:

$$H(0) = \frac{4P}{8R} + \frac{R}{2}Q > 0 \qquad (9.62)$$

The last condition assures that static analyses are well-posed. The translation mode $\varphi^T = [1, 1]$ is dissipative because P is positive definite and $Q\varphi = 0$ so that

$$\text{Im}(H(i\omega)) = \text{Im}\left(\frac{i\omega P}{c} - \frac{1}{8R^2}\frac{cP}{(i\omega + c/R)}\right) = \frac{i\omega P}{c} + \frac{1}{8R^2}\frac{\omega cP}{\omega^2 + (c/R)^2} \geq 0 . \qquad (9.63)$$

On the other hand, the mode $\varphi^T = [1, -1]$ is nondissipative because the term in (8.201) containing the matrix Q gives a negative contribution at low frequencies so that the imaginary part of $H(i\omega)$ related to this mode is negative definite. A detailed analysis shows that $H(i\omega)$ is positive-definite provided that $\omega > c\sqrt{96R^2 - 18h^2}/4Rh$. Because usually $h \ll R$ an accurate approximation of the last equation is $\omega > \sqrt{6}c/h$.

The dissipativity condition gives us a criterion to test the stability of enlarged systems. However, these systems can be stable even if the dissipativity condition is not fulfilled because it is only a sufficient criterion. In fact, many local absorbing boundary conditions turn out to be stable in practice although they don't fulfil this condition.

Remarks:
- A condition similar to (9.59) was proved in [PW96]. However, the proof given in this work is physically much more appealing and more general.
- Because the dimension of the transfer function $H_\infty(s)$ is small compared to that of enlarged systems the condition (9.58) can be generally tested with less numerical effort than the computation of all the eigenvalues of enlarged systems.

Chapter 10

Applications

In this chapter we apply the absorbing boundary conditions developed so far for the finite element analysis of unbounded domains. We shall first analyze finite models of the elastically restrained rod. This gives us the opportunity to analyze the most fundamental features of the novel absorbing boundary conditions. Then, we will examine transient wave propagation problems on infinite solid and solid-fluid strata. Finally, we analyze finite models of dam-reservoir, dam-foundation and dam-reservoir-foundation systems.

10.1 Finite models of the elastically restrained rod

Consider a semi-infinite restrained rod of unit section defined on $x \geq 0$. We assume that the rod is at rest at $t = 0$ and that a loading $f(t)$ acts on the left boundary for $t \geq 0$. The IBVP is defined by the set of equations

$$\frac{1}{c^2}u_{,tt} - u_{,xx} + \frac{2\varepsilon}{c}u_{,t} + \kappa^2 u = 0$$

$$u(x, 0) = 0 \tag{10.1}$$

$$u_{,t}(x, 0) = 0$$

$$\sigma_L = \rho c u_{,x} = -\rho c f(t)$$

10.1.1 The analytical solution

The analytical solution of (10.1) for a loading $f(t)$ can be obtained by first constructing the Green's function $g(x, t)$ associated with the IBVP and then expressing the general solution $u(x, t)$ by way of the convolution of the loading $f(t)$ with the Green's function:

$$u(x, t) = \int_0^t f(y)g(t - y)dy. \tag{10.2}$$

The Green's function $g(t)$ of (10.1) is obtained by replacing the loading $f(t)$ by the Dirac delta function $\delta(t)$ in the last equation. First, we note that the solution of the IBVP is simplified if we consider the variable transformation

$$u(x, t) = v(x, t)\exp(-\varepsilon ct). \tag{10.3}$$

If (10.3) is inserted into (10.1) we obtain a new equation of motion for $v(x, t)$ in which the dissipative term vanish:

$$\frac{1}{c^2}v_{,tt} - v_{,xx} + (\kappa^2 - \varepsilon^2)v = 0. \tag{10.4}$$

Note that the last equation has a stiffness parameter of the distributed springs which is equal to $\kappa^2 - \varepsilon^2$. The initial and boundary conditions are

$$\begin{aligned} v(x, 0) &= 0 \\ v_{,t}(x, 0) &= 0 \\ \sigma_L &= -\rho c\delta(t) \end{aligned} \tag{10.5}$$

Now we apply the Laplace transform to (10.4) and (10.5) and obtain the Green's function

$$g_v(x, s) = \frac{e^{-k(s)x}}{k(s)}, \tag{10.6}$$

with $k(s)$ given by $k(s) = \sqrt{s^2/c^2 + (\kappa^2 - \varepsilon^2)}$. In the time domain, the Green's function is

$$g_v(x, t) = \frac{1}{c}J_0(\sqrt{\kappa^2 - \varepsilon^2}\sqrt{t^2 - (x/c)^2})H(t - (x/c)), \tag{10.7}$$

so that the Green's function of the original IBVP becomes

$$g(x, t) = \frac{1}{c}e^{-\varepsilon ct}J_0(\sqrt{\kappa^2 - \varepsilon^2}\sqrt{t^2 - (x/c)^2})H(t - (x/c)). \tag{10.8}$$

The Green's function is given by a zero order Bessel function of first kind multiplied by an exponential factor and by the Heaviside step function $H(t - (x/c))$. Since the Heaviside step function vanish for $ct < x$ the solution has a sharp wave front which propagates with the velocity c. This wave front decreases proportionally to the exponential factor $e^{-\varepsilon ct}$. If the square-root $\sqrt{\kappa^2 - \varepsilon^2}$ is real ($\kappa > \varepsilon$) the solution is oscillatory behind the wave-front. The amplitudes of these oscillations decrease with the distance to the wave-front. The oscillation periods have variable lengths and increase with decreasing magnitude of the root. For the special case $\kappa = \varepsilon$ the root vanishes and (10.8) becomes

$$g(x, t) = \frac{1}{c}e^{-\varepsilon ct}H(t - (x/c)). \tag{10.9}$$

The solution is a non-oscillatory constant wave-train with a sharp wave-front. If the dissipation parameter ε is increased further such that $\varepsilon > \kappa$, the root in the argument of the Bessel function becomes imaginary and the solution is given by the modified Bessel function of first kind and zero order $I_0(y)$

$$g(x, t) = \frac{1}{c}e^{-\varepsilon ct}I_0(\sqrt{\varepsilon^2 - \kappa^2}\sqrt{t^2 - (x/c)^2})H(t - (x/c)). \tag{10.10}$$

This solution is a decaying non-oscillatory wave-train with a sharp wave-front.

10.1.2   Finite element analysis of the nondissipative restrained rod

In this section we analyse the behaviour of some of the absorbing boundary conditions of the restrained rod. The symmetric realizations of these absorbing boundary conditions are found in section 8.1. The numerical solutions presented in this section are computed using the parameter $c = 1$ m/s and $\kappa = 1$ m. This does not represent a restriction because if the independent variables x and t are transformed to the dimensionless variables x' and t' according to

$$x' = \kappa x \qquad t' = \kappa c t \tag{10.11}$$

the material parameters are transformed to the dimensionless form

$$c \to 1 \qquad \kappa \to 1 \qquad \varepsilon \to \varepsilon' = \varepsilon/\kappa . \tag{10.12}$$

Hence, the solution of a rod with $c \neq 1$ m/s and $\kappa \neq 1$ m are obtained using (10.11).

Let us assume that the rod is exited by a suddenly applied loading on the left boundary:

$$f(t) = f_0 H(t) . \tag{10.13}$$

This excitation has the advantage to exhibit a transient as well as a nonzero stationary response for $t \to \infty$. Hence, with only one loading we are able to study both the short as well as the long-time response of finite models with absorbing boundary conditions. The stationary state is

$$u_\infty(x) = \frac{f_0}{\kappa} e^{-\kappa x} \tag{10.14}$$

as can be easily derived by solving the IBVP (10.1) under the assumption that the term containing derivatives with respect to time vanishes.

We first discuss the result of finite models with the viscous boundary condition at the right boundary (see Figure 10.1, left). The responses of the displacement at the left boundary confirm the results stated in chapter 2. First, the accuracy of the response increases if the length of the near field is increased. Second, the short-time response is accurate and the long-time response not. In fact, the post-peak response of the finite model with a length of 1 m is strongly over-damped and the stationary state is inaccurate. Doubling the length improves considerably the accuracy of the stationary state. However, the accuracy of the post-peak response is still not acceptable. A further doubling of the length reproduces the short-time behaviour with high accuracy up to the first cycle. After this point, the response is more and more out of phase as time increases. The steady state is captured very well and the amplitudes of the post-peak oscillations don't diverge much from the exact solution.

The finite model with the asymptotic boundary condition of second order is more accurate than that with the viscous boundary condition (see Figure 10.1, right). The most significant fact is that the times where the responses diverge appreciably from the exact solution are greater than the ones obtained with the viscous boundary condition. Three-quarter of a cycle for a near field with a length of 1 m, one and a quarter of a cycle for one with a length of 2 m and more than two cycles for one with a length of 4 m. This result is due to the much better approximation of the DtN-map at high frequencies. Nevertheless, finite models with short near field still exhibit a completely wrong post-peak response. However, contrary to the viscous boundary condition, the response is oscillatory.

The responses of the finite model with the doubly-asymptotic boundary condition are given in (Figure 10.2, left). First, notice that the stationary state is reproduced exactly irrespective of the length of the near field. This behaviour is due to the spring term which assures an exact boundary

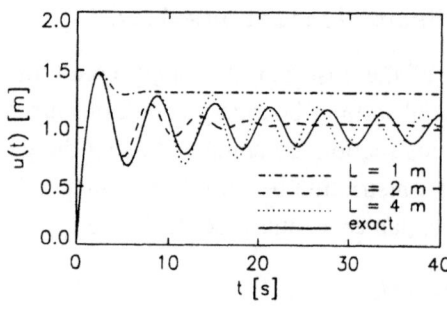 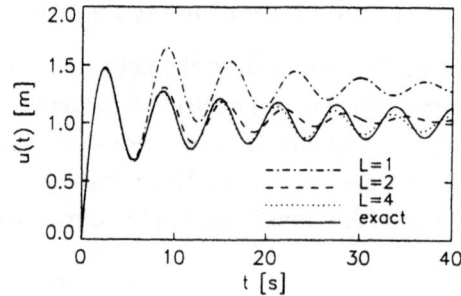

Figure 10.1: Responses of the viscous (left) and asymptotic (right) boundary conditions

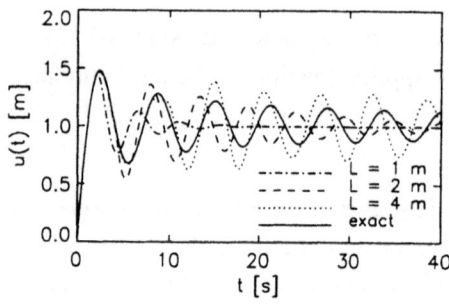 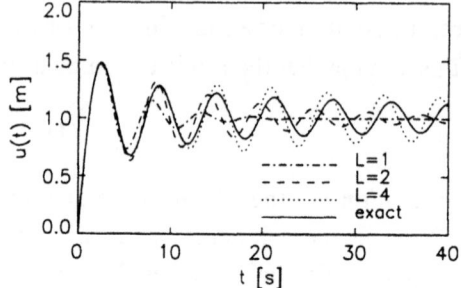

Figure 10.2: Responses of the doubly-asymptotic (left) and Engquist-Halpern (right) boundary conditions

condition for static analyses. In contrast, the finite model with the doubly-asymptotic boundary condition reproduces the short-time response with significantly less accuracy than the model with the viscous boundary condition. The time points where the response diverge from the exact solution are always smaller. In addition, the post-peak oscillations are remarkably inaccurate even for the finite model with a length of 4 m.

The response of the finite model with the Engquist-Halpern boundary condition is significantly better than that with the doubly-asymptotic boundary condition (Figure 10.2, right). The short-time accuracy is similar to that of the model with the viscous boundary condition and the long-time accuracy is better than that of the model with the doubly-asymptotic boundary condition.

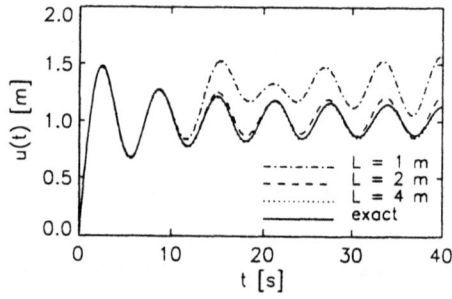

Figure 10.3: Responses of the asymptotic boundary condition of order six

Next, we analyse the responses of a finite model with an asymptotic boundary condition of order six. The points where the responses divergence from the exact solution are significantly shifted to the right when compared to the asymptotic boundary condition of order two (Figure 10.1, right). The response of the finite model with a length of 4 m is very accurate up to 40 seconds.

Finally, we consider the response of a finite model with the novel boundary condition with three auxiliary degrees of freedom. In Figure 10.4 left, we see that the novel boundary condition gives a very accurate response using a model with a length of 1 m. The response is still very accurate, when the length of the near field is reduced to an element (Figure 10.4, right). The novel boundary condition captures both, the short- as well as the long-time response of the rod with high accuracy.

Next, we analyze the behavior of some of these absorbing boundary conditions when the rod is

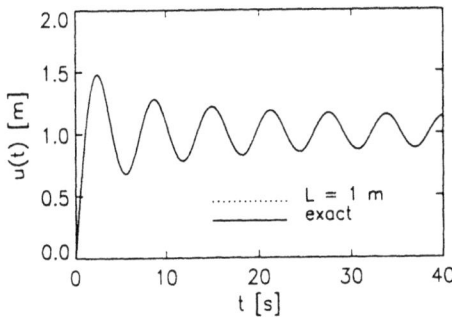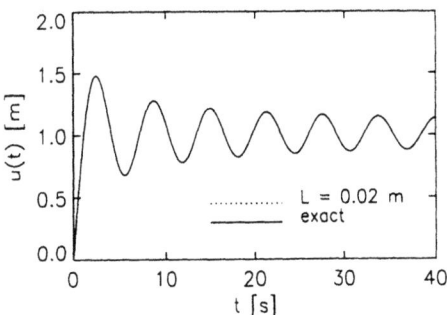

Figure 10.4: Responses of the novel boundary condition

subjected to an impulsive loading. The loading function is

$$f(t) = c\sin(at) \qquad 0 \le t \le 2\pi/a$$
$$f(t) = 0 \qquad\qquad t > 2\pi/a$$

(10.15)

The time history of this loading, it is nonzero for a full period of a sinus function, and its Fourier spectrum is shown in Figure 10.5. The parameters are: $a = 10$ s^{-1} and $c = 5$ N. The loading has only a small frequency content below the cut-off frequency $\omega_c = 1$ Hz. The length of the near field is $L = 1$ m. The responses of the left boundary of the rod with different absorbing boundary conditions are shown in Figure 10.6, left. All boundary conditions capture the short-time peak very accurately. The viscous boundary condition exhibit the most divergent post-peak response. The asymptotic boundary condition of order two captures very well the amplitude of the post-peak response. However, the oscillations are increasingly out of phase as the time advances. The asymptotic boundary condition of order six begins to diverge slightly after 15 s. In contrast, the novel absorbing boundary condition is very accurate up to 40 s (see Figure 10.6, right). If the accurate reproduction of the peak is of concern, then all these absorbing boundary conditions are equivalent. Clearly, the viscous boundary condition is the cheapest one with regard to computational costs. It allows to compute an accurate solution with a moderate far field.

The last computational example considers a sinusoidal loading with circular-frequency equal to the cut-off frequency. In chapter 2, we had shown that the reflection coefficient is always unity at the cut-off frequency no matter what absorbing boundary condition is used. Even the novel boundary condition exhibit complete reflection there. Apparently, this is the weak point of these absorbing boundary conditions. Notice, that the exact solution diverges for $t \to \infty$.

Let us consider the result obtained with a finite model with a length of 1 m (see Figure 10.7). The model with viscous boundary condition reaches quickly a stationary state with finite amplitude. The response increases till the energy dissipation of the dashpot at the right end of the near field balances the energy injected into the system by the loading. Clearly, the solution is over-damped. Similarly, the asymptotic boundary condition of second order reaches a stationary state after several cycles. However, the dissipation rate is smaller than that of the finite model with the viscous boundary condition because the amplitude of the oscillation is significantly larger. The asymptotic boundary condition of order six reproduces very well the exact solution. Therefore, the dissipation of the boundary conditions diminishes with increasing order of the approximation. The novel boundary condition (see Figure 10.7, right) is virtually identical to the exact solution, at least up to 40 s. For this computation the length of the near field was $L = 0.2$ m.

What's about the accuracy if we consider the response in a longer time interval? This result is shown in Figure 10.8. The amplitude of the exact solution still increases as the time advances. However, the amplitudes of the novel boundary conditions lags behind those of the exact solution

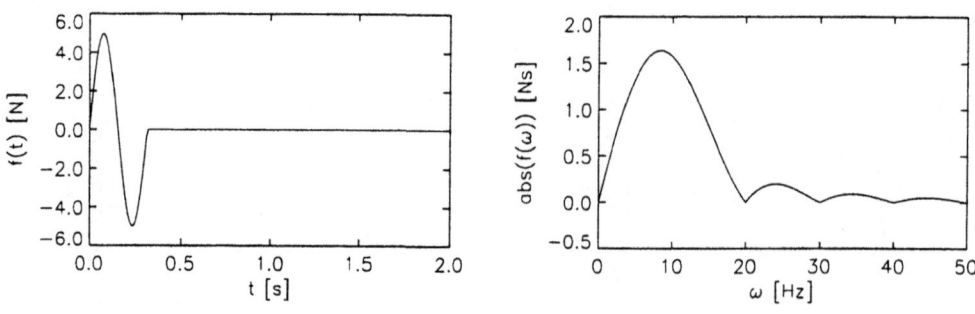

Figure 10.5: Time history and Fourier spectrum of the loading (10.15)

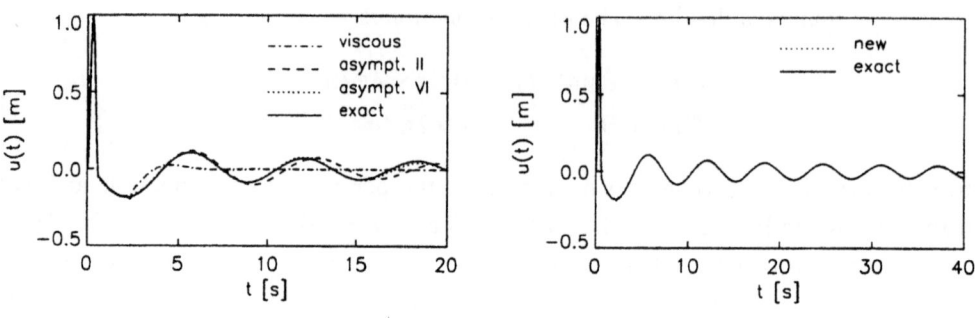

Figure 10.6: Response of the rod due to the loading (10.15)

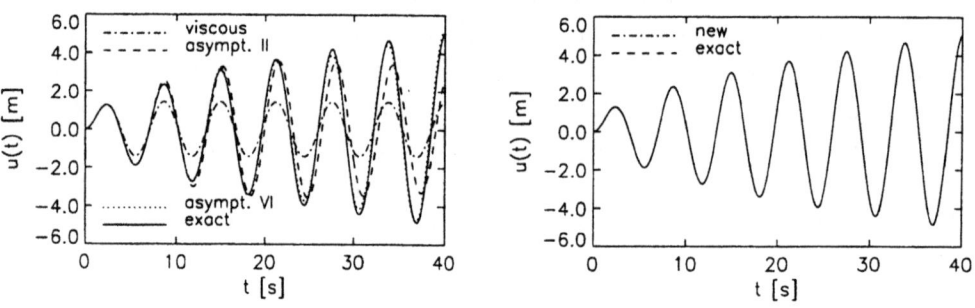

Figure 10.7: Responses of the rod to a harmonic loading with frequency equal to the cut-off frequency

(Figure 10.8, right). Even if the length of the rod is increased to $L = 4.0\,\mathrm{m}$ the response diverge from the exact solution. This fact is related to the error of the approximation in the neighborhood of the cut-off frequency. This error can be made arbitrarily small increasing the order of the approximation. Then, the spurious dissipation is reduced and we obtain a better agreement over a longer time interval. However, this interval will always be bounded.

10.1.3 Finite element analysis of the dissipative restrained rod

Next, let us consider the influence of material damping on the response of the rod with absorbing boundary conditions. Figure 10.10 shows the responses of the viscous and the asymptotic boundary condition of order two. The length of the near field is 2 m (left) and 4 m (right). The material damping is $\varepsilon = 0.1$. The shape of the exact solution exhibits a peak followed by a strongly damped post-peak oscillation. The asymptotic value is equal to that of the nondissipative rod. First, let us consider the responses of the finite model with a near field of 2 m (Figure 10.10, left). The short-time response is captured very well with both boundary conditions. Remarkably, the points where the response of the finite model diverge from the exact solution are approximately

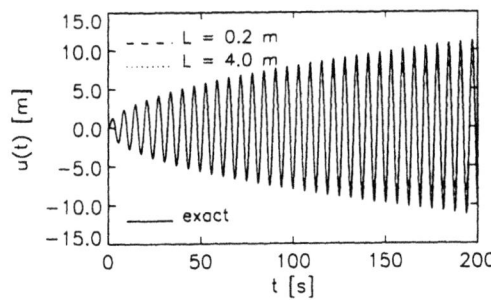 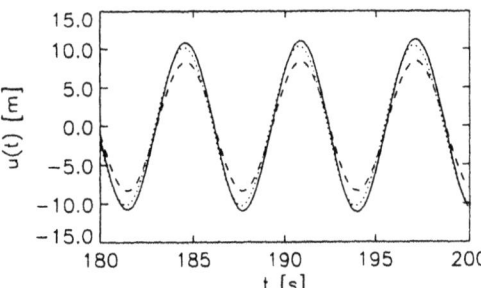

Figure 10.8: Response of the novel boundary condition up to 200 s.

equal to those of the nondissipative rod. Because of the significant damping the responses of the finite model are closer to the exact solution. However, the long-time response is still inaccurate.

The responses of the model with a length of 4 m (Figure 10.10, right) are very good. The finite model with the viscous boundary condition shows still noticeable errors in the post-peak oscillations. However, these are significantly smaller than those obtained with the nondissipative rod. In contrast, the finite model with the asymptotic boundary condition of order two produces a virtually exact solution. Hence, the material damping allows to reduce the size of the near field. Clearly, the larger the damping the smaller can be the size of the near field.

10.2 Finite models of semi-infinite strata

10.2.1 Out-of-plane motion of a semi-infinite solid stratum

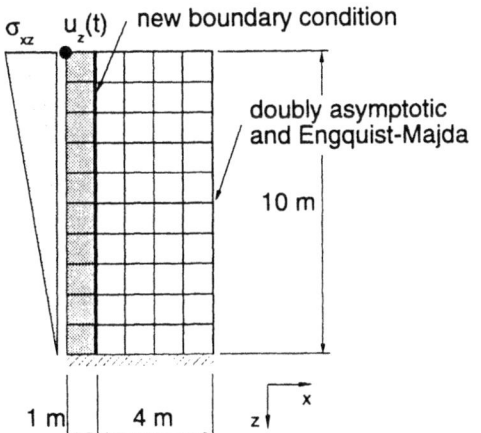

Figure 10.9: Finite models of the semi-infinite stratum subjected to SH-waves

Let us consider the propagation of SH-waves (out-of-plane shear waves) in the semi-infinite solid stratum shown in Figure 10.9. The height of the stratum is 10 m. The bottom of the stratum is fixed and the surface is free of stresses. The mass density of the solid is $\rho_s = 1.0 \, \text{kgm}^{-3}$ and the shear modulus is $G = 2.5$ kPa. The speed of shear waves is $c_s = 50 \text{ms}^{-1}$. In addition, we assume that there is no material damping. The layer is subjected to a step loading at its left boundary. This loading has a triangular stress distribution over the height of the stratum (see Figure 10.9). The maximum value of the stress at the top of the stratum is $\sigma_{xy} = 50$ Pa. The stratum has homogeneous boundary conditions. Therefore, the propagation modes are independent of the frequency. They are given by the set of normalized eigenfunctions

$$u_n(z) = \sqrt{\frac{2}{H}} \cos(\lambda_n z) \qquad \text{with} \qquad \lambda_n = \frac{(2n+1)}{2H}\pi, \, n \geq 0. \qquad (10.16)$$

The DtN-map kernel with respect to the modal basis defined by the set of eigenvectors (10.16) is diagonal and is given by

$$H_{nn} = iG \sqrt{\left(\frac{\omega}{c_s}\right)^2 - \lambda_n^2} \qquad (10.17)$$

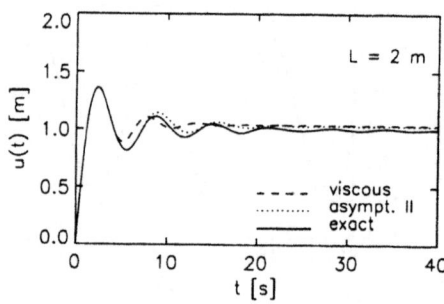
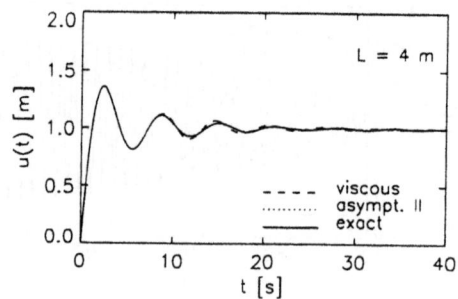

Figure 10.10: Responses of finite models with material damping

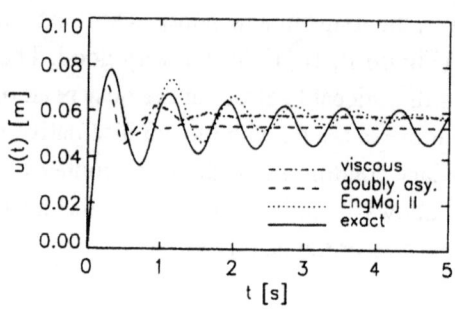
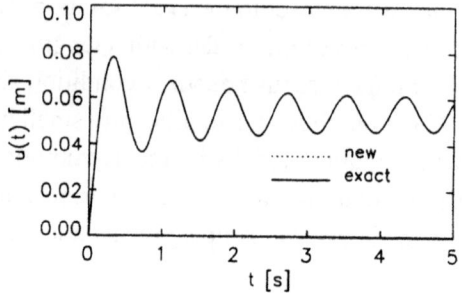

Figure 10.11: Responses of the upper-left node of the solid layer subjected to an out-of-plane loading

The solution may be computed via the superposition of the solutions of restrained rods subjected to a step loading. The parameters defining the restrain of these rods are $G\lambda_n$. The cut-off frequencies are $\omega_c = c_s\lambda_n$. The first two are $f_{\omega_c} = 1.25\,\text{Hz}$ and $f_{\omega_c} = 3.75\,\text{Hz}$. The step loadings acting on these rods are the projection of the step loading shown in Figure 10.9 against the eigenvectors (10.16). The exact solution is computed with the analytical solution developed in section 10.1.1. The contributions of the first three eigenmodes has been considered.

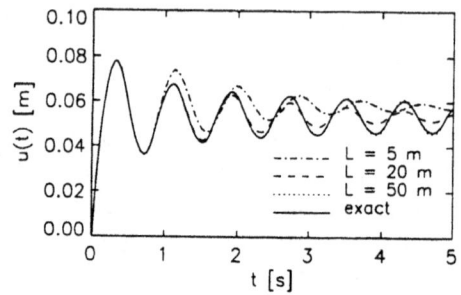

Figure 10.12: Responses of the left-upper node with finite models with the global Engquist-Majda boundary condition of second order

We compare the response of the upper-left node of a finite model with the novel absorbing boundary condition with that of finite models with standard boundary conditions. The near field of the finite model with the novel boundary condition is 1 m long. That with standard boundary conditions has a length of 5 m. The time history plots of the responses of the finite models with standard boundary conditions are shown in Figure 10.11, left. The first three eigenmodes has been considered. These plots show that the responses are dominated by the first propagation mode. The nonlocal Engquist-Majda boundary condition of second order capture very well the peak of the response and the first oscillation period. Afterward, the oscillations are increasingly out of phase and the final displacement is inaccurate. The viscous boundary condition capture very well the peak of the response. However, the post-peak oscillations are very inaccurate. The doubly asymptotic boundary condition captures the final displacement exactly. However, it significantly underestimates the peak of the response and fails completely to reproduce the post peak oscillations. Hence, these results confirm the results obtained with the elastically restrained rod.

In contrast, the novel boundary condition captures very well the exact response in the transient as well as in the post-transient regime (see Figure 10.11, right). This, although this finite model exhibits a very short near field (1 m). The novel boundary condition considers the first three normal modes. The approximation has three degrees of freedom per propagation mode. This result is even more surprising when we consider that the global Engquist-Majda boundary condition of second order needs a near field of more than 50 m to obtain an accurate solution (Figure 10.12).

10.2.2 In-plane motion of an infinite solid stratum

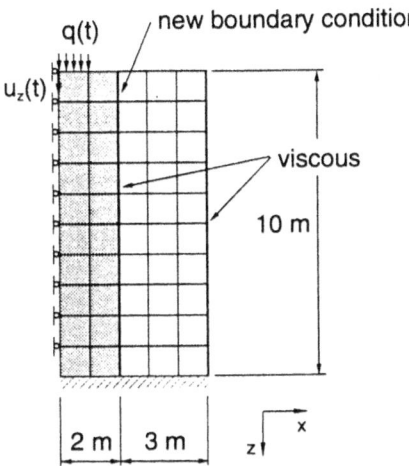

Figure 10.13: Finite models of an infinite stratum subjected to in-plane motion.

Next, let us consider the in-plane motion (plane stress) of an infinite solid stratum of 10 m height (see Figure 10.13). Because the problem is symmetric, only the right half of the stratum is modeled with finite elements. The bottom of the layer is fixed and the surface is free of stresses except along the loaded part. The driving force is a step loading applied at the top of the first column of elements. It has a constant stress distribution of $\sigma_{zz} = 200$ Pa and acts on a length of 1 m on each side of the symmetry axis. The material parameters of the layer are given by the mass density $\rho_s = 1000$ kgm^{-3}, Young's modulus E = 11.111 MPa and Poisson number $v = 0.2$. The speed of compressional waves is $c_p = 100$ ms^{-1} and that of shear waves is $c_s = 86.6$ ms^{-1}. The viscous material damping exhibits a damping ratio of $\zeta = 1.5\%$ at the predominant vibration frequency of the stratum.

Unfortunately, no analytical solution exists of this problem. Therefore, we are obliged to compute an approximation of the exact solution with numerical methods. We consider the finite element solution of a finite model with sufficiently large near field to be equivalent with the exact solution. In order to reduce the size of this near field, we apply the viscous boundary condition at the right boundary. Numerical experiments showed that a near field with a length of 20 m gives a sufficiently accurate approximation of the exact solution.

In Figure 10.14, left, the responses of the vertical displacement of the upper-left node of three finite models are shown. All finite models have viscous boundary condition at the right boundary. The lengths of the near field of these finite models are 2 m, 5 m and 10 m. As expected, the accuracy increases with the size of the near field. All responses are too strongly damped with respect to the exact solution. The response of the finite model with a near field of 2 m length is completely wrong. Neither the short-time peak nor the post-peak oscillations are captured with sufficient accuracy. In contrast, the finite models with near field of 5 m and 10 m captures very well the short-time peak and, with increasing accuracy, even the post-peak oscillations and the static displacement.

The finite model with the novel boundary condition has a near field of 2 m. The novel boundary condition considers six propagation modes: three longitudinal and three transverse modes. The response of this model shows a good accuracy in the transient as well as in the post-transient regime, see Figure 10.14 right. The accuracy of the response is not so good as experienced with the out-of-plane motion. The short-time peak is slightly underestimated and the post-transient oscillations are slightly inaccurate. The short-time peak is better captured if the length of the near field is increased to five meters, see Figure 10.15 left. However, the post-peak oscillations remain substantially unchanged when compared to the response of the finite model with 2 m length. Nevertheless, if this response is compared to that obtained with the viscous boundary condition the accuracy is

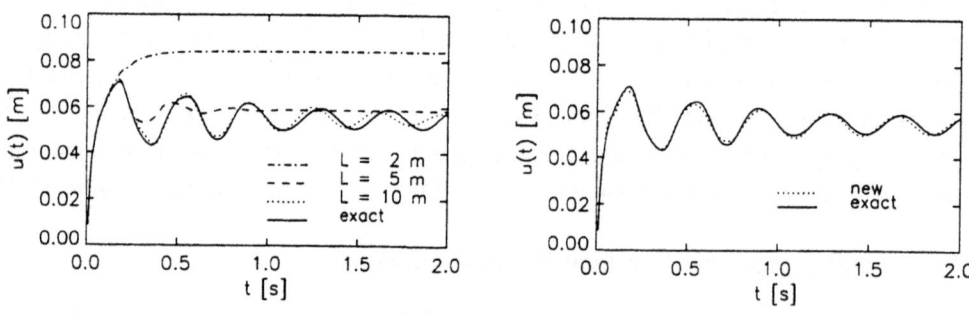

Figure 10.14: Responses of finite models of an infinite solid stratum subjected to an in-plane motion.

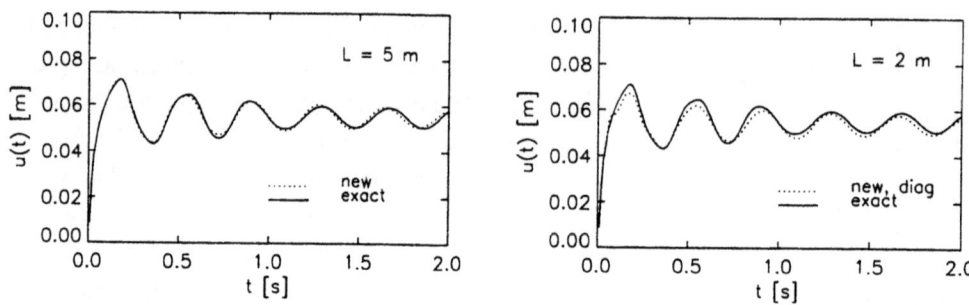

Figure 10.15: Responses of finite models of an infinite solid stratum with the novel boundary condition.

still very good. This inaccuracy is due to the projection of the DtN-map to the modal basis system. In fact, the approximation of the DtN-map is excellent as is shown in Figure 10.16 where the real and imaginary parts of the components of the DtN-map associated with the first two modes are shown. The approximations (dotted line) cannot be distinguished from the components of the DtN-map (full line). The index 1 refers to the first longitudinal and the index 2 to the first transverse mode.

Figure 10.15 right shows the response of a finite model of 2 m length where only the diagonal elements of the DtN-map have been considered to construct the absorbing boundary condition. The accuracy of the response is not as good as that obtained with the absorbing boundary condition developed from the full DtN-map (Figure 10.14 right). Nevertheless, it is still rather good when we consider that the test with a step loading is quite a hard one. Hence, this class of simplified absorbing boundary conditions can be very effective.

10.2.3 Motion of an infinite solid-fluid stratum

The last stratum we are going to analyse is that sketched in Figure 10.17. It is composed of a solid layer of 5 m height. Above this layer, there is a fluid layer of 5 m height. The mechanical properties of the solid layer are identical to those specified in the previous example. The fluid layer has a wave velocity of $c_w = 50$ ms^{-1} and a mass density of $\rho_w = 1000$ kgm^{-3}. Both layers exhibit material damping - Rayleigh damping proportional to the stiffness matrix - such that a damping ratio of 3% is obtained at the predominant vibration frequency. The stratum is subjected to a step loading at the top of the first column of solid elements. The loading has a constant stress distribution of $\sigma_{zz} = 200$ Pa. Because the problem is symmetric the finite model consider only the right half of the stratum. The solid is in a state of plane stress.

We first consider the response of finite models with viscous boundary conditions and near fields having a length of 2 m, 5 m and 20 m. In Figure 10.18 left, time histories of the velocity

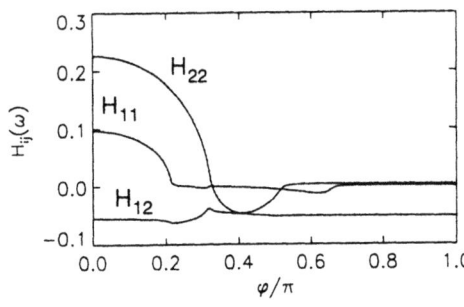 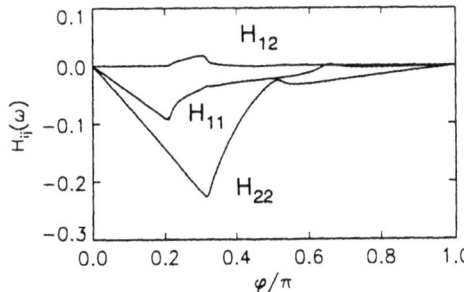

Figure 10.16: Components of the DtN-map kernel of the solid stratum

potential of the interface node on the symmetry axis are shown. The response of the finite model with a length of 20 m is considered to be equivalent with the exact solution. The response of the finite model with a length of 2 m is very inaccurate. It fails to capture even the short-time peak. Afterwards, the response is too strongly damped. Increasing the near field to 5 m we obtain an accurate response up to approximately 0.2 s.

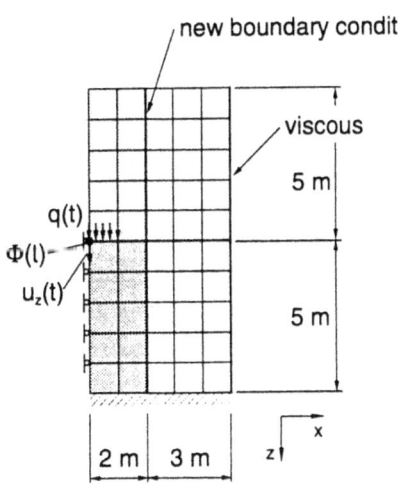

Figure 10.17: Finite models of an infinite solid-fluid stratum.

The finite model with the novel absorbing boundary condition has a near field of 2 m length. The absorbing boundary condition considers 2 propagation modes of the fluid layer and 4 propagation modes of the solid layer (2 longitudinal and 2 transverse modes). The response of the velocity potential is shown in Figure 10.18, right. The accuracy is good. The short-time peak is slightly underestimated. However, the post peak oscillations are well captured. After, 1.5 s the oscillations of the finite model are slightly out of phase. Nevertheless, compared to the responses of the finite models with viscous boundary condition, the result obtained using the novel absorbing boundary condition is very good.

Next we consider the response of the vertical displacement of the interface node on the symmetry axis. Figure 10.19 left shows the time histories of the finite models with the viscous boundary condition. As before, the finite model with a length of two meters is very inaccurate. Neither the short-time peak nor the static displacement is captured. In contrast, the finite model with the novel boundary condition is much more accurate. As before, the short-time peak is slightly underestimated. This inaccuracy is due to the projection of the DtN-map to the modal basis system because the approximation of the DtN-map is excellent. This is shown in Figure 10.20 where the real and imaginary parts of some components of the first row of the DtN-map are shown. The approximations (dotted line) cannot be distinguished from the components of the DtN-map (full line). Note that the coupling between the propagation modes is much stronger than that of the solid stratum. The index 1 refers to the first fluid mode, the index 4 to the first transverse solid mode and the index 5 to the second longitudinal solid mode.

The inaccuracy induced by the projection is showed in Figure 10.21. One model considers the first 2 fluid and the first four solid modes and the second model considers the first four fluid and the first four solid modes. The near fields of both finite models are two meters long. The finite models are subjected to a harmonic loading with the same amplitude of the step loading. The left plot shows the absolute value of the complex response of the velocity potential and the right plot

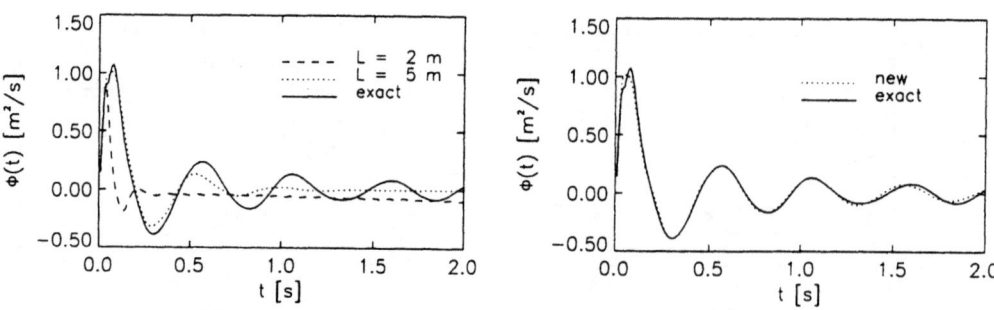

Figure 10.18: Time history of the velocity potential

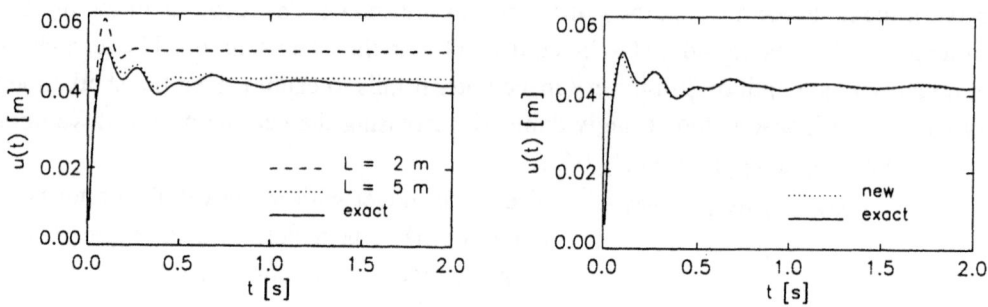

Figure 10.19: Time history of the vertical displacement

that of the vertical displacement. The responses of both finite models are very similar. Both underestimates slightly the peak at 2.08 Hz. This explains the slight excess of damping of these models exhibited in Figure 10.20. The second peak of the vertical displacement is slightly shifted to higher frequencies. The amplitude of the post-peak decay is larger than that of the exact solution in both finite models.

10.3 Applications to interaction problems with concrete gravity dams

10.3.1 Dam-reservoir interaction

We start with the application of the novel absorbing boundary conditions to interaction problems with dams analyzing models considering only the interaction between the dam and the reservoir. This is the standard model used in many investigations addressing the nonlinear seismic behaviour of gravity dams. In our computational examples, we consider a finite element model of the Pine Flat dam (California, USA) [VF89]. The dam has a height of 122.0 m and its basis has a length of 95.87 m. The far field of the reservoir is modeled with a semi-infinite stratum of constant depth. Its height is 116.82 m. We first assume that the dam as well as the reservoir are placed on a rigid foundation. That is, no energy can escape through the base of the dam and the bottom of the reservoir. The dam is modeled with 136 four-node isoparametric finite elements. The concrete has a mass density of $\rho_s = 2.48 \ 10^3$ kg/m^3, a modulus of elasticity of $E = 22.4$ MPa and a Poisson ratio of $v = 0.2$. Viscous material damping is included via Rayleigh-damping with parameters $\alpha = 1.1424$ and $\beta = 7.234 \ 10^{-4}$. This provides for a damping ratio of $\xi = 3.5$ % at the first eigenfrequency of the dam which is $f_1 = 3.19$ Hz. The water has a mass density of $\rho_f = 1.0 \ 10^3$ kg/m^3 and a wave speed of $c_f = 1440$ m/s.

First, let us consider a finite model with a near field of the reservoir consisting only of a column

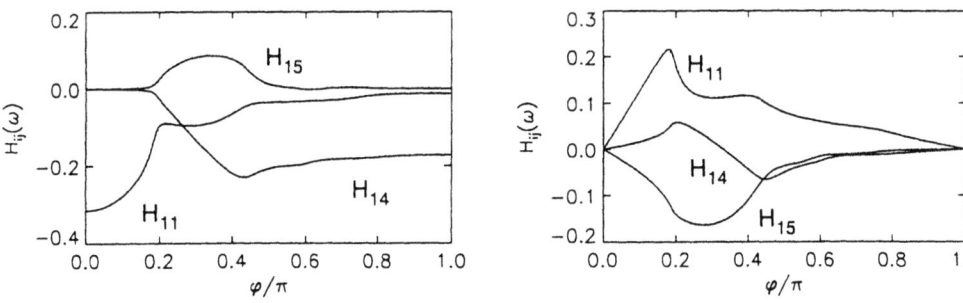

Figure 10.20: Components of the DtN-map kernel of the solid-fluid stratum

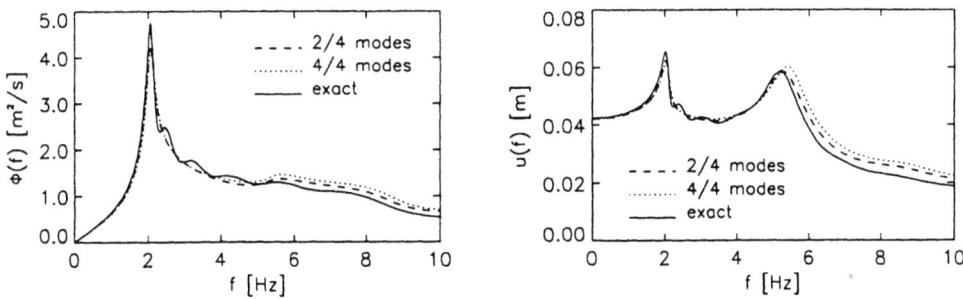

Figure 10.21: Responses of finite models of the solid-fluid stratum

of 16 finite elements (Figure 10.22, left). The response of this finite model is compared to that of the finite model shown in Figure 10.22, right. The near field of the reservoir of this model has a length of 75 m and consists of 112 fluid elements. In both models, the DtN-map of the far field of the reservoir takes into account the first four propagation modes of the semi-infinite stratum. The novel absorbing boundary conditions are modeled with 3 auxiliary degrees of freedom per node. The models are subjected to a transient body loading acting on the dam. It is proportional to the local mass density of the dam times the time history of the horizontal component of the Taft Lincoln School tunnel S69E records (see Figure 10.23).

The responses of the horizontal displacement at the upper-right node of the dam crest are shown in Figure 10.24, left. These are virtually identical. Hence, the finite model with a short near field is very accurate. In Figure 10.24, right, the response of a finite model with long near field but with the viscous boundary condition is compared to that of the finite model with a short near field and the novel boundary condition. The finite model with the viscous boundary condition captures quite well the dominant oscillation. However, the peaks are underestimated. This result is due to the strong energy dissipation of the viscous boundary condition. Hence, in order to obtain an accurate solution with the viscous boundary condition we are obliged to increase the size of the near field.

Next we analyse how many propagation modes are needed to obtain an accurate solution with the finite model with short near field. Figure 10.25, left, shows that with two propagation modes the response is already very accurate. Increasing the size of the near allows to reduce the number of propagation modes to one without any significant loss of accuracy. Figure 10.25, right, shows the response of the finite model with long near field and with an absorbing boundary condition which considers only the first propagation mode. The accuracy is excellent. Clearly, this result can't be generalized because it may depend on the system and on the type of loading. In fact, the first three cut-off frequencies of the reservoir are $f_1 = 3.08$ Hz, $f_2 = 9.25$ Hz and $f_3 = 15.41$ Hz. The largest contributions of the Fourier spectrum of the loading are signifi-

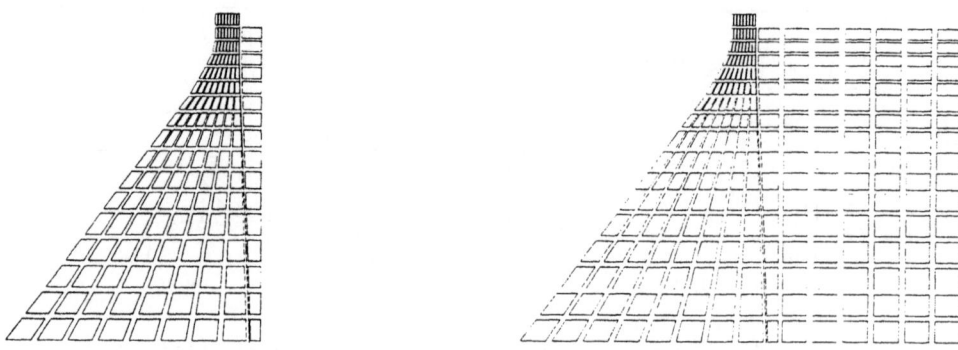

Figure 10.22: Left: Finite model with short near field. Right: Finite model with long near field

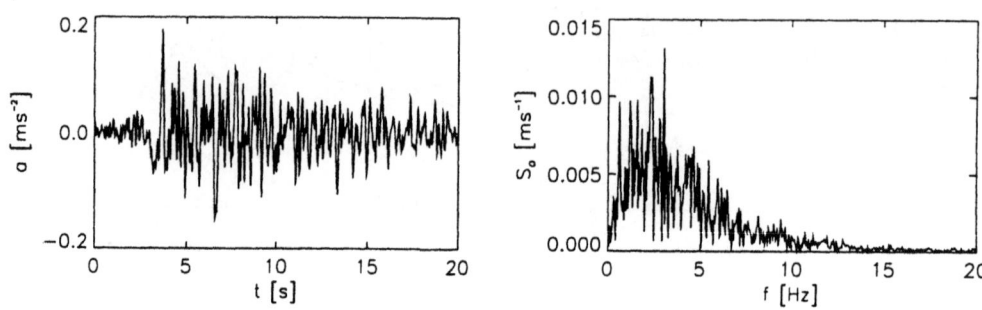

Figure 10.23: Time history and spectral density of the horizontal component of the Taft S96E record

cantly below 10 Hz (see Figure 10.23). Hence, only the first two propagation modes are considerably exited. In addition, most of the energy is transported by the first propagation mode. The second propagation mode is mainly excited below its cut-off frequency so that it reacts as an evanescent mode. This explains why the finite model with long near field needs only one propagation mode to obtain an accurate solution. In fact, the dynamics of the evanescent mode is considered by the finite element model of the near field. The finite model with short near field needs the second propagation mode in the DtN-map because it is too short to properly consider it by the near field alone.

Next we consider a model with a reservoir which is able to dissipate energy at its bottom. There, the boundary condition is a viscous one: $\partial\varphi/\partial n + (q/c)\dot\varphi = 0$. This is a crude model of an interaction with the foundation rock. The parameter q is $q = 1/3$. This is equivalent to a reflection coefficient $\alpha = 0.5$ for waves with a vertical angle of incidence. Note, that the base of the dam is still assumed to be in contact with a rigid foundation. The main effect of this additional dissipation at the bottom is to reduce significantly the maximum response of the dam, see Figure 10.26, left. Figure 10.26, right, shows that even with the additional dissipation, the finite model with long near field and with a viscous boundary condition is still too strongly damped. The error is less pronounced than that obtained with the finite model without dissipation at the bottom (see Figure 10.24, right). This means that the energy dissipation at the bottom is not dominant so that an accurate modeling of the radiation damping is still needed to obtain an accurate solution.

10.3.2 Dam-foundation interaction

In this section, we consider finite models for the analysis of the dam-foundation interaction. The overall model with near and far fields and the discretization of the near field is shown in Figure 10.27. The dam is modeled with 136 finite elements and the foundation by additional 144 ele-

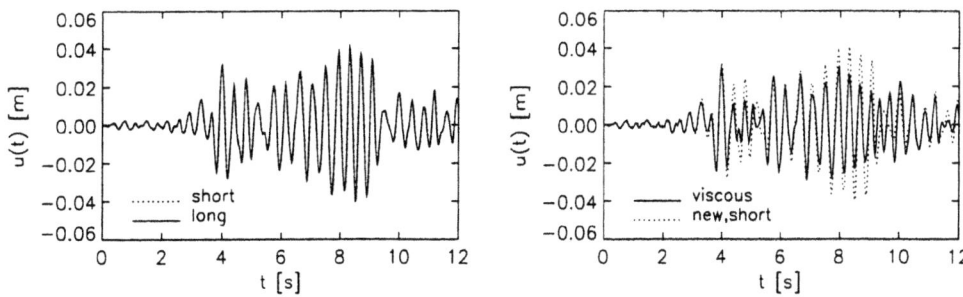

Figure 10.24: Horizontal displacements of the dam crest with different finite models.

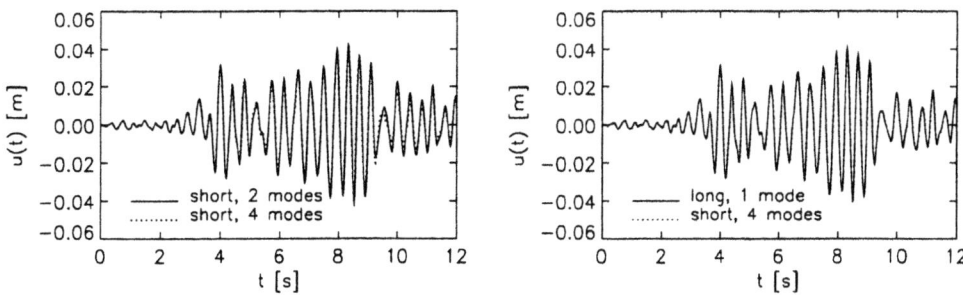

Figure 10.25: Effect of the number of propagation modes on the displacements of the dam crest.

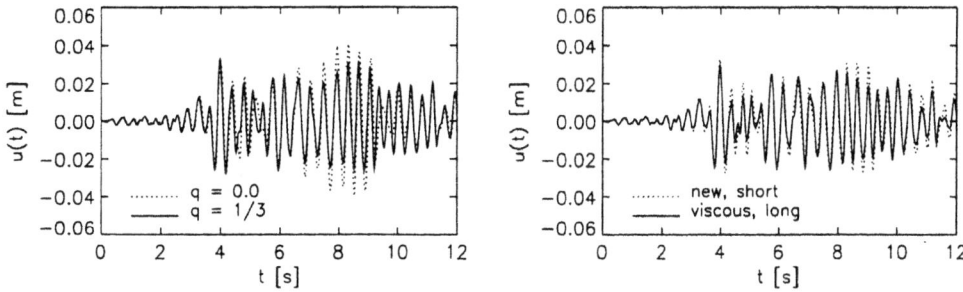

Figure 10.26: Horizontal displacements of the dam crest with energy dissipation at the reservoir bottom

ments. We assume that the displacement at the bottom of the foundation vanish. That is, the foundation is composed by a flexible layer resting on a rigid half space. The height of this layer is 100 m. The absorbing boundary conditions are formulated at the left and right boundaries of this layer. The material parameters of the foundation layer are equal to those of the dam and have been specified in the previous section. The parameters of the Rayleigh damping are $\alpha = 1.361$ and $\beta = 1.608 \ 10^{-3}$. This gives a damping ratio of 5% at the first eigenfrequency of the dam on a rigid foundation (f = 3.19 Hz). The dam as well as the foundation are in a state of plane-stress.

We begin with a steady state analysis of the system. The dam and the foundation are subjected to a harmonic body force proportional to its mass density. One finite model considers DtN-maps formulated with respect to the natural coordinate system. That is, no approximation by a projection to a modal basis of smaller dimension than that of the natural basis is included in these DtN-maps. The second model considers DtN-maps projected to the first four propagation modes (two longitudinal and two transverse modes). These DtN-maps are approximated with rational functions. The absolute value of the horizontal displacements of the upper-right node at the dam crest are shown in Figure 10.28, left. The accuracy of the finite model with the projected DtN-map is

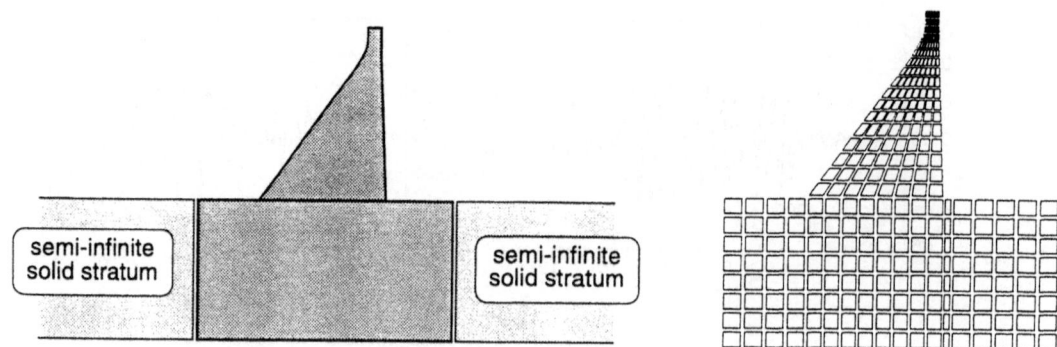

Figure 10.27: Left: Dam-foundation interaction model with two semi-infinite solid strata. Right: Finite element discretization of the near field.

quite good. The first eigenfrequency and the maximum response at this frequency is captured very well. The relative error of the peak is 2%. The second eigenfrequency and its peak is reproduced with less accuracy. The second eigenfrequency is slightly shifted to smaller frequencies and the magnitude of the peak is slightly larger. However, if this result is compared to that obtained with a finite model of the same size but with the viscous boundary condition (Figure 10.28, right) it is still very accurate. The finite model with the viscous boundary condition considerably underestimates the maximum of the peak at the first eigenfrequency and doesn't capture the second eigenfrequency.

The accuracy of the model with the projected DtN-map significantly increases if eight propagation modes are considered (four longitudinal and four transverse modes) instead of four. This is shown in Figure 10.29, left. On the other hand, if only the diagonal components of the projected DtN-maps are considered we obtain a significant reduction of the accuracy. Although the peak of the response at the first eigenfrequency is captured quite well, in the neighborhood of the second eigenfrequency (Figure 10.29, right) the response is significantly different. In particular, the peak is shifted to a larger frequency. This is because the mode associated with the second eigenfrequency is significantly influenced by a horizontal motion of the foundation layer. In fact, the peak at this frequency disappears if the loading acts only on the body of the dam, as is shown in Figure 10.30. Because the foundation layer is not directly subjected to a horizontal loading this mode loose much of its strength.

This result suggests that the accuracy requirements of DtN-maps depend not only on the infinite system being analyzed but also on the type of loading being applied. In fact, if the loading is restricted to act at the body of the dam, the projected DtN-map considering the first four propagation modes (two longitudinal and two transverse modes) gives very accurate results, Figure 10.30, left. Even when only the diagonal elements of the projected DtN-map are used to formulate the absorbing boundary condition, the accuracy of the response is quite good. On the other hand, the finite model with the viscous boundary condition is still too strongly damped.

Figure 10.31 shows how the error induced by considering only the diagonal components of the projected DtN-map (Figure 10.29, right) influences the response of the time-domain analysis. The finite models are subjected to a loading proportional to the mass density of the dam and of the foundation times the acceleration of the horizontal Taft earthquake record. The peaks of the response of the finite model which considers only a diagonal DtN-map are generally smaller than the peaks of the response of the finite model with the full DtN-map. In this case, the accuracy of the finite model with diagonal DtN-maps is unsatisfactory. However, if the loading is restricted to the dam both model exhibit virtually the same time-domain response.

The effect of radiation damping in the time domain is shown in Figure 10.31 (right). Both mod-

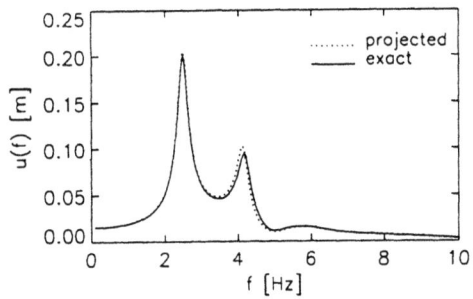
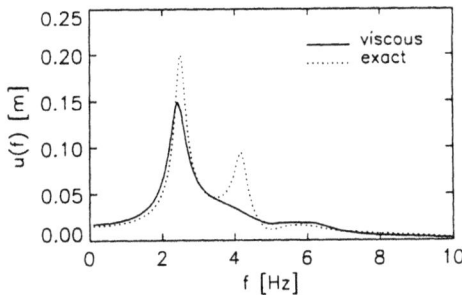

Figure 10.28: Responses obtained with the novel (4 modes) and the viscous boundary condition.

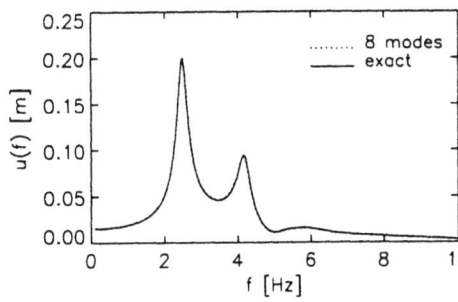
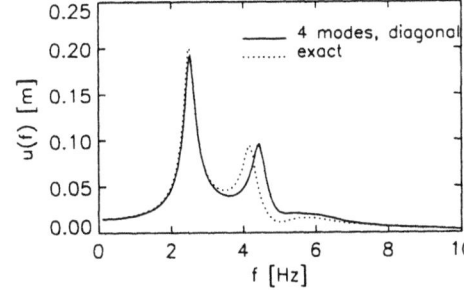

Figure 10.29: Effect of the number of modes (8 modes) and of the diagonalization on the response.

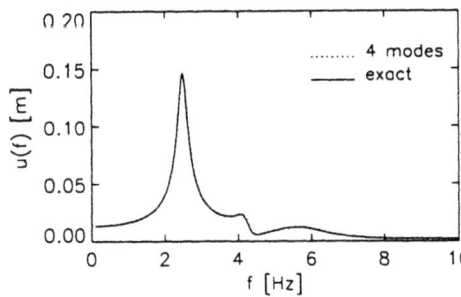
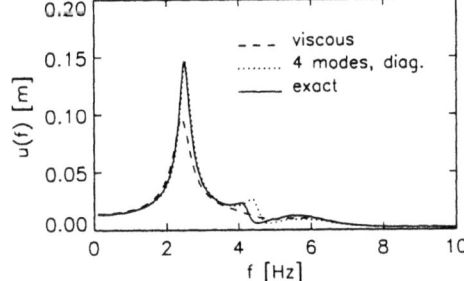

Figure 10.30: Responses when only the dam is excited

els exhibit the same dominant oscillation frequency. However, the peaks of the response of the finite model with no radiation damping are significantly larger than that which considers it.

The last example considers the question how accurate we can model a dam on an elastic half-space with the finite model of Figure 10.27. The assumption of a rigid half-space below the foundation layer has to be replaced. We crudely model the elastic half-space with the viscous boundary condition at the lower horizontal part of the boundary. The dam concrete has a mass density of $\rho_s = 2.60 \ 10^3$ kg/m^3, a modulus of elasticity of $E = 25.13$ MPa and a Poisson ratio of $\nu = 0.1865$. The material parameters of the foundation layer are $\rho_s = 2.60 \ 10^3$ kg/m^3, $E = 22.53$ MPa and $\nu = 1/3$. Viscous material damping is included via Rayleigh-damping proportional to the stiffness-matrix ($\beta = 4.8156 \ 10^{-3}$). This corresponds to a damping ratio of $\xi = 5$ % at the first eigenfrequency of the dam on rigid foundation ($f_1 = 3.3$ Hz). The dam as well as the foundation are assumed to be in a state of plane strain. The viscous boundary condition considers a half space with material parameters equal to those of the foundation layer. The finite model is subjected to unit horizontal and vertical harmonic loadings acting on the upper-right node of the dam crest (see Figure 10.32, left). A second finite model considers the half-space with the viscous boundary condition at the artificial boundary.

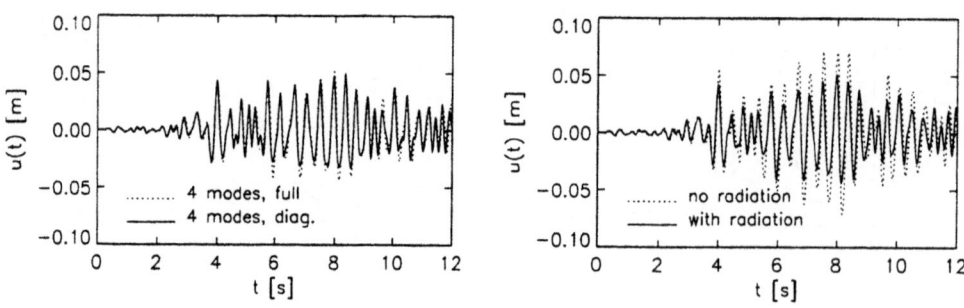

Figure 10.31: Right: Influence of the diagonalization. Right: Influence of the radiation damping

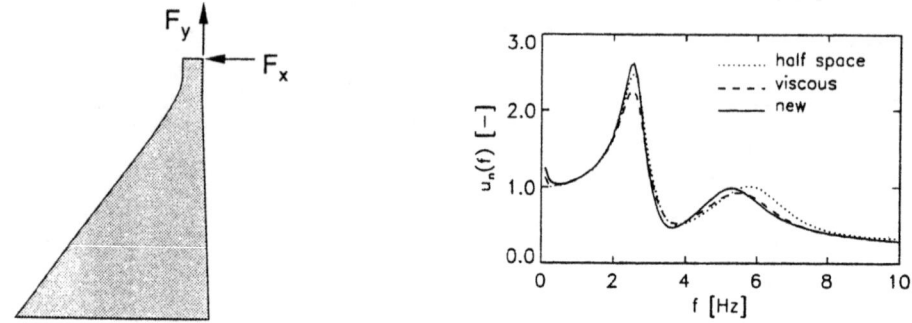

Figure 10.32: Left: Unit forces acting at the dam crest. Right: Normalized horizontal displacements.

The absolute value of the normalized horizontal displacement of the upper-right node of the dam crest is shown in Figure 10.32. The displacement is normalized with respect to the static displacement. The response of this finite model is compared with a more rigorous solution obtained by Szczesiak [Szc96]. The finite model with the novel boundary conditions and that with the viscous boundary condition captures very well the frequency at the first peak of the response. The response of the finite model with the novel boundary condition is moderately larger than that computed with the half-space model. In contrast, the response of the finite model with the viscous boundary condition is significantly smaller. As expected, this model is too strongly damped. The second local maximum of the response is not captured with the same accuracy by the finite model with the novel boundary condition. The magnitude of the response is approximately equal but the frequency at the peak is significantly smaller (5.3 Hz instead of 5.8 Hz). However, the accuracy is rather good considering that the half-space is modeled with two strata and with the viscous boundary condition.

10.3.3 Dam-reservoir-foundation interaction

The last models we shall consider includes all interactions, that is, dam-reservoir, dam-foundation and reservoir-foundation interaction. The overall model with near and far fields and the finite element model of the near field is shown in Figure 10.33. It contains 280 solid elements and 112 fluid elements. The far field on the right consists of a stratum composed of a solid layer and a fluid layer. The solid layer has a height of 100 m and the fluid layer one of 116.2 m. The far field on the left is modeled as a solid stratum of 100 m height. The material parameters of the fluid and solids are equal to those described in section 10.3.1 and section 10.3.2. The dam as well as the foundation are in a state of plane-stress.

We consider finite models with a rigid half-space below the foundation layer. These models exhibit radiation damping only by wave propagation in the two semi-infinite strata. The models

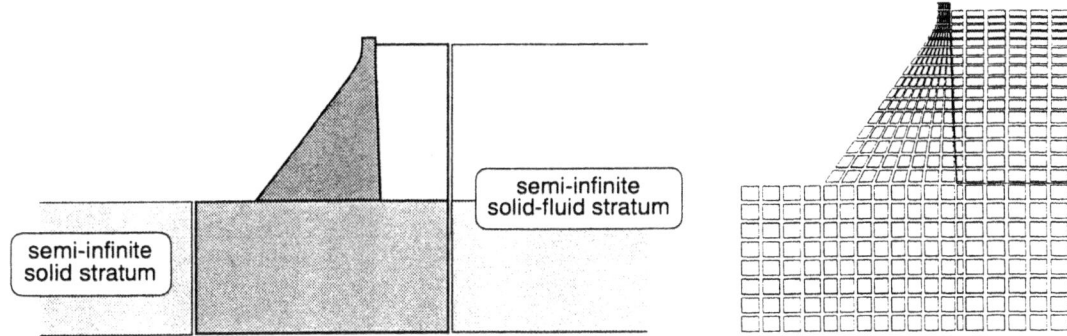

Figure 10.33: Left: Dam-reservoir-foundation interaction model with semi-infinite strata
Right: Discretization of the near field.

are subjected to a harmonic body force which is proportional to the mass density of the dam and the finite part of the foundation. These forces acts in horizontal direction. Figure 10.34 shows the horizontal displacements of the upper-right node at the dam crest of several models. These responses have a dominant peak at 2.05 Hz. The left plot shows the response of three finite models with different absorbing boundary conditions. The first model considers 16 propagation modes (8 fluid and 8 solid) of the right stratum and 8 propagation modes of the left one. The second model considers 8 propagation modes (4 fluid and 4 solid) of the right stratum and 4 propagation modes of the left one. Finally, the third model considers the same number of propagation modes of the second one. However, only the diagonalized DtN-map is taken into account. Figure 10.34, left, shows that the response of the finite models with 8 and 4 propagation modes are virtually identical. The response of the finite model with the diagonalized DtN-map captures very well the dominant peak. However, the second peak at approximately 4.2 Hz is slightly inaccurate. Nevertheless, the response of this simplified model is still very good.

Figure 10.34, right, shows the response of a finite model which considers 8 propagation modes of the right stratum and 4 propagation modes of the left stratum like the second finite model previously described. However, the 4 modes of the fluid and the 4 modes of the solid layer of the right stratum are uncoupled. The fluid layer has a zero velocity boundary condition at its bottom and the solid layer a zero stress boundary condition at its top. The response of this finite model shows slightly increased peaks compared to the finite models taking into account the coupling between the fluid and the solid layer. This suggests that the coupling has little effect on the total amount of energy which is radiated by the stratum. This doesn't mean that there is no energy exchange between the two layers. However, this energy exchange seems to have little effect on the total energy radiated by the stratum. Finally, as usual, the finite model with the viscous boundary condition gives a response which is too strongly damped (Figure 10.34, right).

As last example we compare the response of the finite model shown in Figure 10.33 with that obtained with finite models considering accurately only the interaction between the dam and the reservoir (see Figure 10.22). The last finite model considers the flexibility of the foundation by way of a viscous boundary condition at the bottom of the reservoir stratum. The parameter q is computed using the wave speed in the water and that of longitudinal waves in the foundation. In order to be easily comparable, the loading was applied only to the dam. The frequency-domain responses of the horizontal displacement of the upper-right node at the dam crest are shown in Figure 10.35, left. The response of the fully interacting model (dam-reservoir-foundation interaction) has a dominant response at a frequency which is significantly smaller than those of the partially interacting model (dam-reservoir interaction). In addition, the maximum value of the response of the fully interacting model is larger than that of the partially interacting model. These

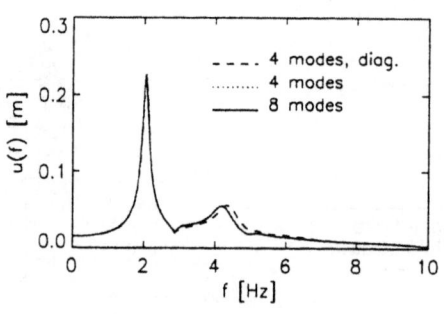 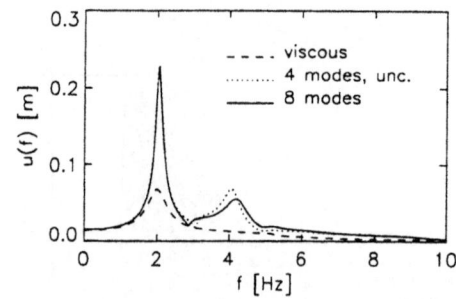

Figure 10.34: Responses of different finite models

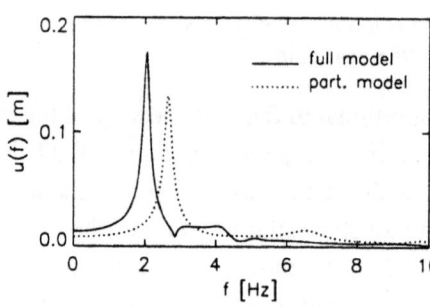 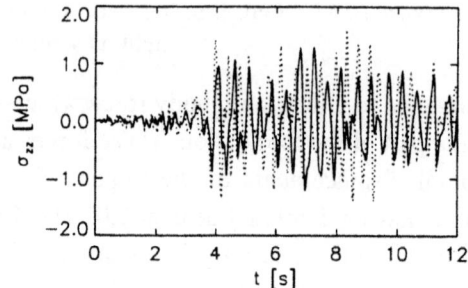

Figure 10.35: Responses of the fully and the partially interacting models

two effects are due to the flexibility of the foundation. A dam on a flexible foundation oscillates with a lower dominant frequency than one on a rigid foundation. In addition, the maximum response of the displacements is generally larger because of the additional flexibility of the foundation.

This is not the case if we consider the stresses. Figure 10.35, right, shows the time history of the stresses near the change of the slope of the downstream face of the dam when it is excited with a horizontal loading being proportional to its mass density and to the time history of the horizontal component of the Taft earthquake ground motion S69E. The maximum stress obtained with the fully interacting model is significantly smaller than that computed with the partially interacting model. Clearly, this result cannot be generalized because the transient response of systems having a dominant peak in the frequency response mainly depends on the intensity of the Fourier spectrum of the loading in the neighborhood of this dominant vibration frequency. However, such a significant difference of the responses suggests that the partially interacting model may not be adequate especially when strong nonlinear phenomena like crack propagation are considered in the transient analysis.

Remark:

• The computational examples shown in the sections 10.2 and 10.3 were computed with the finite element program STRATUM2D written in Fortran 77.

Chapter 11
Closure

11.1 Conclusions

❏ An accurate nonlinear transient finite-element analysis of dam-reservoir-foundation systems has to consider the effects of radiation damping. This can be done using DtN-maps of far field models. The DtN-maps of far fields governed by the acoustic and elastic wave equations are necessarily nonlocal in space and time. When these DtN-maps are coupled to the equations of motion of the near field, a system of Volterra integro-differential equations of convolution type is obtained. The numerical integration of such systems is highly inefficient because of the evaluation of many convolution integrals at every time step and because of the large amount of data necessary to accurately describe DtN-maps in the time domain. Local absorbing boundary conditions are numerically much more efficient. However, these boundary conditions generally require a large near field in order to obtain accurate solutions in the nonlinear earthquake analysis of dam-reservoir-foundation systems. Therefore, this approach is numerically inefficient too.

❏ A technique to eliminate the nonlocality in the time of DtN-maps is to approximate the Fourier transform of DtN-map kernels with rational functions which are analytic in the right complex half-plane C_+. Because DtN-maps are causal the most natural approach is to approximate DtN-maps with Laguerre functions. These are an orthonormal basis of the Hilbert space $L_2(R_+)$ and their Laplace transforms are rational functions which are analytic in the right complex half-plane C_+ and are also an orthonormal basis of the Hardy space $H_2(C_+)$. A Möbius transformation associates the Laplace transform of Laguerre functions expansions to Laurent series which are analytic in the open domain S_∞ defined by $|z| > 1$. This Möbius transformation allows to relate the coefficients of the Laguerre functions expansion of regularized DtN-map kernels to the coefficients of a trigonometric Fourier series expansion of DtN-map kernels mapped on the unit circle T. This allows an efficient evaluation of the coefficients of the Laguerre functions by way of fast Fourier transform techniques. This approach produces a uniformly convergent Laguerre functions expansion of DtN-map kernels if the mapped DtN-map kernels are bounded, continuous and differentiable on the unit circle T.

❏ The Z-transform allows to relate the Laurent series to linear infinite-dimensional time-invariant discrete-time systems. An approximation of these infinite-dimensional systems with stable finite-dimensional systems is obtained by truncating the Laurent series after a finite number of terms. However, these finite-dimensional systems are generally too large for practical applications because the Laguerre functions expansion approach is not optimal when applied to DtN-map kernels which are oscillatory in the time domain. The dimension of these systems can be significantly reduced using model reduction techniques of linear systems theory. This corresponds to an approximation of DtN-map kernels in the time domain with a series of products of trigonometric and exponential functions. The classical Padé approximation method is numerically very efficient but it is not reliable enough for our purpose. Much more reliable are the truncated balanced and Hankel-norm approximation methods. Because Möbius transformations relate discrete-time to continuous-time systems the model reduction can be performed with discrete- as well as with continuous-time systems. Given an a priori error bound with respect to an infinity norm, these methods allow to construct stable rational functions which approximate DtN-map kernels within the prescribed error bound.

❏ In the time domain, the new absorbing boundary condition obtained by the rational approximation of a DtN-map is represented by a linear time-invariant continuous-time system. This is essentially a system of ordinary differential equations of first order. This is local in time and allows to evaluate the convolution integrals much faster than with standard techniques. In addition, the amount of data to describe the DtN-map kernel in the time domain is enormously reduced. If the DtN-map kernel is symmetric it is more convenient to transform the system of ordinary differential equations of first order into a symmetric one of second order (symmetric realization). Then, this system can be coupled with the equations of motion of the near field so that we obtain a symmetric system of ordinary differential equations of second order. This describes the motion of the near as well as of the far field. That is, the system of Volterra integro-differential equations of convolution type is approximated with a system of ordinary differential equations. The time integration of this enlarged system can be handled with standard numerical time integration algorithms of structural dynamics. The theory of passive systems allows to state a sufficient condition for the stability of these enlarged systems. Moreover, this condition has an elegant physical interpretation.

❏ The approximation method developed in this work can be used to construct very accurate absorbing boundary conditions describing far fields modeled by strata of fluids and solids. This allows to compute accurate solutions with very small near fields. Generally, the largest approximation error is not due to the rational approximation but to the projection of the DtN-map to a smaller displacements subspace.

❏ The theory developed for nonlocal DtN-maps can also be used to develop local absorbing boundary conditions for transient finite element analysis. As examples, the symmetric realizations of the Engquist-Majda and Bayliss-Turkel boundary conditions have been developed.

11.2 Future developments

11.2.1 Extensions of the method

❏ The numerically most expensive step of system reduction via truncated balanced realization is the spectral decomposition of the positive definite controllability or observability grammians. This is a significant numerical task if the system is very large because all eigenvalues and eigenvectors have to be computed. On the other hand, we lose the a priori information about the accuracy of the approximation if only a part of the eigenvalues and eigenvectors are computed. A step-by-step approach which constructs a series of reduced systems with an increasing number of degrees of freedom may still be numerically more effective than computing the full spectral decomposition. The accuracy of these reduced systems can be tested a posteriori. This approach works only with truncated balanced approximation methods because Hankel-norm approximation methods require that all eigenvalues of the full system matrix A are computed before the stable part can be extracted. Therefore, the truncated balanced realization methods seem to be numerically more effective than the Hankel-norm approximation methods.

❑ If the system to be approximated becomes so large that a simultaneous approximation of the complete DtN-map kernel would be too time-consuming, the kernel can be partitioned in diagonal blocks which are small enough to be easily approximated. The blocks not on the diagonal could be neglected. This is a generalization of the approximation which results when only the diagonal components of the kernels are considered. The blocks not on the diagonal can be also considered. However, these blocks cannot be represented by way of a symmetric realization so that a staggered integration algorithm is needed to include these blocks in the finite element analysis. An alternative procedure to obtain a simultaneous approximation of the complete DtN-map kernel is to first approximate the diagonal components of the kernel. Then, using the denominator polynomials of this approximation, we could construct the numerator polynomial of the off-diagonal components of the DtN-map kernel using least-squares approximation methods.

❑ The approximation method can be easily extended to approximate causal kernels $h(t)$ which are not in $L_2(R_+) \cap L_1(R_+)$ by simply modifying the kernels according to $h_\alpha(t) = e^{-\alpha t} h(t)$, where $\alpha > 0$. This corresponds to an approximation of the Laplace transform of the kernel $h(s)$ along the vertical line $x = \alpha$ in the right half plane C_+. All results concerning the stability and accuracy of the approximation are still valid for the modified kernel $h_\alpha(t)$. However, they are generally not valid for the kernel $h(t)$ so that it is not known in advance if the rational approximation of $h(s)$ on the imaginary axis iR is stable and uniform. The author has applied this approach to approximate the function $t^{-1/2}$ which is singular at $t = 0$. The approximation was very good except, obviously, at $t = 0$.

❑ Of major concern is the question if it is possible to extend the proof of stability of enlarged systems to absorbing boundary conditions which are not dissipative in the complete frequency domain. Lyapunov's direct method seems to be an adequate instrument to handle this problem. However, the difficulty is to find a Lyapunov function which is simple enough to provide a proof without being obliged to perform large numerical computations. In particular, it should be possible to use the fact that all matrices of the near field are positive definite.

11.2.2 Applications to other problems

❑ Absorbing boundary conditions of three-dimensional wave-guides governed by the acoustic and elastic wave equations should be handled without significant modifications with the theory developed in this work. The main difficulty is probably the construction of DtN-maps of wave-guides governed by elastic wave equations because many large eigenvalue problems have to be solved. The subspace projection technique increases significantly the effectiveness of the construction of these DtN-maps if the induced error remains acceptably small.

❑ The theory developed in this work should also be applicable for constructing absorbing boundary conditions for wave-guides governed by other equations of motion than the acoustic and elastic wave equations. It is essential, however, that these equations of motion are hyperbolic and that the Fourier transform of DtN-map kernels are continuous functions of the frequency. Addressed are, in particular, waves in porous media and electromagnetic waves.

❑ Further research is needed to establish how the theory can be extended to develop absorbing boundary conditions for problems governed by equations of motion of the parabolic type. This encompasses heat conduction and diffusion problems but also wave propagation problems in thin beams (Bernoulli-Euler or Timoshenko theory) and in thin plates (Kirchhoff theory). The Laplace transform of the DtN-map kernels of these types of problem are of order $O(\sqrt{s})$ when $s \to \infty$. That is, these DtN-map kernels have a singularity of the order $O(t^{-3/2})$ at $t = 0$ in the time domain. If the singular term of the DtN-map can be extracted the regularized kernel could be approximated with the methods developed in this work. In addition, the extracted singular term can be reduced to one of order $O(t^{-1/2})$ using the velocity as input variable. The contribution of this term must still be considered by convolution integrals. However, because the kernel is defined analytically by $h(t) = Ct^{-1/2}$ we need only very few data to describe it and the numerical evaluation of the convolution integrals can be computed with fast algorithms.

❏ DtN-maps of elastic half-spaces cannot be constructed using the semi-analytical method because the boundary conditions at the surface of the half-space don't factorize into two functions where the first depends only on the radial coordinate and the second only of the circumferential coordinate. In these cases, the DtN-maps can be constructed using boundary element methods. However, it is not clear if these methods allow to compute DtN-map kernels in the complete frequency range. If not, an asymptotic approximation may fill the gap. However, many of these boundary element methods produce DtN-map kernels which are not symmetric. These can be approximated with the methods developed in this work. A symmetric realization, however, is not possible so that the coupling of the equations of motion of the near field with those defining the absorbing boundary conditions loses much of its appeal. In this case, a staggered time integration procedure may be more convenient.

❏ The approximation method can be used to efficiently solve general linear and nonlinear Volterra integro-differential equations of convolution type by way of systems of differential equations. Clearly, the kernel of the convolution term must be such that it can be approximated with high accuracy. If the Laplace transform of the kernel is a rational function the system of differential equations are exactly equivalent to the associated integro-differential equations. The author has tested this approach with several scalar nonlinear integral equations and it worked very well.

Abstract

The linear and nonlinear behavior of dams subjected to strong earthquake ground motion is strongly affected by the interaction with the impounding reservoir and foundation. However, when such large systems are analyzed using the finite element method only a small part of the reservoir and foundation, usually called the near field, can be directly modeled. Otherwise, the computational costs would be tremendous. In order to avoid the spurious reflection of waves at the artificial boundaries of the finite element model the parts of the reservoir and foundation which are not directly modeled – the far field – have to be considered by so-called absorbing boundary conditions. These take into account the dynamics of the far field and especially the phenomenon of radiation damping.

In frequency domain, nonlocal absorbing boundary conditions are represented by so-called dynamic stiffness functions or DtN-maps. These are numerically very effective for linear problems because the simulation can be performed in the frequency domain. However, when nonlinear effects of the near field should be considered, the simulation has to be carried out in the time domain. Then, the coupling of the equations of motion describing the near field with those describing the far field gives rise to a nonlinear system of Volterra integro-differential equations of the convolution type. Its numerical solution is very cumbersome because generally large data sets are needed to describe DtN-maps in the time domain and because of the huge number of operations necessary to compute the convolution integrals at every time step.

On the other hand, the approximate, local absorbing boundary conditions developed so far for time domain computations generally require a large near field in order to obtain a sufficiently accurate solution so that these are numerically ineffective when applied to earthquake analysis problems. Moreover, no genuine finite element implementation exists for most of these absorbing boundary conditions.

In this work, a new method is developed which allows to construct accurate and numerical effective absorbing boundary conditions. The main idea is to uniformly approximate DtN-maps defined in the frequency domain with rational functions. Within this framework, an DtN-map is considered to be the transfer function of an infinite-dimensional time-invariant linear system. This is approximated by a finite-dimensional linear system. The approximation is performed in two steps. First, the DtN-map kernel is expanded into a series of orthogonal functions (e.g. Laguerre functions). An appropriate Möbius transformation relates this series to a Laurent series. These is truncated after a finite number of terms and identified with a finite-dimensional discrete-time or with a finite-dimensional continuous-time system. These systems, however, contain many degrees of freedom which have little effect on its input-output behavior so that they can be cancelled without reducing significantly the accuracy of the approximation. This system reduction is performed using balanced truncation or Hankel-norm approximation techniques.

The absorbing boundary conditions obtained by this novel method are highly accurate approximations of DtN-maps. Moreover, these absorbing boundary conditions are always stable and causal. Furthermore, the same technique allows to formulate genuine finite element implementations of many of the most popular local absorbing boundary conditions.

251

In time domain, the absorbing boundary conditions are described by systems of differential equations with constant coefficients. These can be easily combined with the differential equations describing the motion of the near field such that the system of integro-differential equations is reduced to one of differential equations. This can be solved using any standard numerical integration algorithm of structural dynamics.

In this work, the method has been applied to simulate the interaction of concrete gravity dams with the impounding reservoir and the foundation in the time domain. The far field is modeled by semi-infinite, two-dimensional strata governed by the acoustic and elastic wave equations. Various examples show the accuracy of the novel method.

Keywords: finite element analysis, fluid-structure interaction, soil-structure interaction, dam-reservoir-foundation interaction, nonlinear transient analysis, earthquake analysis, absorbing boundaries in time domain, linear systems theory, infinite-dimensional linear systems, orthogonal functions, rational approximation, balanced truncation approximation, Hankel-norm approximation.

Zusammenfassung

Das lineare und nichtlineare Verhalten von Gewichtstaumauern unter Erdbebeneinwirkungen wird in bedeutendem Masse von der dynamischen Interaktion der Staumauer mit dem Stausee und dem Untergrund beeinflusst. Werden diese Systeme mit der Methode der Finiten Elementen behandelt, so muss aus Kostengründen die Modellierung auf einen endlichen Bereich, d. h. auf das Nahfeld, beschränkt werden. Die nicht direkt durch die Methode der Finiten Elementen modellierten Teile des Stausees und des Untergrundes – das Fernfeld – werden dann durch sogenannte absorbierende Ränder modelliert. Diese berücksichtigen die Dynamik des Fernfeldes und insbesondere die Abstrahlungsdämpfung.

Im Frequenzbereich können für Fernfelder mit einfacher Geometrie genaue, nichtlokale absorbierende Ränder, sogenannte dynamische Steifigkeitsmatrizen oder DtN-maps, hergeleitet werden. Bei der Behandlung linearer Probleme im Frequenzbereich sind diese dynamische Steifigkeitsmatrizen numerisch sehr effizient. Wenn sich jedoch das Nahfeld nichtlinear Verhalten kann, so muss die Simulation im Zeitbereich durchgeführt werden. Die Kopplung der Bewegungsgleichungen des Nahfeldes mit den dynamischen Steifigkeitsmatrizen im Zeitbereich führt dann zu einem System von nichtlinearen Volterra Integro-Differentialgleichungen vom Konvolutionstyp. Die numerische Lösung dieser Systeme ist sehr aufwendig, da die Beschreibung der dynamischen Steifigkeitsmatrizen im Zeitbereich grosse Datenmengen erfordern und die Berechnung der Konvolutionsintegrale bei jedem Integrationszeitschritt sehr rechenintensiv ist.

Die ungenauen, lokalen absorbierenden Ränder, welche speziell für Simulationen im Zeitbereich entwickelt wurden, erfordern im allgemeinen ein grosses Nahfeld, wenn hohe Genauigkeitsanforderungen gestellt werden, so dass auch sie zu einem numerisch ineffizienten Modell führen. Darüber hinaus existieren von den meisten lokalen absorbierenden Ränder keine Implementationen für die Methode der Finiten Elementen.

Diese Arbeit beschreibt eine neue Methode, um genaue und numerisch effiziente absorbierende Ränder im Zeitbereich zu konstruieren. Die Grundidee ist, die dynamischen Steifigkeitsmatrizen im Frequenzbereich durch rationale Funktionen zu approximieren. Innerhalb dieser Theorie wird eine dynamische Steifigkeitsmatrix als Transferfunktion eines unendlich-dimensionalen, zeitinvarianten, linearen Systems interpretiert. Dieses wird dann durch ein endlich-dimensionales, lineares System approximiert. Die Approximation erfolgt in zwei Schritten. Zuerst wird die dynamische Steifigkeitsmatrix als Reihe orthogonaler Funktionen dargestellt (z.B. Laguerre Funktionen). Mit einer geeigneten Möbius-Transformation wird diese Reihe zu einer Laurentreihe transformiert. Diese wird nach einer endlichen Zahl von Summanden abgebrochen und in Form eines zeit-diskreten oder zeit-kontinuierlichen linearen Systems dargestellt. Diese Systeme enthalten viele Freiheitsgrade, die einen geringen Einfluss auf die Eigenschaften des Systems haben, so dass sie ohne nennenswerten Genauigkeitsverlust gelöscht werden können. Diese Reduktion der Freiheitsgrade wird mit neueren Methoden der Systemtheorie (balanced truncation approximation, Hankel norm approximation) durchgeführt.

Die nach dieser neuen Methode konstruierten absorbierenden Ränder erlauben es, dynamische Steifigkeitsmatrizen, die im Frequenzbereich beschränkt und kontinuierlich sind, gleichmässig

254

und mit hoher Genauigkeit zu approximieren. Darüber hinaus sind die absorbierenden Ränder immer stabil und kausal. Gleichzeitig erlaubt die neue Methode, auf einfache Weise Finite Element Implementationen von lokalen absorbierenden Rändern zu formulieren, die durch Differentialoperatoren dargestellt werden.

Im Zeitbereich werden diese absorbierenden Ränder als Systeme von Differentialgleichungen mit konstanten Koeffizienten dargestellt. Sie können somit einfach mit den Bewegungs-Differentialgleichungen des Nahfeldes gekoppelt werden. Somit wird das ursprüngliche nichtlineare System von Integro-Differentialgleichungen mit einem von Differentialgleichungen approximiert. Dieses kann dann mit den üblichen numerischen Methoden der Strukturdynamik integriert werden.

In dieser Arbeit wird die Methode angewendet, um im Zeitbereich die Interaktion von Gewichtstaumauern mit dem Stausee und dem Untergrund zu simulieren. Die Fernfelder des Stausees und des Untergrundes werden durch zwei-dimensionale Schichtmodelle dargestellt. Die Genauigkeit der absorbierenden Ränder wird anhand verschiedener Beispiele untersucht.

Schlagwörter: Finite Element Analyse, Fluid-Struktur-Interaktion, Boden-Struktur-Interaktion, Staumauer-Stausee-Untergrund-Interaktion, nichtlineare transiente Analyse, Erdbebenberechnung, absorbierende Ränder im Zeitbereich, lineare Systemtheorie, unendlich-dimensionale lineare Systeme, orthogonale Funktionen, rationale Approximation, balanced truncation approximation, Hankel norm approximation.

Bibliography

[AAK71] V. M. Adamjan, D. Z. Arov and M. G. Krein: *Analytic properties of Schmidt pairs for a Hankel operator and the generalized Schur-Tagaki problem.* Math. USSR Sbornik, Vol. 15 (1971), p. 31-73

[AAK78] V. M. Adamjan, D. Z. Arov and M. G. Krein: *Infinite Hankel block matrices and related extension problems.* American Mathematical Society Translations, Vol. 111 (1978), p. 133-156

[AK92] M. T. Ahmadi and Gh. Khoshrang: *Sefidrud dam's dynamic response to the large near-field earthquke of June 1990.* Dam engineering, Vol. 3 (1992), p. 85-113

[AS84] Milton Abramowitz and Irene A. Stegun (ed.): *Pocketbook of Mathematical Functions.* Verlag Harry Deutsch, 1984

[BBK89] S. Boyd, V. Balakrishnan and P. Kalamba: *A bisection method for computing the H infinity norm of a transfer matrix and related problems.* Mathematics of Control, Signals and Systems, Vol. 2 (1989), p. 207-220

[BBM88] A. Barry, J. Bielak and R. C. MacCamy: *On absorbing boundary conditions for wave propagation.* Journal of Computational Physics, Vol. 79 (1988), p. 449-468

[BD94] A. Bedford and D. S. Drumheller: *Introduction to elastic wave propagation.* John Wiley and Sons, Chichester, 1994

[Ben76] A. F. Bennett: *Open boundary conditions for dispersive waves.* Journal of the atmospheric sciences, Vol. 33 (1976), p. 176-182

[BG96] George A. Baker and Peter Graves-Morris: *Padé-approximants.* Vol. 59, (Encyclopedia of mathematics and its applications), Cambridge University Press, 1996

[BO78] Carl M. Bender, Steven A. Orszag: *Advanced mathematical methods for scientists and engineers.* McGraw-Hill, 1978

[BO85] Klaus Jürgen Bathe and Lorraine G. Olson: *Analysis of fluid-structure interactions. A direct symmetric coupled formulation based on the fluid velocity potential.* Computers and structures, Vol. 21 (1985), p. 21-32

[Bro70] R. W. Brockett: *Finite dimensional linear systems.* John Wiley and Sons, New York, 1970

[BS72] R. H. Bartels and G. W. Stewart: *Solution of the equation AX+XB=C.* Communication of the Association for Computing Machinery, Vol. 15 (1972), p. 820-826

[BS90] N. A. Bruisma and M. Steinbuch: *A fast algorithm to compute the H-infinity norm of a transfer function matrix.* Systems and Control Letters, Vol. 14 (1990), p. 287-293

[BT80] A. Bayliss and E. Turkel: *Radiation boundary conditions for wave-like equations.* Communications on Pure and Applied Mathematics, Vol. 33 (1980), p. 707-725

[BZ77] P. Bettess and O. C. Zienkiewicz: *Diffraction and refraction of surface waves using finite and infinite elements.* International Journal for Numerical Methods in Engineering, Vol. 11 (1977), p. 1271-1290

[CC81]   Anil K. Chopra and P. Chakrabarti: *Earthquake analysis of concrete gravity dams including da-water-foundation rock interaction.* Earthquake Engineering and Structural Dynamics, Vol. 9 (1981), p. 363-383

[CC87]   Zhang Chuhan and Zhao Chogbin: *Coupling method of finite and infinite elements for strip foundation wave problems.* Earthquake Engineering and Structural Dynamics, Vol. 15 (1987), p. 839-851

[CC92]   Charles K. Chui, Guanrong Chen: *Signal processing and systems theory: selected topics.* Springer Verlag, 1992

[CF95]   Juan W. Chávez and Gregory L. Fenves: *Earthquake analysis of concrete gravity dams including base sliding.* Earthquake Engineering and Structural Dynamics, Vol. 24 (1995), p. 673-686

[CH93]   R. Chandrashaker and J. L. Humar: *Fluid-foundation interaction in the seismic response of gravity dams.* Earthquake Engineering and Structural Dynamics, Vol. 22 (1993), p. 1067-1084

[Che84]   Chi-Tsong Chen: *Linear system theory and design.* Holt, Rinehart and Winston, 1984

[Cho67]   Anil K. Chopra: *Hydrodynamic pressures on dams during earthquakes.* Journal of Engineering Mechanics Division ASCE, Vol. 93 (1967), p. 205-223

[Cho68]   Anil K. Chopra: *Earthquake behaviour of reservoir-dam systems.* Journal of Engineering Mechanics Division, ASCE, Vol. 94 (1968), p. 1475-1500

[Cho70]   Anil K. Chopra: *Earthquake response of concrete gravity dams.* Journal of Engineering Mechanics Division, ASCE, Vol. 96 (1970), p. 443-454

[CJ83]   Martin Cohen and Paul C. Jennings: *Silent boundary methods for transient analysis.* In: T. Beytschko and T. J. R. Hughes (ed.): Computational Methods for Transient Analysis; Chap. 7, p. 301-357, Elsevier Science Publishers, Amsterdam, 1984

[Cla70]   J. F. Claerbout: *Coarse grid calculations of waves in inhomogeneous media with applications to delination of complicated seismic structure.* Geophysics, Vol. 35 (1970), p. 407-418

[CLW91]   Charles K. Chui, Xin Li and Joseph D. Ward: *Systems reduction vis truncated Hankel matrices.* Mathematics of Control, Signal and Systems, Vol. 4 (1991), p. 161-175

[CR80]   R. R. Coifman and R. Rochberg: *Representation theorems for holomorphic and harmonic functions in Lp.* Asterisque, 1980

[CT90]   Harn C. Chen and Robert L. Taylor: *Vibration analysis of fluid-solid systems using a finite element displacement formulation.* International Journal for Numerical Methods in Engineering, Vol. 29 (1990), p. 683-698

[Dar93]   G. R. Darbre : *Nonlinear reservoir-dam interaction by way of the hybrid frequency-time procedure.* In: Proceedings of the second European Conference in Structural Dynamics, eurodyn'93, Trondheim, Norway; A.A. Balkema, Rotterdam, 1993

[DM90]   Lokenath Debonath and Piotr Mikusinski: *Introduction to Hilbert spaces with applications.* Academic Press, 1990

[DMS+95] W.E. Daniell, R.A. Mir, M.S. Simic and C.A. Taylor: *Seismic behaviour of concrete gravity dams.* In: 10th European Conference on Earthquake Engineering, 28 Aug. - 2 Sep. 1994, Vienna, Austria; p. 1951-1955, A.A. Balkema, Rotterdam, 1995

[EH88]   Bjorn Engquist and Laurence Halpern: *Far field boundary conditions for computation over long time.* Applied Numerical Mathematics, Vol. 4 (1988), p. 21-45

[EH89]   Bahaa El-Aidi and John F. Hall: *Non-linear earthquake response of concrete dams, Part 1: Modelling.* Earthquake Engineering and Structural Dynamics, Vol. 18 (1989), p. 837-851

[Ela96]   Saber N. Elaydi: *An introduction to difference equations*. Springer Verlag, 1996

[EM77]   Bjorn Engquist and Andrew Majda: *Absorbing boundary conditions for the numerical simulation of waves*. Mathematics of Computation, Vol. 31 (1977), p. 629-651

[EM79]   Bjorn Engquist and Andrew Majda: *Radiation boundary conditions for acoustic and elastic wave calculations*. Communications on Pure and Applied Mathematics, Vol. 32 (1977), p. 313-357

[Eve81]   G. C. Everstine: *A symmetric potential formulation for fluid-structure interaction*. Journal of sound and vibration, Vol. 79 (1981), p. 157-160

[FC84b]   Gregory Fenves and Anil K. Chopra: *Earthquake analysis and response of concrete gravity dams*. Report UCB/EERC-84/10, Earthquake Engineering Research Center, University of California, Berkeley, 1984

[FC84c]   Gregory Fenves and Anil K. Chopra: *EAGD-84: A computer program for the Earthquake analysis of concrete gravity dams*. Report UCB/EERC-84/11, Earthquake Engineering Research Center, University of California, Berkeley, 1984

[FC85]   Gregory Fenves and Anil K. Chopra: *Effects of reservoir bottom absorption and dam-water-foundation rock interaction on frequency response functions for concrete gravity dams*. Earthquake Engineering and Structural Dynamics, Vol. 13 (1985), p. 13-31

[Fen83]   Kang Feng: *Finite element method and natural boundary conditions*. In: Proceedings of the International Congress of Mathematicians; p. 1439-1453, 1983

[FHC86]   Ka-Lun Fok, John F. Hall and Anil K. Chopra: *EACD-3D: A computer program for three-dimensional earthquake analysis of concrete dams*. Report UCB/EERC-86/09, Earthquake Engineering Research Center, University of California, Berkeley, 1983

[FN82]   K. V. Fernando and H. Nicholson: *Singular perturbation model reduction of balanced systems*. IEEE Transactions on automatic control, Vol. 27 (1982), p. 466-468

[FNG92]   L. Fortuna, G. Nunnari and A. Gallo: *Model order reduction techniques with applications in electrical engineering*. Springer Verlag, 1992

[Fun65]   Y. C. Fung: *Foundations of solid mechanics*. (Prentice-Hall International Series in Dynamics), Prentice-Hall, Englewood Cliffs, 1965

[FWB90]   G. Feltrin, D. Wepf and H. Bachmann: *Seismic cracking of concrete dams*. Dam Engineering, Vol. 1 (1990), p. 279-289

[Gee78]   T. L. Geers: *Doubly asymptotic approximations for transient motion of submerged structures*. Journal of the Acoustical Society of America, Vol. 64 (1976), p. 1500-1508

[Giv91]   Dan Givoli: *Non-reflecting boundary conditions*. Journal of Computational Physics, 1991, p. 1-29

[Giv92]   Dan Givoli: *Numerical methods for problems in infinite domains*. Vol. 33, (Studies in applied mechanics), Elsevier, 1992

[GK95]   Marcus J. Grote and Joseph B. Keller: *Exact nonreflecting boundary conditions for the time dependent wave equation*. SIAM Journal on Applied Mathematics, Vol. 55 (1995), p. 280-297.

[GKL89]   Guoxiang Gu, Pramod P. Khargonekar and E. Bruce Lee: *Approximation of infinite-dimensional systems*. IEEE Transactions on automatic control, Vol. 34 (1989), p. 610-618

[GL86]   Gene H. Golub and Charles F. van Loan: *Matrix computations*. North Oxford Academic Publisher, 1986

[GL96]   Michael Green and David J. N. Limebeer: *Linear robust control*. (Information and system sciences series), Prentice-Hall, 1996

[Glo84]   Keith Glover: *All optimal Hankel-norm approximations of linear multivariable systems and their Linfinity–error bound*. International Journal of Control, Vol. 39 (1984) / No. 6, p. 1115-1193

[GML94]   F. Guan, I. D. Moore and G. Lin: *A technique for the analysis of the response of dams to earthquakes*. International Journal for Numerical and Analytical Methods in Geomechanics, Vol. 18 (1994), p. 863-880

[GQ95]   Zoran Gajic and Muhammad Tahir Javed Qureshi: *Lyapunov matrix equation in system stability and control*. Vol. 195, (Mathematic in Science and Engeneering), Academic Press, 1995

[Gra75]   Karl F. Graff: *Wave motion in elastic solids*. Clarendon Press, 1975

[GST83]   Martin H. Gutknecht, Julius O. Smith and Lloyd N. Trefethen: *The Carathéodory-Féjer method for recursive digital filter design*. IEEE Transactions on Acoustics, Speech, and Signal Processing, Vol. 31 (1983), p. 1417-1426

[Gut84]   Martin H. Gutknecht: *Rational Carathéodory-Féjer approximation on a disk, a circle and an Interval*. Journal of Approximation Theory, Vol. 41 (1984), p. 257-278

[Hag95]   Thomas Hagstrom: *On the convergence of local approximations to pseudodifferential operators with applications*. In: Gary Cohen (ed.): Proceedings of the Third International Conference on Mathematical and Numerical Aspects of Wave Propagation; p. 474-482, SIAM, 1995

[Hah63]   Wofgang Hahn: *Theory and application of Liapunov's direct method*. Prentice-Hall, 1963

[Ham82]   S. J. Hammarling: *Numerical solution of the stable, non-negative definite Lyapunov equation*. IMA Journal of Numerical Analysis, Vol. 2 (1982), p. 303-323

[HC82a]   John F. Hall and Anil K. Chopra: *Two-dimensional dynamic analysis of concrete gravity and embankment dams including hydrodynamic effects*. Earthquake Engineering and Structural Dynamics, Vol. 10 (1982), p. 305-332

[HC82b]   John F. Hall and Anil K. Chopra: *Hydrodynamic effects in the dynamic response of concrete gravity dams*. Earthquake Engineering and Structural Dynamics, Vol. 10 (1982), p. 333-345

[HC83]   John F. Hall and Anil K. Chopra: *Dynamic analysis of arch dams including hydrodynamic effects*. Journal of Engineering Mechanics, ASCE , Vol. 109 (1983), p. 149-167

[Hen74]   Peter Henrici: *Applied and computational complex analysis vol.1: Power series - integration - conformal mapping - location of zeros*. Wiley, New York, 1974

[Hen86]   Peter Henrici: *Applied and computational complex analysis vol. 3: Discrete fourier analysis - cauchy integrals - construction of conformal maps - univalent functions*. John Wiley and Sons, New York, 1986

[Her63]   Reuben Hersh: *Mixed problems in several variables*. Journal of Mathematics and Mechanics, Vol. 12 (1963), p. 317-334

[Her64]   Reuben Hersh: *Boundary conditions for equations of evolution*. Archive of Ratio- nal Mechanics and Analysis, Vol. 16 (1964), p. 242-264

[Hig86b] R. L. Higdon: *Absorbing boundary conditions for difference approximations to the multi-dimensional wave equation*. Mathematics of Computation, Vol. 47 (1986), p. 437-459

[Hig91]   Robert L. Higdon: *Absorbing boundary conditions for elastic waves*. Geophysics, Vol. 56 (1991), p. 231-241

[Hig92]   Robert L. Higdon: *Absorbing boundary conditions for acoustic and elastic waves in stratified media*. Journal of Computational Physics, Vol. 101 (1992), p. 386-418

[HJ88]   J. L. Humar and A. M. Jablonski: *Boundary element reservoir model for seismic analysis of gravity dams*. Earthquake Engineering and Structural Dynamics, Vol. 16 (1988), p. 1129-1156

[HJ90]   Roger A. Horn and Charles R. Johnson: *Matrix analysis*. Cambridge University Press, 1990

[HJ91]   Roger A. Horn and Charles R. Johnson: *Topics in Matrix analysis*. Cambridge University Press, 1991

[HO78]   Mohamed A. Hamdi and Yves Ousset: *A displacement method for the analysis of vibrations of coupled fluid-structure systems*. International Journal for Numerical Methods in Engineering, Vol. 13 (1978), p. 139-150

[Hof62] Kenneth Hoffman: *Banach spaces of analytic functions*. (Prentice-Hall series in modern analysis), Prentice-Hall, 1962

[HT88]   Laurence Halpern and Lloyd N. Trefethen: *Wide-angle one-way wave equations*. Journal of the Acoustical Society of America, Vol. 84 (1988), p. 1397-1404

[Kai80] Thomas Kailath: *Linear systems*. Prentice-Hall, 1980

[Kau88] Emilio Kausel: *Local transmitting boundaries*. Journal of Engineering Mechanics, ASCE, Vol. 114 (1988), p. 1011-1027

[KB93]   Loukas F. Kallivokas and Jacobo Biellak : *Time-domain analysis of transient structural acoustic problems based on finite element and a novel absorbing boundary element*. Journal of the Acoustical Society of America, Vol. 94 (1993) / No. 6, p. 3480-3492

[Kör88] Thomas W. Körner: *Fourier analysis*. Cambridge University Press, 1988

[Kot59] S. Kotsubo: *Dynamic water pressure on dam due to irregular earthquakes*. Memoirs Faculty of Engineering, Kyushu University, Japan, Vol. 18 (1959) / No. 4

[Kre70] Heinz-Otto Kreiss: *Initial boundary value problems for hyperbolic systems*. Communications on Pure and Applied Mathematics, Vol. 23 (1970), p. 277-298,

[Kre78] Erwin Kreyszig: *Introductory functional analysis with applications*. John Wiley and Sons, 1978

[Kun80] Sun-Yan Kung: *Optimal Hankel norm model reductions – scalar systems*. In: Proceedings of the JAAC Conference in San Francisco; p. FA8-A, August 1980

[Leb72] N. N. Lebedev: *Special functions and their applications*. Dover, New York, 1972

[Leh66] S. H. Lehnigk: *Stability theorems for linear motion with an introduction to Lyapunov' s direct method*. Prentice-Hall, 1966

[LK69]   John Lysmer and Roger. A. Kuhlemeyer: *Finite dynamic model for infinite media*. Journal of the Engineering Mechanics Division, Vol. 95 (1969), p. 859-877

[LL61]   Joseph La Salle and Solomon Lefschetz: *Stability by Liapunov's direct method with applications*. Vol. 4, (Mathematics in Science and Engineering), Academic Press, 1961

[LS82]   William D. Lakin and David A. Sanchez: *Topics in ordinary differential equations*. Dover Publications, 1982

[LW72]   John Lysmer and Günter Waas: *Shear waves in plane structures*. Journal of Engineering Mechanics Division, ASCE, Vol. 98 (1972), p. 85-105

[LW84]   Z. P. Liao and H. L. Wong: *A transmitting boundary for the numerical simulation of elastic wave propagation*. Soil Dynamics and Earthquake Engineering, Vol. 3 (1984), p. 174-183

[Mäk90] Pertti M. Mäkila: *Approximations of stable systems by Laguerre Filters*. Automatica, Vol. 26 (1990) / No. 2, p. 333-345

[Méh76] Bernard Le Méhauté: *An introduction to hydrodynamics and water waves*. Springer Verlag, 1976

[MH87]   Jerrold E. Marsden and Michael J. Hoffman: *Basic complex analysis*. Freeman, New York, 1987

[Mo81]   Bruce. C. Moore: *Principal component analysis in linear systems: controllability, observability and model reduction*. IEEE Transactions on automatic control, Vol. 26 (1981) / No. 1, p. 17-32

[MO92]   Henri J.-P. Morand and Roger Ohayon: *Interactions fluides-structures*. Masson, 1992

[MP82]   Francisco Medina and Joseph Penzien: *Infinite elements for elastodynamics*. Earthquake Engineering and Structural Dynamics, Vol. 10 (1982), p. 699-709

[Mül77] Peter C. Müller: *Stabilität und Matrizen.* Springer-Verlag, 1977

[MV74]   J. W. Meek and A. S. Veletsos: *Simple models for foundations in lateral and rocking motion.* In: Proceedings of the 5th World Conference on Earthquake Engineering, Rome, Italy; p. 2610-2613, 1974

[Neh57] Z. Nehari: *On bounded bilinear forms.* Annals of mathematics, Vol. 15 (1957), p. 153-162

[New66] R. W. Newcomb: *Linear multiport synthesis.* McGraw-Hill, 1966

[NS82]   Arch W. Naylor and George R. Sell: *Linear operator theory in engineering and science.* Vol. 40, (Applied Mathematical Sciences), Springer Verlag, 1982

[Oga87] K. Ogata: *Discrete-time control systems.* Prentice-Hall, 1987

[OV89]   Lorraine Olson and Thomas Valdini: *Eigenproblems from finite element analysis of fluid-structure interactions.* Computers and Structures, Vol. 33 (1989), p. 679-687

[PA89]   Peter M. Pinski and Najib N. Abboud: *Two mixed variational principles for exterior fluid-structure interaction problems.* Computers and structure, Vol. 33 (1989), p. 621-635

[PA90]   Peter M. Pinski and Najib N. Abboud: *Finite element solution and dispersion analysis for the transient structural acoustics problem.* Journal of Vibration and Acoustics, Vol. 112 (1990), p. 245-256

[PA91]   Peter M. Pinski and Najib N. Abboud: *Finite element solution of the transient exterior structural acoustics problem based on the use of asymptotic boundary operators.* Computer Methods in Applied Mechanics and Engineering , Vol. 85 (1991), p. 311-348

[Par90] Jonathan R. Partington: *An introduction to Hankel operators.* Cambridge University Press, 1990

[Pat83] A. R. Paterson: *A first course in fluid dynamics.* Cambridge University Press, 1983

[PS82]   Lars Pernebo and Leonard M. Silverman: *Model reduction via balanced state space representations.* IEEE Transactions on automatic control, Vol. 27 (1982) / No. 2

[PT92]   Peter M. Pinsky and Lonny L. Thompson: *Local high-order radiation boundary conditions for the two dimensional time-dependent structural acoustics problem.* Journal of the Acoustical Society of America, Vol. 91 (1992), p. 1320-1335

[PW94]   A. Paronesso and J. P. Wolf: *Rational approximation and realization of dynamic stiffness of unbounded medium.* In: Proceedings of the Conference for Earthquake Resistant Construction and Design ; p. 303-314, A.A. Balkema, Rotterdam, 1994

[PW95]   A. Paronesso and J. P. Wolf: *Global lumped-parameter model with physical representation for unbounded medium.* In: Proceedings of the 10th European Conference on Earthquake Engineering, Vienna, Austria, 1994; p. 565-570, A.A. Balkema, 1995

[PW96]   A. Paronesso and J. P. Wolf: *Sufficient stability criterion for unbounded soil-structure system.* In: Proceedings of the third European Conference on Structural Dynamics, Florence, Italy; p. 1021-1027, A.A. Balkema, 1996

[SB80]   Leonard M. Silverman and Maamar Bettayeb: *Optimal Hankel approximation of linear systems.* In: Proceedings of the JAAC Conference in San Francisco; p. FA8-A, August 1980

[SBZ78] S. S. Saini, P. Bettess and O. C. Zienkiewicz: *Coupled hydrodinamic response of concrete dams using finite and infinite elements.* Earthquake Engineering and Structural Dynamics, Vol. 6 (1978), p. 363-374

[SG88]   G. Sandberg and G. Göransson: *A symmetric finite element formulation for acoustic fluid-structure interaction problems.* Journal of sound and vibration, Vol. 123 (1988), p. 507-515

[Sze75] Gabor Szegoe: *Orthogonal polynomials.* (Colloquium publications, vol.23), American Mathematical Society, Providence, 1975

[TC95]   Hanchen Tan and Anil K. Chopra: *Earthquake analysis of arch dams including dam-water-foundation rock interaction*. Earthquake Engineering and Structural Dynamics, Vol. 24 (1984), p. 1453-1474

[TH86]   Lloyd N. Trefethen and Laurence Halpern: *Well-posedness of one-way wave equations and absorbing boundary conditions*. Mathematics of Computations, Vol. 47 (1986), p. 421-435

[TL90]   C. S. Tsai, G. C. Lee: *Method for transient analysis of three-dimensional dam-reservoir interactions*. Journal of Engineering Mechanics, ASCE, Vol. 2116 (1990), p. 2151-2172

[TLK90]   C. S. Tsai, G. C. Lee and R.L. Ketter: *A semi-analytical method for time-domain analyses of dam-reservoir interaction*. International Journal on Numerical Methods in Engineering , Vol. 29 (1990), p. 913-933

[Tre81]   Lloyd N. Trefethen: *Rational Chebyshev approximation on the unit disk*. Numerische Mathematik, Vol. 37 (1981), p. 297-320

[UG80]   P. G. Underwood and T. L. Geers: *Doubly asymptotic boundary element analysis of nonlinear soil-structure interaction*. In: R. P. Shaw et al. (ed.): Innovative Numerical Analysis for the Applied Engineering Sciences; p. 413-422, University Press of Virginia, 1980

[VF89]   Luis M. Vargas-Loli and Gregory L. Fenves: *Effects of concrete cracking on the earthquake response of gravity dams*. Earthquake Engineering and Structural Dynamics, Vol. 18 (1989), p. 575-592

[VZ92]   S. Valliappan and C. Zhao: *Dynamic response of concrete dams including dam-water-foundation interaction*. Earthquake Engineering and Structural Dynamics, Vol. 16 (1992), p 79-99

[Waa72]   Günter Waas: *Linear two-dimensional analysis of soil dynamics problems in semi-infinite layered media*. Ph.D. Thesis, University of California, Berkeley, 1972

[Wag85]   L. Wagatha: *On boundary conditions for the numerical simulation of wave propagation*. Applied Numerical Mathematics, Vol. 1 (1985), p. 309-314

[Wal72]   Philip R. Wallace: *Mathematical analysis of physical problems*. Holt, Rinehart and Winston, 1972

[Web90]   Benedikt Weber: *New method for the time domain analysis of dam-reservoir interaction*. In: Proceedings of the first European Conference in Structural Dynamics, eurodyn'90, Bochum, Germany; p. 851-858, A.A. Balkema, Rotterdam, 1990

[Web94]   Benedikt Weber: *Rational transmitting boundaries for time-domain analysis of dam-reservoir interaction*. Ph.D. Thesis, Institute of Structural Engineering, ETH Zurich, 1994

[Wee66]   W. T. Weeks: *Numerical inversion of Laplace transforms using generalized Laguerre polynomials*. Journal of the Association for Computing Machinery, Vol. 13 (1966), p. 419-429

[Wep87]   Dietrich Wepf: *Talsperren-Stausee-Interaktion im Zeitbereich basierend auf der Methode der Randelemente*. Ph.D. Thesis, Institute of Structural Engineering, ETH Zurich, 1987

[Wes33]   H. M. Westergaard : *Water pressure on dams during earthquakes*. Transactions, ASCE, Vol. 98 (1933), p. 418-433

[WHB90]   Benedikt Weber, Jörg-Martin Hohberg and Hugo Bachmann: *Earthquake analysis of arch dams including joint nonlinearity and fluid-structure interaction*. In: Proceedings of the International Conference on Earthquake Resistant Construction and Design, June 1989, Berlin, Germany; p. 349-358, A.A. Balkema, Rotterdam, 1990

[WK83]   Edward L. Wilson and Mehdi Khalvati: *Finite elements for the analysis of fluid-solid systems*. International Journal for Numerical Methods in Engineering, Vol. 19 (1983) , p. 1657-1668

[WM89a]   John P. Wolf and Masato Motosaka: *Recursive evaluation of interaction forces of unbounded soil in the time domain*. Earthquake Engineering and Structural Dynamics, Vol. 18 (1989), p. 345-363

[WM89b]   John P. Wolf and Masato Motosaka: *Recursive evaluation of interaction forces of unbounded soil in the time domain from dynamic-stiffness coefficients in the frequency domain*. Earthquake Engineering and Structural Dynamics, Vol. 18 (1989), p. 365-376

[Wol91a]   John P. Wolf: *Consistent lumped parameter models for unbounded soils: Physical representation*. Earthquake Engineering and Structural Dynamics, Vol. 20 (1991), p. 11-32

[Wol91b]   John P. Wolf: *Consistent lumped parameter models for unbounded soils: frequency-independent stiffness, damping and mass matrices*. Earthquake Engineering and Structural Dynamics, Vol. 20 (1991), p. 33-41

[Won91]   M. W. Wong: *An introduction to pseudo-differential operators*. World Scientific, 1991

[WS86]   John P. Wolf and Dario Somaini: *Approximate dynamic model of embedded foundation in time domain*. Earthquake Engineering and Structural Dynamics, Vol. 14 (1986), p. 683-703

[WVL77]   Weeks White, Somasundaran Valliapan and Ian K. Lee: *Unified boundary for finite dynamic models*. Journal of the Engineering Mechanics Division, Vol. 103 (1977), p. 949-964

[WWB88]   Dieter Wepf, John P. Wolf and Hugo Bachmann: *Hydrodynamic stiffness matrix based on boundary elements for time-domain dam-reservoir-soil analysis*. Earthquake Engineering and Structural Analysis, Vol. 18 (1988), p. 575-592

[ZDG96]   Kemin Zhou, John C. Doyle and Keith Glover: *Robust and optimal control*. Prentice Hall, 1996

[Zem65]   A. H. Zemanian: *Distribution theory and transform analysis*. McGraw-Hill, 1965

[ZT78]   Olgierd C. Zienkiewicz and Robert R. Taylor: *Fluid-structure dynamic interaction and wave forces; an introduction to numerical treatment*. International Journal for Numerical Methods in Engineering, Vol. 13 (1978), p. 1-16

Berichte des IBK beim Birkhäuser Verlag Basel

Die aufgeführten Berichte sind unter Angabe der ISBN-Nr. direkt beim Birkhäuser Verlag Basel (Auslieferungsstelle) zu bestellen.
Adresse: Postfach 205, CH-4105 Biel-Benken (Tel. 061 721 77 84; Fax 061 721 79 50).

The listed publications can be ordered directly from Birkhäuser Publishers Basel by specifying the ISBN No. Dispach address: Birkhäuser Verlag, P.O. Box 205, CH-4105 Biel-Benken/Switzerland Tel. ++ 41 61 721 77 84; Fax. ++ 41 61 721 79 50

Patrick Steffen:
Elastoplastische Dimensionierung von Stahlbetonplatten mittels Finiter Bemessungselemente und Linearer Optimierung
Bericht IBK Nr. 220, Juni 1996, ISBN 3-7643-5478-X, Fr. 78.--

René Steiger:
Mechanische Eigenschaften von Schweizer Fichten-Bauholz bei Biege-, Zug-, Druck- und kombinierter M/N Beanspruchung, Sortierung von Rund- und Schnittholz mittels Ultraschall
Bericht IBK Nr. 221, Juni 1996, ISBN 3-7643-5477-1, Fr. 88.--

Manuel Alvarez, Peter Marti:
Versuche zum Verbundverhalten von Stahlbeton bei plastischen Verformungen
Bericht IBK Nr. 222, September 1996, ISBN 3-7643-5647-2, Fr. 78.--

Leena Eskola:
Zur Ermüdung teilweise vorgespannter Betontragwerke
Bericht IBK Nr. 223, September 1996, ISBN 3-7643-5653-7, Fr. 78.--

Tadeusz Szczesiak:
Die Komplementärmethode: Ein neues Verfahren in der dynamischen Boden-Struktur-Interaktion
Bericht IBK Nr. 224, September 1996, ISBN 3-7643-5655-3, Fr. 68.--

Seyed Mohammad Reza Tabatabai:
Finite Element-based Elasto-Plastic Optimum Reinforcement Dimensioning of Spatial Concrete Panel Structures
Bericht IBK Nr. 225, November 1996, ISBN 3-7643-5684-7, Fr. 78.--

Walter Kaufmann, Peter Marti:
Versuche an Stahlbetonträgern unter Normal- und Querkraft
Bericht IBK Nr. 226, November 1996, ISBN 3-7643-5687-1, Fr. 68.--

Philipp Stoffel, Peter Marti:
Modellversuche Europabrücke
Bericht IBK Nr. 227, März 1997, ISBN 3-7643-5762-2, Fr. 68.--

Marcus Schenkel, Thomas Vogel:
Versuche zum Verbundverhalten von Bewehrung bei mangelhafter Betondeckung
Bericht IBK Nr. 228, Mai 1997, ISBN 3-7643-5769-X, Fr. 58.--

Ralf Martens:
Zum Tragverhalten von Betonplatten mit integrierten Schalungselementen
Bericht IBK Nr. 229, Juni 1997, ISBN 3-7643-5797-5, Fr. 108.--

Matthias Barth, Peter Marti:
Versuche an knirsch vermauertem Backsteinmauerwerk
Bericht IBK Nr. 230, August 1997, ISBN 3-7643-5812-2, Fr. 48.--

Stefan Köppel, Thomas Vogel:
Feldversuch Steilerbachbrücke
Bericht IBK Nr. 231, September 1997, ISBN 3-7643-5838-6, Fr. 48.--